Safety Assessment of Genetically Modified Foods

Kunlun Huang

Safety Assessment of Genetically Modified Foods

Springer

Kunlun Huang
College of Food Science and Nutritional Engineering
China Agricultural University
Beijing, Beijing, China

ISBN 978-981-10-3487-9 ISBN 978-981-10-3488-6 (eBook)
DOI 10.1007/978-981-10-3488-6

Library of Congress Control Number: 2017939578

Printed on acid-free paper

This Springer imprint is published by Springer Nature
The registered company is Springer Nature Singapore Pte Ltd.
The registered company address is: 152 Beach Road, #21-01/04 Gateway East, Singapore 189721, Singapore

Contents

Chapter 1
Introduction

1.1 Generation, Development and Appliance of Genetically Modified Organisms (GMOs) in the World

Nowadays, genetic modification (GM) technology was widely applied into many biology fields. It provides a method to introduce exogenous DNA into the targeted creatures such as batteries, yeasts, plants, and animals, ensuring a specific characteristic existed into the targets. Meanwhile, food issues affect the economy and the livelihood of various countries, and they also affect global development and security. GM crops can be a solution of this problem, because they are modified to contain useful characteristics. Some GM crops are being developed in tolerance of herbicide or insect, and other crops contain new genes that have improved nutritional characteristics for feed or food use. It is no wonder that these modified crops provide more production and can also save our manpower and material resources. This is the reason why GM crops are now applied widely in among 30 countries, making prosperity and development worldwide. Until now, there are many commercial GM crop species including corn, cotton, soybean, canola, and papaya, and they have been grown in hectarage of nearly more than 1.8 million hectares, which was a remarkable 100-fold increase since the start of commercialization in the year of 1996.

However, there also come some notable concerns on the safety of GM crops which mainly focus on the risk to human health and environment. And the media and public are often confused by some unreliable reports, which make the spread of GM crops suffering from a difficult procedure. In fact, before the GM crops will be put on the market, we must make sure that the crops have passed through the safety assessment by national authorities.

Even though the safety assessments of the GM crops differ from countries and authorities, the basic framework and procedure are quite closer to agreements. The safety assessments of GM crops are known to be complex of the subject, including the feeding strategies and periods, animal models (fish, chicken, mice, rat, and pig

K. Huang, *Safety Assessment of Genetically Modified Foods*,
DOI 10.1007/978-981-10-3488-6_1

included), and other parameters including the body weight and food consumption of the animals and some biochemical index related to the study and pathological observation. In these animal studies, a 90-day subchronic toxicology study is thought to be scientific and authoritative to verify if one substance has the potential to threaten animal health.

According to the European Food Safety Authority (EFSA) announcements, safety assessments of GM crops are obeyed by a "substantial equivalence" concept, which is based on the principle, "if a new food is found to be substantially equivalent in composition and nutritional characteristics to an existing food, it can be regarded as being as safe as the conventional food." Moreover, "with different methods, key elements of the assessment procedure such as molecular, compositional, phenotypic, and agronomic traits are analyzed in both the GM line and its near isogenic counterpart". When "molecular, compositional, phenotypic, agronomic and other analyses have demonstrated equivalence of the GM food/feed, animal feeding trials do not add to the safety assessment" (updated in EFSA 2011). Based on the "substantial equivalence" concept, the safety assessments of GM crops are carried out using molecular biology analysis and related animal feeding tests.

A 90-day study is always followed by the principle of OECD Test Guideline No. 408. As for the experimental material of the protocols' in vivo toxicological studies and 90-day rodent study, target animal species, housing, number of doses administered, and gender and number of animals are clearly stipulated in this guideline. In the normal 90-day rodent study, feeds are identical between animal groups (treated or control); usually animals are provided by the GM feeds and non-GM feeds. Using rodent models is essential, because it allows whole GM material to be tested in a 90-day rodent study, usually by the improvement of the feed processing. And such assessment is being motivated by either a modification in the GM plant composition or by indications of potential unintended effects. Additionally, body weight, food consumption, clinical biochemistry, and other phenotypic responses are also needed to be recorded and analyzed. By comparing the data among the GM groups, the non-GM groups, and control groups, researchers can make assessment of the GM crops in this study.

However, there is an adaptation of many specific questions on the strengths and weaknesses of such tests to food safety studies, because an animal study needs much more time and cost compared to the study in vivo, and it is also hard to reproduce perfectly. Ninety-day animal feeding studies are not carried out by one particular effect of a given molecule, but are supposedly designed to detect most of the changes that may occur in the GM foods or feeds, including those potentially generated by the genetic modification as well as those resulting from a compositional change which is directly or not linked to the transgene inserted into the GM crops. Thus, these studies might appear too wide and insufficient to detect weak effects, as EFSA has already stated "It is unlikely that substances present in small amounts and/or with a low toxic potential will result in any observable unintended effects." However, the key point is that in the case of a chemically defined molecule for which human exposure is very low, the dosage of the classical 90-day feeding studies can be increased in further studies, whereas this high dosage could not appear in a food or with a quantitatively important constituent in the one human diet.

However, there is an state that "the subchronic, 90-day rodent feeding study is not designed to detect effects on reproduction or development, other than effects on adult reproductive organ weights and histopathology. Thus, in some cases, testing of the whole food and feed beyond a 90-day rodent feeding study may be needed. In cases where structural alerts, indications from the subchronic study or other information on the whole GM plant derived food and feed are available that suggest the potential for reproductive, developmental or chronic toxicity, the performance of such testing should be considered."

1.2 The Production and Development of GM Crops and Its Planting Situation Throughout the World

The development of genetically modified crops was the outcome of convergence of several discoveries and technological developments, including the discovery of the double-helix structure of DNA, three-base sequences in DNA that specify amino acids, inducing mutations in DNA with chemicals or radiation, recombinant DNA (rDNA) techniques, microprojectile bombardment, and so on(Gould et al. 2016).

Genetically modified crops developed from the 1980s to 2015 depended mainly on the three key technologies: recombinant DNA, tissue culture, and *Agrobacterium*-mediated cell transformation. Another important tool was microprojectile bombardment, which emerged in the latter half of the 1980s, also known as biolistics or the gene gun method. It was developed to increase the plant taxa which could be genetically engineered (Klein et al. 1987).

1.2.1 Global Distribution of Genetically Engineered Crops, by Crop

During the past 20 years of commercialization (1996–2015), the global area of genetically modified crops increased 100-fold, from 1.7 million hectares in 1996 to 179.7 million hectares in 2015. In 2015, about 12% (179.7 million of 1.5 billion hectares) of global cropland produced genetically modified crops; the area of which is equivalent to almost 20% of the total land area of the USA (937 million hectares) or China (956 million hectares) and more than seven times the land area of the UK (24.4 million hectares).

The data of 2015 showed that genetic varieties were commercially available for several food crops, three nonfood crops, and two types of flowers. Among all the genetically modified crops, genetically modified maize and soybean were the most widely grown ones.

Since the first commercial release in 1996, the production of genetically modified maize has substantially increased. In 1996, the global area of genetically modified maize was fewer than 300,000 hectares, which rose to 25.2 million hectares in 2006. In 2015, 53.6 million hectares of genetically modified maize were planted, representing one-third of all land planted to maize all over the world that year. Five countries grew more than one million hectares of biotech maize in 2015, including the USA (33.1 million hectares), Brazil (13.1 million hectares), Argentina (2.9 million hectares), South Africa (1.8 million hectares), and Canada (1.4 million hectares).

As with maize, adoption of GE soybean varieties increased rapidly after their introduction in 1996. In 2001, 33 million hectares of genetically modified soybean were grown worldwide. By 2015, genetically modified varieties dominated the soybean market. The global area of genetically modified soybean in 2015 accounted for more than 80% of the 118 million hectares of soybean harvested in that year and 51% of all the genetically modified crop area worldwide. Among 11 countries which reported planting genetically modified soybean in 2015, the top three countries by far which have been growing the largest area of Bt/HT and herbicide-tolerant soybean were the USA (32.4 million hectares), Brazil (30.3 million hectares), and Argentina (21.1 million hectares).

Other genetically modified food crops grown in 2015 were apple (*Malus domestica*), canola (*Brassica napus*), eggplant (*Solanum melongena*), papaya (*Carica papaya*), potato, squash (*Cucurbita pepo*), sugar beet (*Beta vulgaris*), and sweet corn. The contribution of genetically modified varieties to the production of these crops was small, except for canola. The area of genetically modified canola accounted for 24% of the 36 million hectares of canola planted in 2015. Small areas of PRSV-resistant papaya (1000 hectares) were grown in Hawaii and in the USA in 2015. Previous to this, the papaya industry in Hawaii was knocked out by PRSV and then saved by the PRSV-resistant genetically modified papaya. In 2015, China also planted a total of 6985 hectares of genetically modified papaya, and Guangxi province planted its first genetically modified papaya at 2000 hectares in 2014. Genetically modified sweet corn is conservatively estimated at a minimal nominal area of 1000 hectares of all the sweet corn hectarage of an estimated 300,000 hectares. One thousand hectares of genetically modified virus-resistant squash were grown in Hawaii and in the USA in 2015.

In 2015, genetically modified alfalfa (*Medicago sativa*), genetically modified cotton (*Gossypium hirsutum*), and genetically modified poplar (*Populus* spp.) were entirely or mostly grown for nonfood uses.

Herbicide-tolerant RR® alfalfa was first approved for commercialization in 2005 in the USA. In 2006, 60,000 hectares of RR® alfalfa constituted almost 5% of the about 1.3 million hectares of alfalfa seeded all over the world. Unlike the large genetically modified crops of maize and soybean, genetically modified alfalfa is a perennial plant and may occupy more markets. In 2011, the US area of RR® alfalfa was about 200,000 hectares and up to 1 million hectares in 2015, which represented 13% of total 7.4 million hectares of alfalfa.

In 2015, the global area of genetically modified cotton was 24 million hectares, and a total of 15 countries grew it, among which four countries grew more than 1.0 million hectares, including India (11.6 million hectares), China (3.6 million hectares), the USA (3.4 million hectares), and Pakistan (2.9 million hectares).

A total of 543 hectares of Bt poplar were grown in China (Clive 2015).

1.2.2 Two Types of Flowers

Genetic modified techniques have also been applied to change the color of carnations (*Dianthus caryophyllus*) and roses (*Rosa* spp.) which were sold commercially (S. Chandler, RMIT University, personal communication December 7, 2015) (Gould et al. 2016).

1.2.3 Global Distribution of Genetically Engineered Crops, by Counties

In 2015, a total of 28 countries, 8 industrial and 20 developing countries, grew genetically modified crops. The top ten countries, each of which plant more than one million hectares in 2015, are the USA (70.9 million hectares, 39% of global total), Brazil (44.2 million hectares, 25% of global total), Argentina (24.5 million hectares, 14% of global total), India (11.6 million hectares, 6% of global total), Canada (11.0 million hectares, 6% of global total), China (3.7 million hectares, 2% of global total), Paraguay (3.6 million hectares, 2% of global total), Pakistan (2.9 million hectares, 2% of global total), South Africa (2.3 million hectares, 1% of global total), and Uruguay (1.4 million hectares, 1% of global total) (Clive 2015).

The production of genetically modified crops in 2015 was distributed unevenly worldwide. The USA produced ten genetically modified crops, followed by Canada with four. Besides, genetically modified maize, soybean, and cotton were grown in many countries, whereas genetically modified varieties of alfalfa, apple, eggplant, poplar, potato, and squash were grown in just one country each. Genetically modified varieties of rice, wheat, sorghum (*Sorghum bicolor*), cassava (*Manihot esculenta*), banana (*Musa* spp.), camelina (*Camelina sativa*), citrus (*Citrus* spp.), chickpea (*Cicer arietinum*), cowpea (*Vigna unguiculata*), groundnut (*Arachis hypogaea*), mustard (*Brassica* spp.), pigeon pea (*Cajanus cajan*), and safflower (*Carthamus tinctorius*) were in various stages of development. And a blight-resistant American chestnut (*Castanea dentata*) was also in progress (Gould et al. 2016).

1.3 The Main GMOs Cultivate Around the World

During the 23-year period from 1994 to 2016, a total of 28 different genetically modified plants and 404 transgenic plant events were approved in 40 countries and regions (James 2015). There are 148 transgenic maize events, 58 transgenic cotton events, 46 transgenic potato events, 38 transgenic Argentine canola events, and 34 transgenic soybean events, which have been approved worldwide during the period 1994–2016 (Table 1.1). The first generation of GM plants including traits of herbicide tolerance and insect resistance is still the majority of the GM plants; however more and more new traits such as modified product quality, abiotic stress tolerance,

Table 1.1 Approved transgenic plant events, 1994–2016

Plants	Latin name	Events
Alfalfa	*Medicago sativa*	5
Apple	*Malus x Domestica*	3
Argentine canola	*Brassica napus*	38
Bean	*Phaseolus vulgaris*	1
Carnation	*Dianthus caryophyllus*	19
Chicory	*Cichorium intybus*	3
Cotton	*Gossypium hirsutum* L.	58
Creeping bent grass	*Agrostis stolonifera*	1
Eggplant	*Solanum melongena*	1
Eucalyptus	*Eucalyptus* sp.	1
Flax	*Linum usitatissimum* L.	1
Maize	*Zea mays* L.	148
Melon	*Cucumis melo*	2
Papaya	*Carica papaya*	4
Petunia	*Petunia hybrida*	1
Plum	*Prunus domestica*	1
Polish canola	*Brassica rapa*	4
Poplar	*Populus* sp.	2
Potato	*Solanum tuberosum* L.	46
Rice	*Oryza sativa* L.	7
Rose	*Rosa hybrida*	2
Soybean	*Glycine max* L.	34
Squash	*Cucurbita pepo*	2
Sugar Beet	*Beta vulgaris*	3
Sugarcane	*Saccharum* sp.	3
Tobacco	*Nicotiana tabacum* L.	2
Tomato	*Lycopersicon esculentum*	11
Wheat	*Triticum aestivum*	1
Total plant events		404

James (2015)

Table 1.2 Approved transgenic plant events based on commercial trait

Commercial trait	Events
Abiotic stress tolerance	8
Altered growth/yield	3
Disease resistance	28
Herbicide tolerance	252
Insect resistance	205
Modified product quality	81
Pollination control system	29

James (2015)

and altered growth/yield have been developed and approved (Table 1.2). In 2014, herbicide tolerance, deployed in soybean, maize, canola, cotton, sugar beet, and alfalfa, occupied 102.6 million hectares or 57% of the 181.5 million hectares of biotech crops planted globally. Hectarage featuring insect resistance was 27.4 million hectares in 2014; a decrease in Bt cotton in China contributed to this smaller hectarage. Stacked traits are favored by farmers in all countries for all crops with 51.4 million hectares in 2014. The stacked traits for herbicide tolerance and insect resistance are deployed in cotton and soybean (Bt/HT) and maize (Bt/Bt/IR, Bt/HT, and Bt/Bt/HT) but not in sugar beet and alfalfa. The Bt/Bt/IR stack refers to different Bt or other IR genes that code for different insect-resistant traits, for example, for maize, aboveground pests and belowground pests and herbicide tolerance are all stacked in the same maize product.

The global planted area of GM soybean (50%), maize (30.4%), cotton (13.8%), and Argentine canola (4.96%) accounted for more than 99% of all the biotech crop hectarage in the world in 2014. Another way to provide a global perspective of the status of biotech crops is to characterize the global adoption rates as a percentage of the latest. The data indicate that in 2014, 82% (90.7 million hectares) of the 111 million hectares of soybean planted globally were biotech. Of the 184 million hectares of global maize planted in 2014, almost one-third (30%) or 55.2 million hectares were biotech maize. Of the 37 million hectares of global cotton, 68% or 25.1 million hectares were biotech in 2014. Finally, of the 36 million hectares of canola grown globally in 2014, 25% were herbicide-tolerant biotech canola, equivalent to 9.0 million hectares. If the global areas (conventional plus biotech) of these four crops are aggregated, the total area is 368 million hectares, of which half, 49% or 181.5 million hectares, were biotech in 2014.

The USA continued to be the largest producer of biotech crops in the world, with a global market share of ~40%. The USA planted a record hectarage of 73.1 million hectares featuring eight biotech crops (maize, soybean, cotton, canola, sugar beet, alfalfa, papaya, and squash) in 2014. USDA estimates indicate that the percentage adoption of the three principal biotech crops were at, or close to, optimal adoption: soybean 94%, maize 93%, and cotton 96% with a very high average of 94%; biotech sugar beets are at 98.5% adoption and canola at 94%. In 2014, the total biotech crop hectares in Brazil of 42.2 million hectares comprised 29.1 million hectares of biotech soybean, 12.5 million hectares of biotech maize (summer and winter maize), and 0.6 million hectares of biotech cotton. Argentina maintained its ranking as the third largest producer of biotech crops in the world in 2014, after the USA and Brazil, occupying 13% of global hectarage. In 2014, the 24.3 million hectares comprised 20.8 million hectares of biotech soybean, of which 200,000 hectares were Bt/HT soybean, 3.0 million hectares of biotech maize, and 0.5 million hectares of biotech cotton.

Since 1997, China has been planting large hectarages of Bt cotton and small hectarages of GM papaya, poplar, and other vegetables. In 2009, biosafety certificates were issued for Bt rice and phytase maize. China has successfully planted Bt cotton since 1997, and in 2014, 3.9 million hectares were planted to biotech cotton at adoption rate of 93% compared to 90% adoption in 2013. In addition to cotton, China also grew virus-resistant papaya, which increased in hectarage by 46%, from 5800 hectares in 2013 to 8475 hectares in 2014. Papaya-growing regions Guangdong province and Hainan Island were joined by Guangxi province in 2014. Bt poplar has been cultivated in 2003, and in 2013 and 2014, a total of 543 hectares were planted in China. China's research and commercialization of Bt maize, herbicide-tolerant maize, and phytase maize as well as Bt rice will be very important potential contributions to global food and feed needs and that of China.

1.4 The Issues of Safety of GMOs

However, along with the rapid development of GM foods, there are opposite voices arising around the world. The core topic of the debate on GMOs is whether or not they are safe for the environment and for human health. The supporters claim that GM foods are environment friendly, regulated, and profitable for farmers and bring no risk to human health. Someone even thinks that foods derived from GM crops are more safe due to better evaluation than foods derived from conventionally bred crops according to the additional regulatory requirements. The opponents concern that GM foods may bring unexpected effects and be harmful to human and animal health or be adverse to the environment. The toxicity, allergenicity, unexpected effect of the gene insertion, and adverse effects on nutrition are usually considered to be potential risks to human health (Conner AJ, Jacobs JM 2000).

Among the controversies on GM plants for human health, there are two important affairs, one is the affair Pusztai in 1998 and the other one is the affair Seralini in 2012. In 1998, a senior nutrition scientist in the UK, Dr. Pusztai, announced on television that a kind of pest-resistant GM potatoes with *Galanthus nivalis* lectin gene was harmful to rats and affected their immune systems (Ewen and Pusztai 1999). The premature release of information to the mass media resulted in persistent opposition on GMOs in EU until now. The other affair was that a toxicologist in France, Dr. Seralini, published a paper about the two-year carcinogenic test for the roundup-tolerant GM maize on rats (Séralini et al. 2012). Even if the result of this paper was widely queried by scientists and governmental organizations, it was popularly cited worldwide as one of the evidence to oppose the GM foods.

In recent years, based on thousands of published studies, many international panels of experts claimed that there are no confirmed scientific evidences showing that recombinant DNA technologies are harmful to human health. It is based on the fact that all DNA is chemically and structurally the same and the transferred genes and their products are digested and degraded in human gastrointestinal tract as same as other genes and proteins. Meanwhile, the transfer of genetic material across species barriers occurs not only in laboratories but also in natural world as a major driving force in evolution.

However, the risk of GMOs is still needed to be well evaluated based on the precautionary principle required by international organizations and governments. Hazards related to GM crops usually include (1) the presence and characteristics of newly gene-expressing products, usually proteins; (2) the possible changes in the level of natural constituents beyond normal variation, especially natural toxins and antinutrients and allergens; and (3) the possible occurrence of unintended effects by gene insertion.

While it is concerned that the safety assessment of whole foods is mainly based on the guidelines for low-molecular-weight chemicals, such as industrial chemicals, pharmaceuticals, food additive pesticides, and contaminants, these protocols have limitations for testing of whole food and feed. The bulky foodstuffs cannot be included in the diet at very large multiples of the expected human exposure as same as low-molecular-weight chemicals. And the traditional safety assessment methods cannot follow the dramatic development of GM techniques, such as RNA interference (RNAi) and gene targeting, CRISPR/Cas9.

Fortunately, the characterization of recombinant DNA introduced into GM crops is being better investigated by the increasing understanding of whole genomes of plants and the development of new molecular biology tools. Furthermore, the ways of gene transfer are being improved to minimize recombinant DNA transferred into GM crops and decrease the burden of work for safety assessment. It is believed that the debates on the safety of GM foods will be proved by time.

1.5 The Principle and Content of Safety Assessment for Genetically Modified Food

The safety assessment for genetically modified food (GMF) is based on the following six principles: scientific research, concept of substantial equivalence, precaution, case-by-case basis, step-by-step evaluation, and concept of familiarity.

1.5.1 Scientific Research

In the process of safety assessment of GMF, scientific attitude, scientific methods, data from scientific research, and result should be applied on it.

1.5.2 Concept of Substantial Equivalence

Concept of substantial equivalence is the basis in safety assessment of GMF. FAO and WHO proposed that it was a key point to compare GMF with traditional food in 1990. Then OECD raised the concept of substantial equivalence which became the basic rule and practical approach of safety assessment of GMF in the global world. At the current time, this concept occupies an important position in the safety assessment of GMF. The comparison between GMF and traditional food includes several aspects: agronomic traits, morphological traits, main nutrients, trace nutrients, toxins, and key antinutritional factors. If these components are substantial equivalent between GMF and traditional food, they are considered the equivalent safe. If they are different, further assessment and analysis are needed.

1.5.3 Precaution

Some measures should be taken in advance to prevent some risks from GMF. Biological safety risks probably happen because of unintended effects. So we should take some precautions to avoid the potential risks.

1.5.4 Case-by-Case Basis

Case-by-case basis is considered basic rule and approach to evaluate safety for genetically modified crops. There are different receptors, different sources and functions of exogenous gene, different methods to transfer exogenous genes, different sites, and different impacts on animals and human beings; as a consequence, they are regarded as different events based on case-by-case principle.

1.5.5 Step-by-Step Evaluation

A series of experiments needs to be done to assess safety of GMF executed in different times and different locations such as laboratories, trial lab, environment release, and production plant, which would take a long time. It is requested that analysis for GMF should be conducted step by step. The data in each step is analyzed before it was turned to next step.

1.6 Concept of Familiarity

Only when we are familiar with biological receptor, biological characters, and food history of genetically modified crops and interact with other biological environment, we can make the reasonable and effective prediction in the safety assessment procedure.

The content of food safety assessment of transgenic plants refers to the following aspects: nutritional components and antinutritional factors, toxicity, allergenicity, and antibiotic resistance. Nutritional components include protein, fat, fiber, ash content, moisture, carbohydrates, amino acids, fatty acid, mineral elements, vitamins, natural toxins, harmful substances, and antinutritional factors. Animal experiments are conducted to assess toxicity of GMF, which include evaluation of foreign proteins and the whole genetically modified food. In addition to toxicity of foreign proteins, allergenicity assessment is also requested. The assessment of antibiotic resistance is mainly due to two reasons. First, the expression product of a marker gene may be toxic or allergenic; second, the resistant gene horizontal transfer may occur.

The book was fulfilled through the joint efforts of all contributor: Xiaoyun He, Wentao Xu,Huilian Che,Yao Sheng,Shujuan Zheng,Junyu Liu,Jianbing Ge, Xiaoqun Ou,Tianqi Lang,Haoyu Li,Shiying Zou,Zixin Xie, Xiaozhe Qi, Haiyan Liu.I am very grateful for their help and suggestions in preparing the manuscripts.

Part I

Chapter 2
Nutrient Assessment of GMOs

Genetically modified (GM) plants and derived foods and feeds have been on the market since 1994. The worldwide cultivation of GM plants increased from 1.7 million to about 180 million ha/year (about 12% of total arable land) from 1996 to 2015, with soybean, maize, rapeseed, and cotton as the predominant crops (James 2015). Most of the GM plants are characterized by herbicide tolerance or insect resistance or both. Such plants can be considered as substantially equivalent to their isogenic counterpart (OECD 1993a) and are termed first-generation GM plants (EFSA 2004; ILSI 2004). The purpose of substantial equivalence is to compare transgenic crops of the first generation with their isogenic counterparts. Some of these crops are used directly, such as papaya. Some are preserved in animal nutrition, such as sugar beet pulp. Some are used as by-products from the processing industry, such as extracted oil meal.

GM plants are now under development in which significant intended alterations in composition have been achieved to improve the agronomic properties or to enhance the nutritional properties or health benefits. GM plants of the so-called second generation (ILSI 2004) are characterized by output traits such as increased content of valuable components (amino acids, fatty acids, vitamins, etc.), improved availability of nutrients, or a lower concentration of undesirable substances (e.g., phytate, lignin, allergenic substances).

Whether GM plants and derived foods/feeds have equivalent or improved nutritional value is always a concern for the public and developers. Nutrient assessment is the one of most important aspects of the food safety determination of GMOs. Safety assessment of GM plants and derived foods/feeds follows the "substantial equivalence" concept; that is, the food and feed are compared with their non-GM counterparts to identify intended and unintended differences that subsequently are assessed with respect to their potential impact on the environment, safety for humans and animals, and nutritional quality (OECD 1993a). This idea is based on this principle: "If a new food is found to be substantially equivalent in composition and nutritional characteristics to an existing food, it can be regarded as being as safe as

K. Huang, *Safety Assessment of Genetically Modified Foods*,
DOI 10.1007/978-981-10-3488-6_2

the conventional food" (Anonymous 2003). The principle of "substantial equivalence" may cover the nutrient assessment for GM plants of the first generation; however, for second-generation GM plants with changed components, the "case-by-case" principle should be introduced to assess nutritive value and food safety.

In the future, therefore, assessing the nutritive value and also the safety of foods and feeds from GMOs will be a real challenge for animal nutritionists. Various types of animal feeding studies are required to answer all the scientific and consumer questions and to improve the public acceptance of such foods/feeds. The current state of the nutritional and safety assessment of foods from GMOs and the future challenges are analyzed in this chapter.

2.1 Nutritional Components Analysis of GMOs

2.1.1 Contents of Nutritional Components Analysis of GMOs

According to the Organisation for Economic Co-operation and Development (OECD) (1993a), a 'new' food or a new food ingredient is regarded as substantially equivalent if no significant differences appear in comparison with an appropriate traditional source. A food/feed crop is thus substantially equivalent if it corresponds to a conventional variety in its agronomy, composition, metabolic processes, and content of undesirable substances. If substantial equivalence can be established then by analogy, the novel food can be assumed to be as safe as the material to which it was compared.

In practice, substantial equivalence is assessed mainly by comparing the agronomic characteristics of the plant and its composition. However, in determining the degree of equivalence, it should not be overlooked that conventional feedstuffs also exhibit considerable biological variability in their growth characteristics and constituents. The consensus documents prepared by OECD (2001, 2002) on the compositional analyses proposed for new varieties of soybean, maize, potatoes, rapeseed, and sugar beet provide excellent guidance for the analyses needed as part of the nutritional assessment of GM crops modified for agronomic traits and improved nutritional characteristics.

The principle of "substantial equivalence" is regarded as a suitable robust framework for the nutritional and safety assessment of first-generation GM crops. Although it is not a statutory requirement, various bodies have also undertaken the nutritional assessment of first-generation GM crops in a series of extensive digestion and feeding experiments with various species and categories of animals, in addition to laboratory studies. However, the concept of substantial equivalence is much more difficult to apply to second-generation GM crops (Clark and Ipharraguerre 2001). In such plants, modifications may be intended to change the composition and biological value. With such plants, nutritional studies such as conversion and feeding experiments with laboratory or farm animals using in vitro or in sacco measurements may assume a much more important role in confirming that the changes produced were those, and only those, intended. With some constructs, studies of this sort could replace the concept of substantial equivalence with one of nutritional equivalence.

Nutritional evaluation of transgenic crops used for human food and animal feed is an important aspect of safety evaluations in the determination of substantial equivalence. The evaluation includes compositional analysis, nutrient utilization ratio, dietary status in different population. Compositional analysis of genetically modified (GM) crops has continued to be an important part of the overall evaluation in the safety assessment program to identify similarities and potential differences between the GM product and its conventional counterpart (Brune et al. 2013).

The nutrient content in crops is influenced by many factors such as cultivar, origin, and picking time. Therefore, in addition to comparison with non-GMF foods, data on nutrients in GMF are also compared with those in OECD and ILSI. The evaluation encompasses proximate analysis of macro-nutrients (protein, starch, fat, ash, carbohydrates, moisture, fiber, amino acids, fatty acids), analysis of micronutrients (minerals, vitamins), and analysis of inherent toxins such as phytic acid, erucic acid, gossypol, glucosinolates, endogenous allergens, anti-nutrients such as phytic acid, lectin, and trypsin inhibitors, and secondary metabolites. For specific crop plants, Consensus Documents with lists of parameters to be measured have been developed by the OECD. Validated analytical methods for each compound (targeted analysis) should be used. Studies using compositional analysis have been noticed by Tang, Li, and Xu (Li et al. 2007, 2008; Tang et al. 2006, 2013; Xu et al. 2009). Compositional analysis is the cornerstone for the nutritional assessment of plants modified for improved nutritional characteristics and has been used widely both at home and abroad. Numerous papers comparing the composition of GM plants to their near-isogenic counterparts have been published (Gayen et al. 2013, 2016; Hamilton et al. 2004; Herman et al. 2004, 2007, 2010; Jiao et al. 2010; Jiménez et al. 2009; Junhua et al. 2005). Results emphasize that the GM and non-GM comparators are of similar composition. The work conducted by Ridley et al. provides an excellent example of the extensive compositional analyses conducted when comparing the grain and forage component of HT maize (NK603) with its near-isogenic counterpart and a number of commercially grown varieties (Ridley et al. 2002). Compositional equivalence between the GM and non-GM plants was clearly demonstrated. Even though some differences between the GM material and its near-isogenic counterpart were statistically significant, the values fell within the range of currently available commercial varieties and those reported in the literature. Examples of GM crops for which compositional equivalence to a non-GM comparator has been investigated are also presented elsewhere. A crop composition database for conventional varieties of soybean, maize, and cottonseed has been developed by ILSI (http://www.cropcomposition.org/) (Ridley et al. 2004). One of the primary focuses of comparative assessment has been the thorough evaluation of crop composition (Herman et al. 2009). Typically, 50 to 100 or more individual components are determined for each crop variety; the selection of analytes follows the recommendations of OECD Consensus Documents that describe key nutrients, anti-nutrients, and other compounds of biological interest for many crops. This evaluation of crop composition is sometimes referred to as "targeted analysis." It should be noted that the "targets" of the analysis typically comprise more than 95% of the crop composition (Herman et al. 2009).

2.1.2 *Nutritional Components Analysis of GMOs*

2.1.2.1 Test Basis for Compositional Analyses

Compositional analyses, including protein, fat, fiber, ash, moisture, vitamins B_1, B_2, and E, fatty acids, total carotene, and anti-nutrients (phytate, phosphorus), were conducted in accordance with methods that the Association of Analytical Communities (AOAC) and the American Oil Chemists Society (AOCS) recommended abroad. However, detection methods of food components in the national standard were commonly used in China. For a method without formal rules, use of the OECD file reference method could be recommended. Table 2.1 (following) shows some test bases.

Evaluation of Main Nutrients in GMOs

Protein

Proteins are of special concern in nutritional assessments because they could act as toxins (e.g., phytohemagglutinin), anti-nutrients (e.g., protease inhibitors), or allergens or may be involved in their synthesis. Moreover, in contrast to the genome, which is constant for an organism, the proteome is highly dynamic and depends on the cell cycle, environmental influences, and tissue or cell types (Pastorello et al. 2000). Detection and quantification of novel proteins expressed in GM crops are accomplished by immunoassays. The two most applicable immunoassays concerning large-scale GM detection are lateral flow strip tests and enzyme-linked immunoassay (ELISA). Commercial varieties of both tests are available for the detection of numerous GM crops expressing proteins that confer insect resistance or herbicide tolerance. Although Western blots are highly sensitive and have limits of detection (LODs) of 2.5 g/kg for Roundup Ready soybean seeds and 10 g/kg for toasted soybean meal (Rogan et al. 1999), the time and cost requirements of this type of assay limit its usefulness. Lateral flow strips utilize a double-antibody system to detect novel plant proteins in bulk and some processed feeds. The strips contain excess antibodies (Ab) coupled to a color reagent and are specific to a recombinant protein. Lateral flow strips are highly sensitive and some commercial varieties can detect as little as 1 GM maize kernel in 800 or 1.25 g GM/kg. Although they are qualitative in nature and provide only a positive or negative outcome, results can be obtained in less than 10 min and can be performed easily in the field. They can be used for

Table 2.1 National standard methods in China

Test parameter	Reference standard	Standard name
Moisture	GB/T 5009.3-2010	Determination of moisture in foods
Ash	GB/T 5009.4-2010	Determination of ash in foods
Fat	GB/T 5009.6-2003	Determination of fat in foods
Protein	GB/T 5009.5-2010	Determination of protein in foods
Crude fiber	GB/T 5009.10-2003	Determination of crude fiber in vegetable foods
Fatty acid	GB/T 17377-2008	Animal and vegetable fats and oils: analysis by gas chromatography of methyl esters of fatty acids
Amino acid	GB/T 5009.124-2003	Determination of amino acids in foods
Fe, Mg, Mn	GB/T 5009.90-2003	Determination of iron, magnesium, and manganese in foods
K, Na	GB/T 5009.91-2003	Determination of potassium and sodium in foods
Ca	GB/T 5009.92-2003	Determination of calcium in foods
Zn	GB/T 5009.14-2003	Determination of zinc in foods
Se	GB/T 5009.93-2003	Determination of selenium in foods
Cu	GB/T 5009.13-2003	Determination of copper in foods
P	GB/T 5009.87-2003	Determination of phosphorus in foods
Phytic acid	GB/T5009.153-2003	Determination of phytic acid in vegetable foods
Retinol	GB/T 5009.82-2003	Determination of retinol and tocopherol in foods
Tocopherol	GB/T 5009.82-2003	Determination of retinol and tocopherol in foods
Thiamine	GB/T 5009.84-2003	Determination of thiamine (vitamin B_1) in foods
Riboflavin	GB/T 5009.85-2003	Determination of riboflavin in foods
Vitamin B_6	GB/T 5009.154-2003	Determination of vitamin B_6 in foods
Vitamin B_{12}	GB/T 5413.14-2010	Determination of vitamin B_{12} in foods for infants and young children; milk and milk products

quality control to quickly ascertain which shipments of plant materials contain an engineered trait. However, test strips do appear to be subject to user error. A study conducted to examine the ability of test strips to detect GM soybean in a grain-handling facility reported substantial variation in detection, depending on the concentration of GM soybean in the sample (Fagan et al. 2001). When unknown samples contained 0–10 g GM/kg, the frequency of false positives was 0.067, whereas the frequency increased to 0.223 when the samples contained 0–100 g GM/kg. The higher rate of false-positives for the 100 g/kg samples was likely the result of cross-contamination. On the other hand, false-negatives occurred at frequencies of 0.677 and 0.682 when the actual GM content was 5 and 10 g/kg, respectively. However, at a content of 100 g GM soybean/kg, all samples were correctly identified as GM. The study highlights the importance of training field personnel in the proper techniques to avoid contamination, which are usually in place within laboratory settings. In contrast to lateral flow strips, ELISA can provide quantitative data. An international study involving 38 laboratories showed the capability of ELISA to detect GM soybean expressing the recombinant protein CP4 EPSPS (Lipp et al. 2000). The experiment was designed to determine if the Roundup Ready soybean could be detected above or below a threshold set at 20 mg/g. The test samples contained GM soybean powder at 0 to 20 mg/g (in conventional soybean powder). The ELISA had a LOD of approximately 3.5 mg/g. When the sample concentration was less than 20 mg GM plant/g, the ELISA identified the sample as being less than 20 mg/g with a 99% confidence level. In addition, quantitative use of the assay resulted in good repeatability (R.S.D.r = 7%) and reproducibility (R.S.D.R = 10%). Both R.S.D.r and R.S.D.R are measurements of precision within the same laboratory and between laboratories, respectively. Assays for quantification of molecules within biological matrices are generally given an acceptable limit of 15% R.S.D. at a particular concentration level unless the concentration is at the lower limit of quantification, in which case it should not exceed 20% R.S.D. Although these immunochemical tests can provide useful data on identity preservation and screening for the presence of GM crops in test samples, their usefulness for detecting specific GM events is limited. For example, Bt176, Bt11, and MON810 events all result in expression of the Cry1Ab protein, to varying degrees. Thus, protein (Cry1Ab) detection will not indicate the concentrations of these individual GM crops within a mixture. In addition, the rates of expression of novel proteins will vary among tissues of a plant and with the maturity of the plant (Stave 2002). In certain applications, therefore, use of protein standards diluted on a w/w basis will not yield useful information.

However, it is not enough to know only the protein content. For nutritional evaluation, other information about the protein is also needed. Proteomics has been used as a complementary analytical tool for improving the concept of substantial equivalence.

Cao et al. evaluated the safety of Cry1C protein from genetically modified rice using proteomics according to the National Standards of P.R. China for a New Food Resource. The Cry1C fusion protein was expressed and identified by Xu et al. (2009). The protein was expressed in *Escherichia coli* after incubation at 37 °C and a 4-h induction. The bacteria were fragmented by ultrasound and then purified using a His TrapTM FF crude column (Amersham Biosciences, Uppsala, Sweden). After purification and refolding, the protein was identified by Western blot and liquid chromatography–mass spectrometry (LC-MS/MS). The bioinformatics search for allergen sequence matches was conducted according to Goodman et al. (2007). Cao et al. searched for 80 amino acid alignments, looking for identities greater than 35% on AllergenOnline (http://www.allergenonline.org/, version 9.0). Cao et al. also searched for eight exact amino acid matches on the same web site. A sequence similarity search was conducted of the amino acid sequence of the Cry1C protein using BLASTP 2.2.20 with an E-score cutoff of 1.0, according to Delaney et al. (2008b). A total of 8,996,415 sequences were searched. Sequence annotations of proteins with similarity to the Cry1C protein sequence were inspected manually to identify known allergens or toxic proteins. The Cry1C protein was dissolved in 20 mM Tris–HCl and 5 mM EDTA buffer at a concentration of 0.1 mg ml^{-1} in 1.5 ml microcentrifuge tubes. The stability was assessed at a temperature of 100 °C for periods of 10, 30, and 60 min. The assay was terminated by placing the sample tubes on ice and adding SDS-sample buffer [50 mM Tris–Cl, 8% sucrose, 2% sodium dodecyl sulfate (SDS), with 5% 2-mercaptoethanol, and 0.02% bromophenol blue]. Control samples subjected to no incubation (kept at 4 °C) were also prepared (Hérouet et al. 2005). The resulting proteins were analyzed by SDS–polyacrylamide gel electrophoresis (PAGE) and Western blot (anti-Cry1C rabbit polyclonal antibody was used in Western blot: 1:10,000). The results showed that the Cry1C protein had no sequence homology with any known allergens or toxins by bioinformatics analysis and had no mutagenicity potential.

Comparative proteomics has been used as a tool to analyze differences in food (Jin et al. 2014). Comparisons based only on the centesimal composition are not sufficient to answer questions about the differences between foods. The proteome is defined as the entire complement of a genome and the result of genetic expression, ribosomal synthesis, and proteolytic degradation. Barbosa et al. (2012; Lehesranta et al.2005) have found differences between GM and non-GM lines in potato and soybean seeds at the proteomic level, considering only genetic modifications.

Wang et al. analyzed the proteomic profiles and nutritional composition of transgenic rice seeds containing the Cry1Ab/Ac protein to assess the safety of these transgenic seeds (Wang et al. 2012). Cry1 is a group of relatively highly homologous genes that exhibit specific toxicity toward Lepidoptera. The Cry protein encoded by the Cry1ab/ac gene: a fusion of cry1ab (GenBank accession no. X54939) and cry1ac (GenBank accession no. Y09787) into a single gene is highly toxic to *Chilo suppressalis*, *Scirpophaga incertulas*, and *Cnaphalocrocis medinalis*, three important lepidopteran pests of rice (Xu et al. 2009). The increasing use of transgenic plants has led to a greater number of animal and human transgenic consumables, provoking strong demand for appropriate risk assessment for the safety of transgenic plants (Davies 2010; Llorente et al. 2011). However, concerns are seldom reported regarding the analytical procedures used to detect the unintended effects of transgene. Thus, to assess the existing potential safety hazards caused by unintended effects, proteomics was used as a complementary analytical tool. The proteomic profiles of a transgenic rice variety (cry1ab/ac) and its isogenic controls (WT01 and WT02) in two different growth environments (T01 and T02) were analyzed. To obtain the protein, rice samples were extracted with 10% TCA. The samples were kept at −20 °C for 3 h and then centrifuged. The pellets were then washed three times with acetone. The dried pellets were dissolved in lysis buffer containing 7 M urea, 2 M thiourea, 4% (w/v) CHAPS, 1% (w/v) IPG buffer (pH 3–10), 40 mM Tris–HCl (pH 8.8), 65 mM DTT, 1 mM PMSF, and 0.5% (v/v) protease inhibitor cocktail. The protein solution was sonicated intermittently for 100 s and then centrifuged to remove the insoluble material. To analyze the proteomic profiles, the protein concentrations were measured by Bradford assay. SDS-PAGE and two-dimensional (2D) electrophoresis were used to separate proteins, and MALDI-TOF/TOF-MS/MS experiments were carried out for protein identification. The results showed that a total of 900 spots were obtained in each sample. Most of the proteins had molecular weights between 14 and 94 kDa. In each sample, albumins and globulins were mostly distributed in the pH range 4–7. By comparing proteomic profiles, 21 proteins were up- or downregulated as a consequence of environmental influence (WT01 vs. WT02). Similarly, 20 to 22 protein levels were differentially modulated in transgenic rice seeds in comparison to their nontransgenic counterparts (T01 vs. WT01; T02 vs. WT02).

Vidal et al. (2015) also used comparative proteomics to evaluate differences between two types of crop samples: transgenic (MON810 event with the Cry1Ab gene, which confers resistance to insects) and nontransgenic maize flour commercialized in Brazil because these flours are consumed at high rates around the world. Protein extraction was carried out according to the method proposed by Sussulini et al. (2007). To separate the protein, 2D electrophoresis was used, which allows the analysis of protein maps with a high protein resolution. To ensure repeatability and reproducibility (and, thus, the homogeneity of the flour samples), six 2D–PAGE gels were run for each sample before the difference gel electrophoresis (DIGE) procedure. After image analysis, 2D gel electrophoresis was performed as previously described: the gel obtained for a maize flour sample was stained with colloidal Coomassie, and the spots of interest were excised manually and subjected to the identification of protein species by nESI-QTOF MS/MS analysis. The results showed that the 2D DIGE technique revealed 99 differentially expressed spots. The abundance of protein differences between the transgenic and nontransgenic samples could arise from genetic modification or as a result of an environmental influence pertaining to the commercial sample. The major functional category of proteins identified was related to disease or defense and, although differences were observed between samples, no toxins or allergenic proteins were found. This study showed that there are differences between T and NT maize flours commercialized in Brazil at the proteomic level. These differences can be attributed to genetic modification or environmental influences.

Lipids and Fatty Acids

A large body of evidence suggests that there is a significant underconsumption of omega-3, long-chain, polyunsaturated fatty acids (LC-PUFAs) and that this is the cause of multiple chronic diseases and developmental aberrations. The scope for increasing omega-3 LC-PUFA consumption from seafood is limited because global wild fisheries are unable to increase their harvests, and aquaculture fisheries currently rely on wild fisheries as a source of LC-PUFAs. Agricultural production of oils is highly efficient and has the potential to be sustainable. The transfer of genes from marine microalgae and other microorganisms into oilseed crops has shown that the production of terrestrial omega-3 LC-PUFA oils is indeed possible. The specifications of these oils or whole seeds for use in human nutrition and in Atlantic salmon (*Salmo salar*) aquaculture have been discussed, and it was found that the nutrient value of omega-3 LC-PUFA from seaweed was consistent with the result of that from genetically modified oil crops (Robert 2006).

Forty-four fatty acids, which were analyzed as methyl esters by GC/MS in scan mode, have been determined in genetically modified corn and soybean seeds by Jimnez et al. (Jiménez et al. 2009). Their relative concentrations have been compared with those of isogenic lines grown in the same conditions. Studied compounds comprised saturated and unsaturated fatty acids, including cis/trans isomers and minor fatty acids. The profile of major and minor fatty acids is similar in the isogenic and its transgenic counterpart; it had not been found as a specific isomer (cis/trans, double-bond position) or a compound that denotes specifically the genetic origin of the crop, but for some sample pairs, significant differences have been found in relationship to the percentage of certain acids. With irrelevant exceptions, trans fatty acids have not been detected in the samples. The concentration of some acids is significantly different for some transgenic lines. The highest observed differences for major acids were 4.1% in corn and 4.8% in soybean; the variations were relatively small and within tolerable limits. The concentrations of certain long-chain acids such as 25:0, 26:0, 27:0, and 28:0 were higher in some isogenic lines, whereas the concentrations of some short-chain acids such as 6:0, 8:0, 9:0, 10:0, and 12:0 were higher in transgenic lines.

Inositol polyphosphate kinase (IPK; EC 2.7.1.151) participates in inositol metabolism, calcium signaling, stress response, gene transcription, and other physiological and biochemical processes. In the present work, ThIPK2 (an inositol polyphosphate kinase from *Thellungiella halophila*) was transferred into soybean through an *Agrobacterium tumefaciens*-mediated transformation. Transgenic plants with the ThIPK2 gene displayed tolerance to water deficit, salt, and oxidative balance compared to nontransformed controls. Furthermore, the expression of ThIPK2 altered the ratio of fatty acid components in soybean seeds, resulting in an increase of oleic acid (C18:1). Also, seed size was increased in transgenic plants. These findings might be available to improve soybean seed quality and soybean yield (Liu et al. 2012b).

Minerals

The mineral content will differ between some transgenic and nontransgenic crops, but the difference has no significance. Genetically modified, low phytic acid strains of maize were developed to enhance mineral absorption. Mendoza et al. evaluated the mineral and phytic acid content of a low phytic acid "flint" maize (LPM, the lpa-1-1 mutant) and its parent, wild-type strain (WTM) and measured iron absorption from tortillas prepared with each type of maize and from a reference dose of ferrous ascorbate. Mineral and phytic acid content was measured by standard techniques. Iron absorption from the tortillas was evaluated by using the extrinsic tag method and was measured as the incorporation of radiolabeled iron into the red blood cells of 14 non-anemic men 2 weeks after intake. The results showed that the phytic acid content of LPM was 3.48 mg/g, approximately 35% of the phytic acid content of WTM; concentrations of macronutrients and most minerals were not significantly different between strains. Iron absorption results were adjusted to 40% absorption of ferrous ascorbate. Iron absorption was 49% greater from LPM (8.2% of intake) than from WTM (5.5% of intake) tortillas ($p < 0.001$, repeated-measures analysis of variance). In conclusion, consumption of genetically modified, low phytic acid strains of maize may improve iron absorption in human populations that consume maize-based diets (Mendoza et al. 1998).

Mendoza et al. also assessed iron absorption from porridges prepared from the same LPM and WTM, both of which were fortified with either ferrous sulfate or sodium iron EDTA. Results showed that no significant effect of phytate content on iron absorption was found when porridge was fortified with either sodium iron EDTA or ferrous sulfate. Fractional absorption of iron from WTM porridge fortified with sodium iron EDTA (5.73%) was 3.39 times greater than that from the same porridge fortified with ferrous sulfate (1.69%). Fractional absorption of iron from the sodium iron EDTA-fortified LPM porridge (5.40%) was 2.82 times greater than that from LPM porridge fortified with ferrous sulfate (1.91%) ($p < 0.0001$ for both comparisons, repeated-measures analysis of variance). Thus, the previously identified benefit of LPM was no longer detectable when maize porridge was fortified with additional iron. It may be concluded that iron was absorbed more efficiently when the fortificant was sodium iron EDTA rather than ferrous sulfate, regardless of the type of maize (Mendoza et al. 2001).

Iron deficiency is estimated to affect about 30% of the world population. Iron supplementation in the form of tablets and food fortification has not been successful in developing countries, and iron deficiency is still the most important deficiency related to malnutrition. Drakakaki et al. have generated transgenic maize plants expressing *Aspergillus* phytase either alone or in combination with the iron-binding protein ferritin to increase iron availability. They found that phytase in the maize seeds was associated with increased cellular iron uptake, and that the rate of iron uptake correlated with the level of phytase expression regardless of the total iron content of the seeds. They also investigated iron bioavailability under more complex meal conditions by adding ascorbic acid, which promotes iron uptake, to all samples. This step resulted in a further increase in iron absorption, but the effects of phytase and ascorbic acid were not additive. In conclusion, the expression of recombinant ferritin and phytase could help to increase iron availability and enhance the absorption of iron, particularly in cereal-based diets that lack other nutritional components (Drakakaki et al. 2005).

Lucca et al. also presented experiments that aim to increase the iron content in rice endosperm and to improve its absorption in the human intestine by means of genetic engineering. We first introduced a ferritin gene from *Phaseolus vulgaris* into rice grains, increasing their iron content up to 2 fold. To increase iron bioavailability, we introduced a thermotolerant phytase from *Aspergillus fumigatus* into the rice endosperm. In addition, as cysteine peptides are considered major enhancers of iron absorption, we overexpressed the endogenous cysteine-rich metallothionein-like protein. The content of cysteine residues increased about 7 fold and the phytase level in the grains about 130 fold, giving a phytase activity sufficient to completely degrade phytic acid in a simulated digestion experiment. This rice, with higher iron content, rich in phytase and in cysteine peptide, has great potential to substantially improve iron nutrition in those populations where iron deficiency is so widespread (Lucca et al. 2002).

Shireen et al. investigated the bioavailability of calcium from two varieties of sweet potatoes, and supplementation of sweet potatoes with soy flour was investigated in hamsters using plasma calcium concentration and femur calcium content as indicators. Five different diets were fed to five groups of animals for 28 days. There was no significant difference in plasma calcium concentrations of hamsters in all the diet groups. However, the femur calcium content of hamsters with transgenic sweet potato flour (TSPF) and parent nontransgenic (from which transgenic was produced) sweet potato flour (NTSPF) diets was significantly higher than that of the transgenic sweet potato flour supplemented with soy flour (TSPF + SF) and parent nontransgenic sweet potato flour supplemented with soy flour (NTSPF + SF) diets. The relative bioavailability of calcium from the control (100%), TSPF + SF (30%), NTSPF + SF (23%), TSPF (57%), and NTSPF (46%) indicated that sweet potatoes could be the better source of calcium; however, supplementation with soy flour might reduce the bioavailability of calcium (Shireen et al. 2002).

Numerous papers that reported evaluation of the contents of the main nutrients of GM plants compared to their near-isogenic counterparts have shown no significance between GM plants and non-GM comparators (Gayen et al. 2013, 2016; Herman et al. 2004, 2007, 2010; Jiao et al. 2010; Jiménez et al. 2009).

2.1.3 *Research Review of Nutritional Evaluation of Common GM Crops*

Following the first commercial planting of a transgenic crop in 1996, agricultural biotechnology has been rapidly adopted in many countries of the world (James 2015). The majority of existing commercial genetically modified (GM) crops have been designed to express transgenic proteins with a limited spectrum of biological activity (e.g., insect resistance, herbicide tolerance). To date, most GM crops exhibit traits such as insect resistance and herbicide tolerance, as either single or stacked traits. Now, maize, soybeans, oilseed rape, and cotton have become widely genetically modified crops throughout the world.

2.1.3.1 Corn/Maize

In China, maize (*Zea mays*) is an important food crop and feed stuff. Because of the lack of essential amino acids lysine and tryptophan, which human beings and monogastric animals cannot synthesize, the nutritional value of maize is for malnutrition as well as other cereals. (Tang et al. 2013). It is predicted by the Food and Agriculture Organization (FAO) that maize products will be needed to increase 60 million tons in the world by 2030 to satisfy the requirements. Especially in Asia, such as corn for food, the demand will reach 40 tons. The transgenic technology applied to maize breeding may make the traditional maize varieties to obtain insect resistance, herbicide resistance, disease resistance, high lysine expression, high phytase expression, and delayed ripening and other traits, which can improve the nutritional value of corn, increase corn production, and meet yearly corn demand. Genetically modified corn is one of the most widely grown GM crops. The European Commission (EC) has authorized a total of 30 varieties, including 14 stacked GM corn. Up to 2015, for corn alone, about 24 GM may have been commercialized, and this may be the equivalent of more than 2,000 triple-stacked lines and more than 12,000 quadruple-stacked lines (Stein and Rodriguez-Cerezo 2009).

Recently, much plant biotechnology has been applied to improve the lysine content of maize seeds (Altenbach et al. 1992; Ohtani et al. 1991; Torrent et al. 1997; Yu et al. 2004). In a recent study, the sb401 gene that exists in potato (*Solanum berthaultii*) was successfully integrated into the maize genome, leading to the increase of lysine content and total protein level in maize seeds. The sb401 gene encodes a 240-amino-acid pollen-specific protein with 40 lysine residues (Yu et al. 2004).

As the starting point of safety assessment for genetically modified (GM) crops, compositional analysis, in accordance with the concept of substantial equivalence (FAO/WHO 1996; OECD 1993a) between transgenic lines and their counterparts, showed that the content of most key nutrients was not changed significantly with the insertion of foreign genes (Chrenková et al. 2002; Hammond et al. 1996; Kuiper et al. 1999; Noteborn et al. 2000; Ridley et al. 2002). For the nutritionally improved genetically modified food (GMF) modified in nutrients or flavor, however, the nutritional profiles may be affected compared with their near-isogenic line.

In the study by Tang et al. (2013), the transgenic lysine-rich maize lines (an inbred line 642 and a hybrid line Y642) were compared with a conventional quality protein maize Nongda 108 for their similar characteristics of lysine and protein level on maize, and the history of safe use (more than 20 years) of Nongda 108 in China (Fan et al. 2004). Subsequently, the composition data of GM maize were compared with the data set from the ILSI crops composition database (International Life Sciences 2004) to provide a scientific comprehension of the comparison results. The study was focused on the effect of the insertion of the lysine-rich gene into maize on nutrients, amino acid balance, and the distribution of protein solubility fractions to demonstrate whether the introduced lysine-rich protein was nutritionally available to human and mono-gastric animals.

The experimental maize under study was grown in the experimental field of Beijing farm. The transgenic hybrid Y642, its parental inbred line 642, and conventional maize Nongda 108 were planted in adjoining fields under identical climate conditions. After harvest, samples were collected from five points on each diagonal of the fields. The seeds were dried, milled into fine flour to pass a 40-mesh sieve, and stored in polyethylene bags at 4°C. For the nutrient composition analyses, three random samples were selected from each type of maize, and each sample was analyzed in triplicate.

Compositional analyses including protein, fat, fiber, ash, moisture, vitamins (vitamin B_1, vitamin B_2, and vitamin E), fatty acids, total carotene, and anti-nutrient (phytate phosphorus) were conducted following the relative methods of the Chinese Standard GB/T 5009–2003. Amino acids were analyzed according to the method of Herman. Selected minerals (sodium, magnesium, calcium, phosphorus, potassium, copper, zinc, manganese, iron) were determined by inductively coupled plasma optical emission spectrometry (ICP-OES) according to AOAC method 999.1117. All compositional analyses were performed at the Testing Center for Agricultural Products Quality (Ministry of Agriculture of China). Different solubility fractions of maize protein (albumin, globulin, zein, G1-glutelin, G2-glutelin, G3-glutelin) of 642, Y642, and Nongda 108 were extracted stepwise by a series of solvents according to the method of Aisha et al. with some modifications. The proximates of transgenic inbred maize (642) and hybrid line (Y642) and the conventional QPM (Nongda 108) indicated that the moisture and fiber levels of the two transgenic varieties were not significantly different from Nongda 108. The protein, lipid, ash, and carbohydrate content of 642 was significant different from those of Nongda 108, but for the hybrid transgenic line Y642, only protein content was significantly higher than the QPM Nongda 108. This result, in agreement with prior research (Fufa et al. 2003; Yu et al. 2004), indicated that the protein content of maize with introduction of the sb401 gene was higher than conventional quality protein maize (QPM). The statistic changes of transgenic maize did not go beyond the range of the ILSI composition database apart from the carbohydrate content of the three varieties, all lower than the minimum data.

The amino acids analysis indicated that, compared to Nongda 108, the 17 tested amino acids of 642 and Y642 all showed higher levels, resulting in higher content of total essential and nonessential amino acids, which suggested a greater nutritive value of the transgenic maize. There were nine kinds of amino acids of 642 maize and eight kinds of Y642 maize significantly different from those of Nongda 108. Most of the contents of amino acid of transgenic maize lines fell within the range of the database; nevertheless, three of them (threonine, glycine, and phenylalanine) were higher than the reported range. According to the proposed amino acids scoring pattern of the WHO Technical Report 935, the first three limiting amino acids of all the maize varieties are lysine, leucine, and valine. The amino acid score of Nongda 108 was 0.61. The amino acid scores of high-lysine transgenic maize 642 and Y642 were improved to 0.76 and 0.70, respectively, which were notably improved compared with Nongda 108.

The results of minerals analysis showed that compared to Nongda 108 the content of iron, zinc, and manganese of inbred line 642 and of zinc of the hybrid line Y642 was significantly higher. All the data of transgenic maize were covered by the range of the ILSI database, except for the lower content of phytate phosphorus, which may be the result of different measurement methods.

The vitamin B_1, vitamin B_2, vitamin E, and total carotene content of 642 were all higher than Nongda 108; especially, the content of vitamin B_2 and total carotene of 642 were significantly increased. The vitamin B_2 of Y642 was higher than the conventional QPM, but other nutritional compositions were comparable to Nongda 108.

Maize is a protein source for both human beings and livestock. Grain protein can be separated into six fractions according to their solubility, named albumin, globulin, zein, G1-glutelin (or zein-like protein), G2-glutelin (glutelin-like protein), and G3-glutelin (true glutelin). In comparison with the conventional QPM, the water-soluble protein of transgenic maize lines were significantly increased by 60%. But as one of the easily digested protein fractions, the salt-soluble protein of transgenic maize was just about 20% of the content of conventional QPM. As a result, the sum content of the albumin and globulin of transgenic maize was comparable to that of the conventional one. The other kind of nutritionally available protein G2-glutelin (glutelin-like protein) of Y642 was about 12 fold that of Nongda 108, which was statistically significantly higher. The G3-glutelin (true glutelin) of the transgenic lines was slightly higher than that of the conventional line. The sum of the nutritional available glutelin of transgenic maize was 20.8%, which was higher than that of the conventional one. The alcohol-soluble proteins including zein and G1-glutelin of the two cultivars were comparable. The higher lysine content appeared to be a result of a generally higher content of total protein or free amino acids.

The comparator applied in this study was in accordance with a history of safe use and a similar compositional characteristic to the GM maize. Also, the review of comparison results against the background of values of maize showed that most of the composition values of the transgenic maize fell within the range of the ILSI database. The nutritional composition assessment is the starting step of the GM safety assessment, but not the endpoint. In the next step, the protein digestibility corrected amino acids score (PDCAAS) is detected to evaluate the bioavailability of the increased protein and amino acids.

The transgenic lysine-rich maize showed a good balance of amino acids with improved total content of essential amino acids and will have a protein-sparing effect (Brandt et al. 2000). Also, because of the higher protein level, every kind of amino acid of the transgenic maize was higher than that of conventional QPM Nongda 108. It was demonstrated that the insertion of the lysine-rich gene did not affect the amino acid compositional balance, and the transgenic lysine-rich maize lines were proved to be superior, with higher content of lysine and amino acid score, to the conventional QPM widely cultivated in China. The analysis of mineral elements demonstrated that transgenic lysine-rich maize was higher in phosphorus than Nongda 108 but lower in phytate phosphorus, which might indicate more usable phosphorus in transgenic lysine-rich maize. This property could promote the utilization of phosphorus, iron, and zinc (Mendoza et al. 2001; Spencer et al. 2000a). These results are meaningful for the nutritional improvement of maize.

In the study by Rayan et al., three different genetically modified types of corn, namely, NK603 (RR corn), MON88017 × MON810 (VT3 corn), and MON89034 × MON88017 (VT3 PRO corn), were compared with a non-GM hybrid corn control (B73 × MO17). The compositional analyses were evaluated through a series of chemical analyses following the guidelines of the OECD consensus document. This analysis functions to assess whether any significant changes in the compositional and nutritional profile of the crop resulted from the insertion of new traits or a combination between several traits into the corn genome or from the presence of the new proteins. Proximate composition of the corn samples such as moisture content, ash, crude fat, protein, crude fiber, carbohydrate, and minerals was determined according to the standard Association of Official Analytical Chemists (AOAC) procedures. The fatty acid analysis was based on the American Oil Chemists Society (AOCS) method Ce 2–66. The results showed that moisture content ranged from

10.5% to 11.17%. It is well known that the corn grain should be at stable (equilibrium) moisture content of 12%. The protein content of the analyzed samples differed significantly ($p < 0.05$). The VT3 PRO sample had the highest protein content (12.05%) and the RR sample had the lowest (8.55%). The results fell within the range of 6% to 13.6% protein, which has been reported for several corn varieties (Ridley et al. 2002). Similarly, there was a significant increase in the crude fat content of the VT3 PRO sample compared with the other samples. The crude fat content values were 3.5%, 3.55%, 3.7%, and 4.3% for non-GM, RR, VT3, and VT3 PRO corn samples, respectively. These results were similar to those of Wu et al. (2006), who found that fat content ranged from 3.2% to 6.3% for different corn varieties. It is important to note that high fat content may pose a problem for the possible shelf-life of milled grains, resulting in auto-oxidation and a rancid flavor (Cuevas-Rodríguez et al. 2006). Crude fiber content ranged between 2.3% and 2.65% for the investigated corn samples. These results were within the range (1–2.54%) reported in the literature for corn lines with a similar genetic background (Marzok 2004; Wu et al. 2006). Ash levels reported in this study were 1.26%, 0.989%, 0.989%, and 0.989% for non-GM, RR, VT3, and VT3 PRO corn samples, respectively. These results were similar to those observed by Marzok (2004), who found that ash content for isogenic and transgenic corn was 1.30% and 1.36%, respectively. Carbohydrates are a class of energy-yielding substances that also influence the utilization of corn. Carbohydrate content, which was calculated by difference, was 84.1%, 84.4%, 82.1%, and 80.0% for non-GM, RR, VT3, and VT3 PRO corn samples, respectively. These results were higher than that reported by He et al. (2008), who reported that carbohydrate content for transgenic and nontransgenic corn samples was 77.75% and 77.41%, respectively. Also, Appenzeller et al. (2009) reported that the carbohydrate content for the control and a genetically modified maize hybrid was 75.9% and 73.8%, respectively. However, these results were within the literature range reported by Ridley et al. (2002) and OECD (2002). No significant differences were observed in Ca, P, Na, Zn, Fe, Mn, and Mo values among the investigated samples, whereas small but significant differences in Mg, S, and Cu content were observed between the samples. The values of all amino acids were generally similar, although there are some significant differences in leucine, glutamic, aspartic, glycine, and alanine values. Other amino acids were not significantly different. All fatty acid values were either within ranges found in the published literature or within the range of historical conventional control values determined from previous studies (Ridley et al. 2002). Statistically significant differences between GM corn samples and the non-GM control were observed with respect to levels of palmitic, stearic, arachidic, palmitoleic, oleic, linoleic, linolenic, and gadoleic acids. It is worth mentioning that there was a clear increase in the content of all fatty acids in the GM corn varieties when compared with the control sample, which may result from the synergistic effect of introducing new traits into varieties of corn as a result of the genetic modifications.

In conclusion, although there are some differences between the GM corn samples and non-GM control in some biochemical components, they are unlikely to be biologically significant, as they are well within the range of values reported in the literature. Therefore, on the basis of the principle of substantial equivalence as proposed by the World Health Organization (WHO) and the United Nations Food and Agricultural Organization (FAO), all data indicate that the GM corn samples analyzed in this study are compositionally equivalent to, and as nutritious as, non-GM corn.

Rice

Rice (*Oryza sativa* L.) is the most important staple crop on which half of the world's population, more than 3 billion inhabitants of the planet, depend, providing 21% of the energy for human beings. More than 90% of the world's rice is grown and eaten in Asia, providing at least 30% of daily caloric intake. China is the world's largest rice-producing country in Asia with 186 million tons, accounting for 30% of world output (Hareau 2006; OECD 2004). Advances in rice biotechnology have opened up new ways for the improvement of agronomic traits, including pest and disease control, abiotic stress resistance (drought, herbicide, low temperature, salt, high temperature), and increased nutritional value of rice variety development (vitamins, fatty acids, and other micronutrients) (Bajaj and Mohanty 2005). The application of rice breeding biotechnology in easing the pressure caused by the expansion of the world's population, which reduces arable land, water, and other resources, to strengthen rice production provides such an effective and promising way. As a staple food, the safety assessment of transgenic rice should be conducted before full commercialization. Compositional analysis has been done to determine whether the insertion of the foreign gene in rice affects the quality of food and feed (Han et al. 2005; Momma et al. 1999; Vasconcelos et al. 2003). These authors found that the nontransgenic and transgenic lines are essentially nutritionally equivalent, and their results agree with those of our current studies.

A herbicide-tolerant rice nutritional composition (Bar 68-1) and its nontransgenic counterpart were compared by Li et al. (2008). The current development of Bar 68-1 in China is generating two copies of the genome bar gene by particle bombardment by inserting a conventional isolate (D68) isolated from *Streptomyces* strain HP632. The field tests results revealed that it has steady tolerance to glufosinate-ammonium herbicide, which offered an efficient way to control identical physiological characteristics and agronomic performance compared to the nontransgenic control. The usual form of rice for food is brown and milled grain; thus, following the guidelines of the OECD consensus document, much work was conducted to assess the effect of new gene insertion on anti-nutrient content and nutritional composition of brown and milled transgenic Bar 68-1 rice.

To determine the chemical components, the rice samples were ground in a cyclone mill with an 80-mesh sieve. Compositional analyses including fat, protein, ash, fiber, fatty acids, vitamins (vitamins E, B_1, B_2), moisture, and other vitamins (VE, riboflavin, thiamine) were conducted following the related methods of the Chinese Standard GB/T 5009–2003. Dietary fiber (total) was analyzed according to AOAC method 991.43. Selected minerals were determined by inductively coupled plasma optical emission spectrometry (ICP-OES) according to AOAC method 999.11. The data obtained from the detection were analyzed with the proof of safety approach (Hothorn and Oberdoerfer 2006), a modified form of the common *t* test. In practice, first the nontransgenic group means were calculated for each nutritional component. Then, 20% of the means was set as equivalence boundaries according to the Nordic Council of Ministers recommendation. If the 90% confidence interval of the difference was within 20% of the mean value of the respective reference treatment, the two treatments would be considered equivalent in these studies (nontransgenic plants). Given the natural variation of values not in the same lines, the results of Bar 68-1 grains were compared not only to the D68 line used in the experiment but also to values from other reported commercial lines.

No significant differences were found in the levels of lipid, protein, ash, fiber, and carbohydrate between the D68 and Bar 68-1 rice samples. The content of 17 amino acids was determined by an automatic amino acid analyzer. The percent fraction of a particular amino acid in the total protein was also calculated for convenient comparison with the reference range. The results revealed that the amino acid profiles in the milled and brown rice samples were analogous between the Bar 68-1and D68 both in ratio (% protein) and in content (g/100 g). However, significant differences of methionine and glutamic acid were found in brown rice between Bar 68-1 and D68 in content (g/100 g) but not in ratio (% to protein). The main fatty acids in both D68 and Bar 68-1 are oleic acid (C18:1) and palmitic acid (C16:0). Fatty acid profiles were evaluated in brown and milled rice for Bar 68-1, and no statistically significant differences were found compared to D68 rice in the content of palmitic (C16:0), myristic (14:0), oleic (C18:1), palmitoleic (C16:1), gamma-linoleic (C18:3), arachidic (C20:0), and linolenic (C18:2) acid. Between D68 rice and Bar 68-1, statistically significant differences were observed for iron in both milled and brown rice samples. The other eight mineral contents in D68 and Bar 68-1 rice are similar. In both milled and brown rice, the levels of vitamin B_1 and vitamin B_2 were similar, but the level of VE of D68 rice was significant lower than that of Bar 68-1. All significantly different means for Bar 68-1 rice were not greater than commercial rice reference ranges, and thus these differences were thought to indicate relative safety.

Thus, these results show that the transgenic rice grain nutritional quality was equal to that of the nontransgenic control. After insertion of the bar gene improving weed control, no harmful changes in the nutritional composition of the transgenic rice grain were observed.

Nutritional composition of transgenic rice seeds with Cry1ab/ac genes have been compared with their nontransgenic counterparts (Wang et al. 2012). *Bacillus thuringiensis* (Bt) is an already useful supplement or alternative to synthetic chemical pesticide applications in commercial agriculture (Kuiper et al. 2001). The Cry protein encoded by the Cry1ab/ac gene, a fusion of cry1ac (GenBank accession no. Y09787) and cry1ab (GenBank accession no. X54939) into a single gene, is highly toxic to *Scirpophaga incertulas*, *Cnaphalocrocis medinalis*, and *Chilo suppressalis*, three significant lepidopteran pests of rice (Xu et al. 2009c). Introducing exogenous DNA sequences into the plant genome may influence the regulation of biochemical pathways or other genes. A chief matter of these genetic modifications is the possible appearance of unintended results; for example, the modified metabolism may be affected by interruption of an important regulatory sequences or open reading frames, novel fusion proteins, or other pleiotropic effects that may compromise the safety of the product (Ruebelt et al. 2006). There are other toxic or harmful side effects that include intentional and unintentional toxic new allergens or toxins. It is therefore necessary to evaluate transgenically improved crop safety to address the expected and unexpected effects of GM (Herman et al. 2009; Lehesranta et al. 2005).

Essentially equivalent is an important aspect; component analysis was used in this study. Four random samples [a transgenic rice variety (cry1ab/ac) and its isogenic controls (WT01 and WT02) in two different growth environments (T01 and T02)] were selected for each kind of rice. The fiber, protein, carbohydrate, moisture content, and ash were comparable among all the samples. Although the lipid content in rice harvested at Jingzhou (WT02 and T02) was slightly higher than that at Xuancheng (WT01 and T01), there were no significant differences between the nontransgenic and transgenic groups. No statistically significant differences in the levels of fatty acid, amino acids, anti-nutrients, and micro-nutrients, such as lectin, trypsin inhibitors, and phytic acid, were revealed. The gamma linoleic (C18:3) and oleic acid (C18:1) contents of the WT02 and T02 samples were slightly higher than in WT01 and T01.

Nutritional and anti-nutritive components examined in this study suggest that the nutritional quality of transgenic rice was essentially the same as nontransgenic. When comparing the proteomic profiles, many horizontal and vertical position shifts were observed. A number of seed storage proteins, such as putative globulin proteins and globulins, showed shortened forms with molecular weights significantly lower than the native ones. A possible explanation is that frequent mutations lead to changes in protein molecular weight (Zolla et al. 2008). However, the foregoing results will not be observed in traditional substantial equivalence evaluations. Seed storage proteins that are very rich, including gliadin, albumin, globulin, and glutenin, are important in determining the nutritional value of rice seeds (Wang et al. 2010). The shortened forms of storage proteins may affect intestinal digestion or nutrient value and assimilation of animals and humans in an undetermined manner. Zolla and others (Zolla et al. 2008) found frequent molecular weight changes in the major storage proteins of transgenic maize lines.

In this study, a large number of proteins have been shown to be regulated both by different environments and being genetically modified. Approximately 21 different levels of protein were modulated as a result of environmental effects (WT01 vs. WT02), whereas 20 and 22 differentially expressed proteins were found in transgenic seeds compared to their nontransgenic lines (T01 vs. WT01; T02 vs. WT02). These differentially expressed protein spots were related to glycolysis, starch synthesis, the glyoxylate cycle, and gluconeogenesis. However, no significant differences were observed between the different genes and environmental issues. Thus, we suggest that the presence of differentially expressed proteins between transgenic and nontransgenic lines falls within the natural range of variation. Amazingly, MS/

MS identification of modulated spots revealed that the proteins belonged to several different metabolic pathways. No toxic or new protein unique to individual transgenic lines is observed. The expression of the target Bt gene is not over the limit of detection (LOD) of the method used here. This finding is expected because proteomics is a valid method for comprehensive analyses, but if the level of a target protein is very low it would not detect the content of the protein. Through the work of Corpillo and coworkers, the product of their marker gene (nptII) was only detectable by a highly sensitive immunodetection assay (Corpillo et al. 2004).

We conclude that there is little difference in differential proteomics between transgenic rice seeds and the nontransgenic counterparts. This review also included a comparison of different environmental groups, suggesting that these two single-gene insertions and slightly different growth environments had only a modest effect on the rice proteome. Many environmental elements can seriously affect the growth of rice, including soil, plants, pressure, and climate (weeds and insect pests); these factors then affect the nutritional value and the quality of the crop (Barros et al. 2010). Genetically modified rice and non-GM counterparts are grown in the same local environment, such as soil, water, and climate. However, because the transgenic rice Cry1ab/ac gene is highly toxic to three important lepidopteran pests, it can grow in an ideal environment, nontransgenically slightly different. In other words, transgenic rice grows in different growth conditions, even if they are in the same location. Differences in results of genetic modification can also be affected by environmental changes. More specifically, the effect of environmental growth conditions on the rice proteome was not significantly different from the effect of single-gene insertion. Single-gene insertions may affect only a limited number of proteins; by contrast, more potential protein changes may lead to significant environmental changes.

Because all significant differences imply that transgenic rice is within the reference range of rice, these differences are not considered to be biologically relevant. In addition, differential protein spots have more than one-half abundance monitoring; however, differentially expressed proteins identify storage proteins or proteins involved in essential metabolism, not allergens or toxic proteins. Environmental effects were also not significantly different for transgenic effects. These results thus confirm that the nutritional quality of transgenic rice seeds is virtually equivalent to that of nontransgenic counterparts and that no deleterious changes in the potential unintended effects of the transgene are observed with Cry1Ab/Ac insertion of genes to improve control of insects. In addition, our study approach could provide a more comprehensive assessment of the safety of transgenic crops.

The insect-resistant genes of genetic improvement provide an effective and promising new method to manage insects. Nutrition of insect-resistant rice (IRR) (*Oryza sativa* L., CV two optimal divisions feng 6) was compared with Indica rice (*Oryza sativa* L., CV two optimal 2186) by Li et al. (2007). Insect-resistant rice was developed by Chinese rice scientists with a gene that is a modification of the cowpea trypsin inhibitor gene and *Bacillus thuringiensis* gene (Deng et al. 2003). Live demonstration shows that the total of expressed genes and the cry1Ac gene have different insecticidal mechanisms to improve the rice insecticidal spectrum width and resistance, leading to higher yields, and can reduce pesticide use for an extended period compared to the conventional variety (Zhu 2001). Analysis of IRR and control rice samples showed that in the protein, fat, and fiber content of the main nutrients there were no significant differences between control and IRR rice samples in ash and carbohydrates. Through the use of an automatic amino acid analyzer, the content of 17 kinds of amino acids were determined, for the convenience of comparison and reference range, more than the protein content of each amino acid in the form of expression. The results showed that the amino acid spectrum in brown and white rice samples is very similar between control and transgenic rice content (g/100 g) and the ratio of protein (%). In yield and control of fatty acids, mainly palmitic acid (C16:0), oleic acid (C18:1), and linolenic acid (C18:2) were present. Fatty acid profiles were assessed for brown rice and IRR brown rice: no significant difference was found in the level of myristic acid (14:0) and palmitic acid (0), sixteen carbene (C16:1), eighteen (C18:1), linolenic acid (C18:2), and gamma linolenic acid (C18:3), peanut (C20:0) phase ratio. Statistically significant differences were observed between the Fe IRR rice and the control of the brown rice samples. These differences are not considered to be biologically relevant because all significant differences in the mean of IRR rice are within the reference range of commercial rice. The other eight minerals (sodium, magnesium, calcium, phosphorus, potassium, copper, zinc, manganese) contained in the control and IRR rice are similar. Mineral content did not differ significantly between the two types of rice.

The level of vitamin B_2 in brown rice is the control of the rice, but the content of vitamin B_1 and vitamin E is significantly higher than control genetically modified rice. All want to use a different average to rice trade reference value, so these differences are thought to be not related to the biological phenomenon.

The levels of vitamin B_1 and vitamin K are comparable to that of the parents of all the different grains and the average rice grain of the framework. Such changes could help to improve the usual nutritional value of rice, but the trend requires review. In addition to natural factors, the nutrition of rice is an integral part of the plants (and a concrete value), speaking also for the yidanbaimei inhibitor, IRR parental authority, and the comparison of repeated equivalence, which confirm the quality of the genetically modified rice cereal to be essentially the same as the non-genetically modified food. The genetically modified rice has no changes in the composition of grain shipments.

We have made great progress in the nutritional improvement of genetically modified (GM) crops, including rice. Development of the iron level of rice has been reported with soybean ferritin gene expression by endosperm-specific gluten promoter. Studies show that the levels of iron and zinc are still high in milled and brown rice compared to non-GM rice (IR68144) after different levels of milling (Vasconcelos et al. 2003). Ferritin genes were expressed in the analysis of the components of a genetically modified (GM) event of rice seed endosperm of rice protein (iron) and its correspondence with the same genetic background (non-GMO) evaluation through a series of chemical analyses after the OECD guidelines document to assess the effect of new gene insertion of nutrients and the brown rice content of anti-nutritional factors (Gayen et al. 2013). The value of the direct component ferritin rice and grain control infrared was 68,144 people within the OECD published literature. Results show that the content of iron protein in brown rice seeds increased 1.28 times compared to the control rice seed. In the case of polished rice protein and iron, the ferritin gene consists of an endosperm specificity promoter; iron content was reduced by 21.0%, only keeping about 2.5 fold the iron accumulation as the control IR68144 rice seed. Sodium, potassium, and magnesium in transgenic rice showed no significant difference from control. However, a statistically significant difference was found between copper and manganese in brown IR68144 ferritin and genetically modified rice, but within the scope of a function of the number of references (OECD 2004), so biological analysis does not consider the difference. The data showed no statistically significant difference in the amino acid content of control and transgenic rice in addition to glutamic acid. Significant difference ($p < 0.05$) was observed in the content of glutamic acid grinding IR68144 (19.49%) and ferritin of transgenic rice (20.0%), respectively, and the number of different references. No statistical differences were found at the levels of palmitic acid (0), 9–16 carbon olefin acid (C16:1), stearic acid (C18:0), oleic acid (C18:1), linoleic acid (C18:2), and linolenic acid (C18:3). Statistically significant differences appeared between vitamin E values for IR68144 milled ferritin rice. In addition, these values are within the scope of the reference (OECD 2004). Composition analysis results showed that the nutrient composition of rice grain ferritin was comparable in composition with non-GM IR68144 control seeds. Nutrition and anti-nutritional component values indicated that there was no difference between GM rice ferritin and non-GM rice IR68144 in iron, zinc, copper, manganese, glutamic acid, and vitamin E content resulting from the insertion of ferritin gene expression in endosperm and all mean values in the range of reference values. Therefore, on the basis of the principle of substantial equivalence, proposed by the WHO and the United Nations FAO, studies support the conclusion that genetically modified (GM) rice nutrition with ferritin as reported is within the scope of the value of other business lines.

Recently, genetically modified rice contains Agb0103 CaMsrB2 genes, showing resistance to drought by the South Korea Rural Development Agency (RDA). The pepper (*Capsicum annuum*) methionine sulfoxide B_2 CaMsrB2 reductase gene may be important as a novel regulator of defense to oxidative stress and pathogen attack under drought stress conditions. Cao et al. conducted a study to investigate nutrition of transgenic rice lines containing the CaMsrB2 gene HV23 from hot pepper (*Capsicum annuum*), and compared with parent HV23 rice varieties (Cao et al. 2012). The results showed no significant difference ($p > 0.05$) in the observed levels of protein, fat, fiber, ash content, and total carbohydrate of Agb0103 between the rice and rice parents from the aspect of growth. Average of moisture, protein, fat, ash, and carbohydrates in the content of this research from rice samples analysis of organizational consensus document reference range (OECD 2004) was as the other commercial rice varieties, and the fiber content of genetically modified rice is higher than the OECD report reference range of 0.6% to 1.0%. No significant difference ($p > 0.05$) was observed between the fatty acid composition of Agb0103 rice and non-transgenic rice, in addition to which the oleic acid content is significantly different ($p < 0.05$). Fatty acid composition analysis is similar to the six reference values observed for typical varieties of South Korea by Jung et al. (Jung et al. 2002); the level of long-chain (eicosenoic), mountain, 24 fatty acid values is slightly higher than the report. Agb0103 amino acid content in rice and profile is similar to control in 100 g/g. All amino acids levels, except alanine and glycine, by analysis report agreement with the reference range (Choi et al. 2012). Alanine and glycine level were slightly below the reference range (Choi et al. 2012). However, the amino acid profile is similar to the value previously reported (Oberdoerfer et al. 2005). In Agb0103 the sodium content of rice increased by 1.24 fold, compared with the control level of rice. No significant difference ($p > 0.05$) was observed in other minerals in Agb0103 rice and nontransgenic content between peers. Mineral content reports are within the scope of all rice samples (OECD 2004). Similar to vitamin content observation, comparing Agb0103 with non-GM rice, vitamin B_2 content is relatively lower than reported values (OECD 2004). Low levels of vitamin B_2 at similar levels were recently reported for insect-tolerant transgenic rice (Park et al 2013). The content of vitamin E in the scope of this study also was reported slightly below the level (OECD 2004), but the level was in the range of previously reported values (Choi et al. 2012). In short, the nutrition and anti-nutritional components of Agb0103 showed no difference between rice and non-GM rice, other than some small differences of nutrients. Genetically modified ingredients of drought-tolerant Agb0103 rice previously reported values within the scope of other commercial rice varieties.

The development of disease-resistant transgenic rice is considered as the best option to manage plant diseases that are directly linked with the productivity and quality of grains. In Chesapeake, a disease-resistant rice variety (OsCK1) was developed by inserting a choline kinase (CK1) gene into the rice genome that is related to blast diseases and bacterial leaf blight of rice. Park et al. analyzed the development of disease-resistant transgenic rice, considered to be the best choice of plant disease management, directly with grain productivity and quality. In South Korea, a disease-resistant variety (OsCK1) is composed of a choline kinase (t) gene into rice genome diseases related to blast and bacterial leaf blight of rice. Analysis of the grains of main nutrients and anti-nutritional factors in OsCK1, its parent varieties, and other commercial rice varieties grown in the same field in South Korea, then evaluated the nutritional quality of rice whether to change the insert on the control of the gene (Park et al. 2015). The results found no significant differences of average protein, ash, fiber, insoluble dietary fiber, starch, and carbohydrates in comparing OsCK1 to a separate non-GMO. OsCK1 amino acid values agreed with the corresponding brown rice, excepting alanine, arginine, glycine, lysine, methionine, proline, or valine. For the value of the fatty acid composition, stearic acid (C18:0) and linoleic acid (C18:2) were slightly higher but oleic acid (C18:1) was lower in OsCK1 than its parent varieties. However, all the amino acids were in the range value of the four commercial varieties provided by the OECD for rice and reference. Except for zinc, sulfur, and phosphorus, the gap between these minerals was significant for OsCK1 and nontransgenics. The values (24.50 and 30.49 g/g), zinc sulfide (0.97 mg/g), and phosphorus (2.32 and 2.54 mg/g) were provided in the OsCK1 of the OECD range of values (7–33 g/g, 2.2 and 0.3 mg/g, 2–5 mg/g, respectively). Nine minerals in the value of brown rice from OsCK1 were analyzed to comply with a good range of four commercial rice varieties planted in the same field. No significant difference in vitamin B_1 content was found between OsCK1 and its corresponding counterpart. The content of vitamin B_2 (0.73 and 0.69 g/g) and E (9.48 and 8.51 g/g) in OsCK1 was lower than that of the parent varieties, but the level of measurement range values in four commercial rice varieties (1.01 g/g vitamin B_2, 0.62; vitamin E, 6.50–11.08 g/g). OsCK1 g/riboflavin levels to 100 g (4.31 and 4.24) were in the range of 2–7.84 g/100 g in four commercial rice varieties. In conclusion, the key components of nutrition and OsCK1 anti-nutrient content are related to nontransgenic counterparts and commercial rice varieties grown in the same field. These results suggest that the nutritional quality of rice is not affected by the insertion of CK1 genes.

Xanthomonas oryzae pv. *oryzae* causes bacterial blight. *Aspergillus niger*, one of the most devastating diseases of rice, can be reduced by up to 80% of the total initial production of crops in tropical and temperate regions (Kumar et al. 2013). In rice varieties resistant to bacterial blight, the Xa21 gene was inserted into rice IR72 by particle bombardment (Gayen et al. 2016). Nutritional quality assessment of transgenic rice with Xa21 resistance to bacterial blight and nontransgenic parental rice IR72 showed that the lipid content of the transgenic rice seeds (2.03%) was also almost indistinguishable from that of the nontransgenic rice seeds (2.02%). Transgenic and nontransgenic rice seeds were 75.84% and 73.66%, respectively, in carbohydrate content: these two values are located in a suitable reference range as reported by the OECD. Together, these results suggest that the approximate composition and quality of Xa21 rice seed are essentially identical to that of the nontransgenic control, as also shown in previous studies (Gayen et al. 2013). Mineral content of genetically modified and unmodified rice seeds showed only small changes in the composition of sodium, iron, and zinc in each seed type. All the data were found to be statistically insignificant in the report of the OECD and any of the observed changes in the desired reference range, as was also shown recently in other crop plants (Rayan and Abbott 2015). The study found that nicotinic acid (6.64 and 6.60 mg/100 g) and thiamine (0.45 and 0.45 mg/100 g), respectively, control the content of transgenic rice. The study also supports Gayen et al. (2013) in that the local rice genetically modified protein and vitamin content resembled the nontransgenic rice IR68144 by statistical distinction. In conclusion, the transgenic Xa21 rice seeds revealed that because of the integration of exogenous genes into the nutritional components of the results, no deleterious changes were observed in the transgenic seeds.

Soybeans

Soybean was the first genetically modified crop in a large-scale commercial operation. GM soy retains its position as the largest food, nutritional, industrial, and pharmaceutical use of genetically modified crops worldwide. From 1996 to 2012, with a total of 763,100,000 hectares (ha) planted with genetically modified soybeans, nearly $37 billion of income was brought to farmers. In 2013, about 845,000 hectares were planted in GM soybeans in 11 countries in the world, accounting for 48%

of all GM crops grown and 79% in the world. Although the herbicide wear-resistant transgenic soybean market share is leading in the world, the soybean crop has been regarded as a recalcitrant unchanged for many years.

In the Zhang et al. study (Zhang et al. 2014c), with the development of antiviral composition of genetically modified soybean seeds into three short IRS, containing specific, highly conserved sequences from a virus source, using the principle of "substantial equivalence" IR changes, investigated whether serious adverse biological insertion compared with controls occurred. The results showed that, in the near component, the confidence level was 95%, with no statistically significant differences observed between the transgenic soybean seeds and controls, and also in the different transgenic lines. Individual values are determined within the tolerance limits of the commercial variety and within the published literature. Most amino acid levels were measured in the range with the database, except for threonine and tryptophan; these levels were higher than the maximum values. However, there was no significant difference in the determination of the 18 amino acids between the virus-resistant and the corresponding nontransgenic soybean seeds. These results indicated that the amino acid content of GM soybean varieties was essentially equivalent to that of nontransgenic plants. Palmitic acid (C16:0), oleic acid (C18:1), and linoleic acid (C18: 3) are the most abundant fatty acids in the antiviral transgenic soybean and nontransgenic. The percentage of most of the measured fatty acids decreased in the ILSI database, except for palm oil and 17 carbon; the two fatty acids were all lower than the minimum. However, there was no statistical difference in the confidence level of 95% to be found in fatty acid composition. In conclusion, the results of component analysis showed no significant differences in the levels of confidence that had been found in transgenic soybean lines with glyphosate tolerance and virus resistance from the traditional correspondence to 95%. On the basis of the principle of substantial equivalence (FAO/WHO 2001; OECD 1993a), genetically modified soybean plants resistant to glyphosate have resistance to the virus essentially equivalent to those not genetically modified. The results showed that no significant change occurred in RNAi mediated by the Internal Revenue Service of the internal state administration.

To determine the nutrition and das-444 o 6-6 soybean feed and grain composition of anti-nutritional factors and nontransgenic soybeans, Lepping (Lepping et al. 2013) analyzed the approximate fiber and mineral content; particle analysis also included vitamins and amino acid and fatty acid profiles, anti-nutritional factors, and active ingredients (lectin, phytic acid, raffinose, stachyose, trypsin inhibitor, isoflavones). The results show that the proximate cause, fiber, minerals, and feed samples is between das-444 o 6-6 entries and contour similarity, and das-444 o 6–6 [77.2–77.6% items of fresh weight had a slightly lower level of humidity anomaly (FWT) compared with the contour (78.7%)]. From the das-444 o 6-6 entry-level particles were similar in contour difference with small differences in carbohydrate abnormalities and isolated protein and fat, in some das-444 o 6-6 entry contains more protein [2,4-D spraying (38.5% dry weight, DWT, or herbicide (38.6% tons)] [19.5% (do not spray or fat DWT) or spraying glufosinate (19.3% tons) and fewer carbohydrates (all items, 37–37.4% DWT) than the contour (protein, fat, carbohydrate: 37.8%, 18.9%, 38.1% tons, respectively). For the industry and fiber of das-444 o 6-6 analysis of the items of value in the observation range of non-GMO soybeans, calcium is three, five o das-444 6-6 entry slightly higher concentration [not currently spraying (324 mg/100 g DWT) and 2,4-D spraying (318 mg/100 GDWT) or glyphosate (320 mg/100 g DWT)], and potassium status in all das-444 o 6-6 entries in a slightly higher concentration (1770–1790 mg 100 g/DWT) ratio contour (calcium, 301 mg/g 100 tons; potassium, 1730 mg/100 g DWT).However, calcium and potassium values are similar to those observed in non-GMO soybeans. From the das-444 o 6-6 entry food, in only 18 amino acid cystine levels (not spraying, spraying herbicide, or all applications) were the differences statistically significant in the contour (Table 2.1). Cystine levels were slightly higher in the das-444 o 6-6 entry (0.6% tons each) compared with the equivalent line (0.5% tons); however, all the data are similar to the level observed in non-GMO soybeans. Grain fatty acid levels were similar between das-444 and 6–6 to check items of fatty acid isoline. Most of the vitamin and tocopherol levels in the checked food found between das-444 o 6-6 entries and contour are similar. The effect of vitamin B_9 is an isolated exception (folic acid), which is considered to be slightly lower in das-444 o 6-6 plots of all herbicides (3.9 mg/kg ton) compared with the contour (4.3 mg/kg ton). The observation is not supported by the remaining four o das-444 6-6 projects, and the average values were similar to those of nontransgenic soybean, showing that a difference is not observed as a result of genotype or herbicide treatment. In the gamma-tocopherol, alpha-tocopherol levels slightly higher das-444 o 6-6 entries (179 and 185 mg/kg DWT) compared to the equivalent line (174 mg/kg, reflecting tons) observed in total tocopherol in trend, namely, the das-444 o 6-6 entry values similar to those in similar commercial soybean lines in the commercial soybean varieties in the detection of previous years. The results show that the das-444 o 6-6 is equivalent to the non-GM soybean soybean ingredients. The results were similar to those of other input features. The results were consistent with the endogenous metabolic pathway, which was expected to be less affected by the transgenic and traditional plant breeding methods.

Soybean products MON 87701 * Mon 89,788 expression from *Bacillus thuringiensis* EPSPS gene and CP4 gene (EPSP synthase) gene from *Agrobacterium* strain CP4. Each biotechnology-derived trait confers specific benefits on pest resistance and glyphosate resistance, respectively, compared to the seed and feed composition of this combination of traits of the product than the traditional soybean geographic and climatic different regions of the plant (Berman et al. 2011). Field trials were conducted in the United States during the 2007 growing season, Argentina during 2007–2008, and in northern Brazil and southern regions of 2007–2008 and 2008–2009 during the growing season of soybean. The results showed that the average value of seed protein was slightly higher than that of the control group on Monday 87,701 to 9788 on Monday. The mean difference was 0.24% DW (northern Brazil, 2008–2009) 1.95% data warehouse. The relative magnitudes of the differences, relative to the control group, ranged from 0.64% to 5.16%, and the statistical significance ($p < 0.05$) was observed in only three of the six regions. Monica 87,701 and the control group had no significant difference in the mean protein values in the different production of this study. In general, for all production, seed protein value range for Monday to 87,701 and the differences between the control and observed the extensive overlap in the size of the average Monday to Monica 89,788 and control protein were small. Differences in average protein amino acid Mon 87,701 × MON 89788 and controls were also relatively small in magnitude of the test, control of a wide range of overlap, and commercial value. There were no consistent differences in the mean values of seed fat between all production on Monday 87,701–Monday 89,788 and control. The average difference was a Monday 87,701 × Mon 89,788 DW down 0.90% (southern Brazil, 2008–2009) to an increase of 0.17% DW (southern Brazil, 2007–2008).The relative magnitude of the difference, with respect to the control, decreased from 4.44% to an increase of 0.88%. Statistical significance ($p < 0.05$) was not observed in any of the productions. Overall, the range and control of seed fat values from Monday to 89,788 were wide, and the difference between the Monica 87,701 and the control group was smaller than the difference between the two groups. As for the protein, genetic and environmental contributions from each of the changes in the level of production of fat were also evident, as observed in the values observed in the commercial reference literature. Monica 87,701 and the United States, Argentina, and Brazil produced no consistent differences in the distribution of fatty acids, and their production. There were greater differences in the levels of fatty acid production in the conventional control group of harvested seeds than in the 87,701 to Monica 89,788 and the control of any production on Monday. The average value of seed vitamin E was slightly lower than that of 87,701 on Monday, with all six production controls. No consistent differences were observed between the mean values for the approximate values (no seed protein above), seed, feed fiber Mon 87,701 * MON 89788 and control. The results showed that the components of the * MON87701 * * MON * 89788 were equivalent to the traditional soybean. In addition, the insect protection of glyphosate-tolerant soybeans, on Monday to Monica 89,788, is equivalent to the composition of conventional soybeans, in different world regions and multiple growing seasons, of the traditional soybean.

Other Crops

Cotton is a source of fiber in the textile industry. The genetically engineered *Gossypum hirsutum* L., which is planted mainly all over the world, has an important role in the cotton industry to control target insect pests, increase production, and decrease costs. Bt (*Bacillus thuringiensis* gene) transgenic cotton is dominating cotton production over the world at an amazing speed (Pray et al. 2002). In 2003, there were 6.9 million acres of transgenic cotton growing in China, which accounted for 58% of the cotton crop in China, more than 33% growth in 2002 and 4% of biotech crop worldwide. As a survey shows, growing the genetically modified cotton led to nearly 1 billion dollars in gross economic benefits in China over the past 5 years. By evaluating cottonseed after the introduction of the *Bacillus thuringiensis* gene for effects of nutrition (Tang et al. 2006), Tang et al. analyzed chemical component of

Bollgard II cotton event 15,985, DPH37B, and 5415. Cotton DPH37B contains *cry1Ac* gene, and Bollgard II 15985 contains *cry2Ab* and *cry1Ac* genes, while making Cotton 5415 as a control. The results showed that, compared with 5415, the fiber, fat, ash, and carbohydrate of DPH37B had significant differences ($p < 0.05$), while all were in the commercial reference range. The ash of 15,985 was statistically different with 5415 ($p < 0.05$), but in the commercial reference range. Mineral content analysis found that there was no statistical differences between DPH37B, 15,985, and 5415. All results were within the scope of the commercial reference range, in addition to the iron content of 5415. Arginine, glutamic acid, and aspartic acid is the main amino acid in the cotton seeds. Glutamic acid and aspartic acid of 37B was statistically different ($p < 0.05$), although in the commercial scope of reference; 15,985 had no statistical differences of three kinds of amino acids. Aspartic acid, glutamic acid, glycine, arginine, cysteine, proline, valine, methionine, isoleucine, aspartic acid, cysteine, methionine, lysine, and leucine content had no significant difference. Phenylalanine levels of 37B had significant differences, but the means of the transgenic cottonseeds and nontransgenic 1 were within the scope of all reported values (OECD, 2004). As a result, the comparative analysis of DPH37B, 15,985 and non-GM heterozygote amino acid content had acceptable results. The results showed that the transgenic cottonseed composition is comparable with nontransgenic cottonseed. Many qualitative and quantitative analysis results showed that the transgenic cottonseed and traditional cottonseed are identical in composition.

In previous studies (Tohidfar et al. 2008, 2012), two genes, cry1Ab and Chi, were transferred into cotton by a carcinoma called *Agrobacterium tumefaciens*. The achieved lines revealed resistance both to cotton bollworm (*Helicoverpa armigera*) and to verticillium wilt disease (*Verticillium dahliae*) in laboratory tests and greenhouse bioassay to observe the stability of the gene and its expression in the T_2 generation. The present study concluded that in comparison with the nontransgenic counterpart (Modirroosta et al. 2013), two T_2 transgenic cotton lines included chitinase (Line 11/57) and Bt lines (Line 61). To achieve the results by using GC/MS, high performance liquid chromatography (HPLC), and ion chromatography (IC) analysis, respectively, the biochemical characteristics of leaves and seeds, such as the content of amino acids, fatty acids, carbohydrates, ion, and cations, are analyzed. The results showed that the chitinase and Bt lines of the content of fatty acid were significantly different ($p \leq 0.05$), compared with their non-GM counterpart. In addition to stearic acid content, all observed differences in the study were covered within the International Life Sciences Institute (ILSI 2006) guidelines for cotton seed (crop composition database; can be found at http://www.cropcomposition.org). Thus, they are considered to be biologically nonsignificant. These data suggest that there is no obvious difference between the Bt cotton line and its nontransgenic counterpart. The Bt and chitinase transgenic and nontransgenic counterpart cotton leaf extracts showed no significant differences in sucrose and glucose content. However, the evidence shows that glucose content of chitinase lines, fructose content of transgenic lines (Bt and chitinase), and asparagine and glutamine of chitinase lines were significantly higher than the nontransgenic counterpart plants. In all the

20 measured amino acids, 18 amino acids had no significant differences when compared with the leaf extracts of the non-GM counterpart, other than asparagine and glutamine content that was obviously changed in the chitinase plants. However, there was no significant difference in amino acids between the Bt lines and the non-transgenic counterpart. Compared to the nontransgenic counterpart, there are no considerable changes in either Bt or chitinase transgenic lines in all anions, including acetic acid, chloride, nitrate, phosphate, sulfate, succinic acid, and oxalic acid. The amount of sodium in transgenic lines (both Bt and chitinase) was significantly reduced compared to the large amount of sodium in the transgenic plants. The experimental results on the seeds show that histidine, isoleucine, leucine, and phenylalanine content of transgenic and nontransgenic lines are all the same, whereas other amino acids in transgenic lines increased significantly. Surprisingly, it is found that compared with both the Bt line and its nontransgenic counterpart, the concentrations of stearic acid, myristic acid, oleic acid, and linoleic acid in the chitinase line were the same, but were obviously different than those of non-GM counterpart plants. Differing from the changes of leaves, changes in the content of seeds is more obvious, possibly because the seeds are metabolite storage organs, so they show greater changes in metabolites than in concentrations of the leaves. Various data of Bt and chitinase cotton show that in addition to lepidopteran insect resistance and tolerance of fungi diseases, respectively, the transgenic lines in essence are equal to the nontransgenic counterparts substantially, and are as safe as non-GM cotton.

The Bollgard II cotton event 15,985 produces the Cry1Ac and Cry2Ab2 proteins. It has been genetically engineered to develop a plant spectrum for insects, which is an insect resistance, has an insect tolerance, and was regarded as management tools to prevent the emergence of resistance. To assess cotton Bollgard II ingredients and nutrition, compared to the traditional cotton varieties, Hamilton et al. studied the composition analysis (Hamilton et al. 2004). This analysis shows that in the approximate level there were no statistically significant differences between the seed Bollgard II and the control. Furthermore, these values are within the scope determined for the commercial varieties analyzed at the same time. No statistically significant differences in any Bollgard II cotton fiber value and the control was found. The contents of 18 kinds of amino acids in Bollgard II cotton seeds and the control line are comparable. No amino acid results for Bollgard II cotton statistically significantly differs from the parental control, DP50 cotton varieties. Fatty acids of Bollgard II have no significant statistical differences between cotton and the control observed in the analysis of six of ten fatty acids. There are statistically significant differences in fatty acid content between Bollgard II cotton and the control, which were noted as myristic (14:0), stearic (18:0), linolenic and γ-linolenic (18:3), and arachidonic (22:0) fatty acids. Compared with the control line, the contents of myristic, stearic, linolenic and γ-linolenic, and arachidonic fatty acids in Bollgard II cotton were higher in the data combined from 1998 and 1999 field trials, which were 16.8%, 14.3%, 13.3%, and 7.4%, respectively. A range of values for the stearic and flax and gamma linolenic acid 99% tolerance range of commercial varieties shows the difference between two groups in the same population and confirms it is unlikely to be of biological significance. The cyclopropenoid fatty acids sterculic, malvalic, and dihydrosterculics are unique fatty acids that exist in cotton. The dihydrosterculic acid value range is within the 99% tolerance range of the commercial varieties, showing the difference in the same population and confirming that of the two groups observed is unlikely to have biological significance. There are statistically significant differences in mineral content between Bollgard II cotton and the control, which were noted as copper, iron, and phosphorus. For these small ions, however, significant differences are unlikely because the average value is within the scope of the commercial, and the value range is within the 99% tolerance range of commercial varieties, set at the 95% confidence level, in the analysis of the two groups in the same population. For antioxidant content, vitamin E is similar in cottonseed oil from Bollgard II cotton and the control, and similar levels (136–660 mg/kg) were previously reported in the literature. Thus, the conventional control cotton line and other commercial varieties of the cottonseed and cottonseed Bollgard II were comparable in their composition. These data support the conclusion that Bollgard II cotton conventional cotton is as safe and nutritious as conventional cotton for food and feed use.

Potato (*Solanum tuberosum* L.) is fourth after the three most important food crops, which are wheat, rice, and corn. Potatoes are a good source of carbohydrates, with more dry matter and high-quality protein than the main grain crops (Al-Saikhan, 2000). In addition, there is preliminary evidence that potato contains many important antioxidants, including phenolic compounds, flavonoids, and carotenoids (Brown, 2005). Because of the importance of potato all over the world, great effort has been made to develop new varieties of potato that are tolerant to different pathogens. One of these pathogens is the potato virus Y (PVY), a potyviruses member. An efficient and safe conversion system of transgenic potato plants harboring potato virus Y-coat protein gene (CP–PVY) conferring resistance against potyvirus Y has been developed (Saker 2003). Saker et al. (2011) analyzed a transgenic potato microtuber (*Solanum tuberosum* L. Desiree) line (which harbors potato virus Y coat protein gene) in composition and genetic variation, and compared this with the traditional line to further understand the metabolic and genetic changes related to the conversion process. Microtuber lines of mineral content display the CPPVY gene, causing the conversion process of minerals (phosphorus, nitrogen, iron, zinc) to increase significantly compared with the nontransformed line. The content of reducing sugar, vitamin C, phenols, flavonoids, and beta carotene is similar in the evaluation of the two microtubers. The variation in content of reducing sugar line height (6.52 mg/g DW) was more than the conversion line of content (5.09 mg/g DW). Vitamin C content changed in value from 22.76 mg/100 g DW (transformed line) to 18.63 mg/100 g DW (nontransformed line); this difference is statistically significant. Two lines of total phenol content changed from 7.76 to 8.33 mg/g DW; however, the total flavonoids ranged from 5.43 to 5.26 mg/g DW. No significant differences were found between the two lines ($p = 0.05$) for less total phenol, total content of flavonoids, and beta carotene. In sum, the data showed that there were no significant differences between the conversion and the amount of amino acid in a line, some minerals, total phenol, total flavonoids, and beta carotene content. In general, it is concluded that conversion microtubers proved to be similar to the nontransformed microtubers line.

The traditional bean (*Phaseolus vulgaris* L.) is the main food legume in human diets. It is significant for providing protein, complex carbohydrates, iron, calcium, and vitamins for a billion people worldwide, especially in Latin America and sub-Saharan Africa. A golden mosaic disease severely reduces bean production in Latin America. A transgenic bean event called Embrapa 5.1 (EMB-PV051-1) that can prevent the bean golden mosaic virus was planted widespread for the purpose of commerce in Brazil in 2011. Carvalho et al. (2015) measured and evaluated the nutritional components of the beans, as well as the anti-nutrient levels in the primary transgenic line and its derived near-isogenic lines after crosses and back-crosses with two commercial cultivars. In the amounts of cysteine and methionine in grain, statistically meaningful differences ($p < 0.05$) were found in Londrina in the year 2008. However, differences were not nutritionally meaningful for the numbers of cysteine and methionine were within the values observed in the databank of bean nutritional composition. In any of the three locations in the year 2009, no statistical differences ($p < 0.05$) were found in grains gained. In Santo Antonio de Goias in the year 2009, the amount of vitamin B_1 was statistically different in grain gained. In Santo Antonio de Goias in the year 2009,there were no statistical differences ($p < 0.05$) in the amount of vitamin B_2 in grains gained. Most importantly, results described that the transgenic bean event, cultivated under field conditions, was adequately equal to that of the nontransgenic bean plants. Also, the amounts of the nutritional components are within the range of values found for some bean market-oriented varieties grown across a range of environments and seasons.

Arachis hypogaea L. (peanut) is an important food crop worldwide for both commercial and subsistence production. Peanut is susceptible to many fungal and viral diseases. Sclerotinia blight, a result of the soil-borne fungus *Sclerotinia minor* Jagger, is a dominating and economically important disease of peanut in the mid-Atlantic and southwestern regions of the United States (Beute et al. 1975). Transgenic lines provide a sustainable and profitable tool to combat the devastating disease. Three Virginia-type peanut cultivars (NC 7, Wilson and Perry) have been successfully modified with a barley oxalate oxidase gene to provide resistance to Sclerotinia blight (Partridge-Telenko et al. 2011; Livingstone et al. 2005). These three cultivars and their corresponding transgenic lines (N70, W73, and P39) with a barley oxalate oxidase gene were analyzed by Hu et al. for differences in key mineral nutrients, fatty acid components, hay constituents, and grade characteristics (Hu et al. 2014). The results showed that no significant differences in five major mineral elements were observed across all 3 years between transgenic line N70 and the nontransgenic parent NC 7, indicating that they had a very similar mineral composition. For transgenic line W73 and nontransgenic parent Wilson, magnesium and phosphorus levels in P39 were similar to those of their corresponding nontransgenic parent cultivars in other years, suggesting that the elevated and/or reduced values in a field season were likely caused by environmental variations among three growing seasons. Moreover, the values were always within the reported ranges of conventional peanut cultivars (Jonnala et al. 2005) and were not sufficient to cause any nutritional or health concerns. Statistical differences were observed only in sulfur in 2007 and in phosphorus in 2006 and 2008. The fatty acid composition of transgenic lines was similar to that of their nontransgenic counterparts. Furthermore, transgenic lines W73 and P39 showed no statistical differences in all eight fatty acid components tested from those of their corresponding parents. In conclusion, this study provided multiple-year data on chemical composition in different climates and agronomic conditions. No biologically meaningful unintended compositional or quality effects were found between the transgenic lines and their paired nontransgenic counterparts, and the composition of transgenic lines was within the known range of the nontransgenic peanut. Results clearly indicated that insertion and expression of an oxalate oxidase gene in peanut did not cause substantial unintended changes in chemical composition in seeds or hay constituents. Therefore, this study provides strong evidence for concluding substantial equivalence in quality and compositional traits among transgenic lines expressing oxalate oxidase and their nontransformed parent cultivars.

2.2 Anti-nutritional Factors

Anti-nutrients inhibit or prevent important paths of human metabolism, or cause damage to the digestion. Anti-nutritional factors can decrease nutrient utilization; usually these are a protein, vitamins, or minerals, thereby reducing the nutritional value of food. So it is important to undertake that no new anti-nutrition factor is inadvertently introduced, causing differences between the results of genetic modification. So it is necessary to make sure the transformation does not introduce new compounds of this type or influence the content or characteristics of endogenous compounds. The negative changes, which may affect human health, already exist in the crop plants. There are many examples of anti-nutritional factors in a certain scope of feed, including alkaloids, glucoside, sulfur, glycoside, lectin, and phenol derivatives, such as tannin and gossypol and protease inhibitors and six inositol phosphates (OECD 2001, 2004; ILSI 2004). The concentration of the GM lines significantly reduces these negative materials; low-phytate corn is one example. Composition analysis is not sufficient to provide an accurate and extensive nutrition assessment for the feed, and animal feeding studies with target species is needed. In studies with fattening pigs, Spencer et al. (Spencer et al. 2000a, b) compared diet containing corn grain from a commercial variety or a modified phytic acid content at a low level. The author suggests that although similar intake and weight records the traditional diet of daily life for conventional and GM diet with the low-phytate maize, it eliminates the need to supplement the use of phosphorus and significantly decreases phosphorus excretion, with potential impact in decreasing the environmental footprint of livestock production systems of the stomach.

2.2.1 Anti-nutrients in Rice

The content of anti-nutritional factors in rice grains, such as phytic acid, lectin, and trypsin inhibitors, has been proposed to be analyzed in the OECD consensus documents (OECD 2004). Li et al. compared the contents of anti-nutrients between Bar 68-1 and D68 (Li et al. 2008). Hemagglutination activity of lectin activity has been determined (based on Keller, Field, and Li 1987). Cheever Lipinski detected phytic acid by the use of AOAC 986.11 ion-exchange analysis. The activity trypsin inhibitor was analyzed by solvent extracted with an alteration of the American Association of Cereal Chemists Method 71–10 (AACC 1983). The content of phytic acid in Bar 68-1 and of D68 was comparable in both brown and milled rice grain. Lectin activity did not reach the detection limit for either group. The inhibition activity of trypsin inhibitors was at a low level both in brown rice and milled rice samples of both D68 and Bar 68-1.To report this as similar with previous observations, trypsin inhibition was only found in rice bran (Oberdoerfer et al. 2005; Tashiro and Maki 1979), but not in brown or milled rice samples. Determination of inhibitory activity against trypsin inhibitors was based on and did not distinguish between D68 and Bar 68-1 in brown rice and milled rice.

Gayen et al. sequenced phytic acid in homozygous genetically modified IR68144 rice and corresponding nontransgenic rice (Gayen et al. 2013). Phytic acid extract of 2.4% hydrochloric acid was approximated by spectrophotometer at 500 nm after reaction with 0.03% $FeCl_3/6H_2O$ solution containing 0.3% sulfosalicylic acid (Bhandari and Kawabata 2006). No significant differences were observed, according to the results of the content of phytic acid between GM and non-GM rice seeds. Research shows that the content of phytic acid in all samples of seeds is consistent with reported values (Oberdoerfer et al. 2005). Gayen et al. tested transgenic Xa21 phytic acid content (1.15%) and non-GM rice seeds (1.13%) by spectrophotometer (Bhandari and Kawabata 2006) but found no difference that was statistically significant.

Li et al. sequenced anti-nutritional factors such as insect-resistant rice (IRR) and their nontransgenic counterparts (Li et al. 2007). Phytic acid ion exchange was evaluated using the method recognized by AOAC 986.11. The way to measure trypsin inhibitor activity determined by the solvent extraction uses a modified method of the American Association of Cereal Chemists, Method 71–10 (AACC 1983). Pectin activity measurement is based on erythrocyte agglutination activity (Klurfeld and Kritchevsky 1987). As results, the level of phytic acid IRR can be comparable to those of control in both brown and milled rice grain. In the two groups, the activity of pectin was not found. Based on the inhibitory activity of trypsin inhibitor measurement, the control in both brown and milled rice grain has no different in IRR.

Wang et al. evaluated phytic acid, trypsin inhibitor activity, and lectin activity in transgenic rice seed with Cry1ab/ac genes and their nontransgenic counterparts (Wang et al. 2012). The analysis method is similar to Li et al. (2007). The result shows that inhibition of trypsin inhibitor activity and the content of phytic acid is at a very low level in the two groups. Lectin activity was not found.

Cho et al. evaluated the ant-inutritional factors in transgenic rice harboring CaMsrB2 genes and traditional counterparts (Cho et al. 2016). Phytic acid extraction rice with 2.4% hydrochloric acid and analysis in the 500-nm spectrophotometer to determine the activity of trypsin inhibitor in alkali solvent-extracted rice samples modification method using a modification of AACC method 71–10 (AACC 1983). Phytic acid in food reducefs the nutrient absorption of metal ions, and trypsin inhibitor nutritional characteristics inhibit the activity of protease. Results show that there was no significant difference ($p > 0.05$) between the phytic acid content of Agb0103 rice and non-GM rice. Inhibitory activity of trypsin inhibitor was at a low level in all rice samples, which is consistent with previous research (OECD 2004; Oberdoerfer et al. 2005; Li et al. 2008).

The levels of the significant anti-nutrients were compared between the disease-resistant transgenic rice (OsCK1) and its parent cultivar (cv. Nakdongbyeo), and four types of traditional rice (cv. Chucheongbyeo, Ilmibyeo, Ilpumbyeo, and Shindongjinbyeo) by evaluating the contents of trypsin inhibitors and phytic acid. Phytic acid content was analyzed by the basis of the modification of Latta and Eskin (Latta and Eskin 1980) by following polyprep-prefilled chromatographic columns (Bio-Rad Laboratories, Richmond, CA, USA) including an AG-1-X8 anion-exchange resin (100–200 mesh chloride form, 0.8 × 4 cm) so that the segregation of phytate from the rice extract (Park et al. 2012). Trypsin inhibitor activity in the grains was evaluating by using AOCS method Ba 12–75. The study revealed that the inhibitory activity of trypsin inhibitors was at a very low level in all the samples, which is consistent with the results from old studies performed using transgenic rice as well as traditional rice. In old studies, the component contents were less than 0.1 TIU/mg (24, 32). The phytic acid content (1.26% and 1.29%) of OsCK1 was lower than that of its non-GM counterpart (1.91% and 2.41%). Nevertheless, the levels were within the ranges of values (1.18–2.68%) that were evaluated in the four commercial rice kinds.

Oh et al. (Oh et al. 2016) analyzed the phytic acid level and the activity of trypsin inhibitor synthesis of resveratrol Agb0102 (*Oryza sativa* L. cv. Dongjin rice), drought-tolerant Agb0103 (*O. sativa* L. cv. Ilmi rice) and 23 commercial reference varieties of rice (*O. sativa* L.). Phytic acid content was analyzed based on the modification of Latta and Eskin (Latta and Eskin 1980). Trypsin inhibitor activity was decided using AOCS method Ba 12–75. Phytic acid and the concentration of trypsin inhibitor were evaluated, and the absorbance was measured with a spectrophotometer, to evaluate the safety of the two kinds of transgenic rice (Agb0102 and Agb0103) and check substantial equivalence with a non-GM comparator. The results show that the phytic acid and trypsin inhibitor content of GM rice is the same under environmental conditions similar to the traditional control. Observation of GM rice (Agb0102 and Agb010) showed no difference from the normal control under the same conditions. Phytic acid content and trypsin inhibitor of the two genetically modified rice varieties is within the scope of the reference rice and reference rice TI. Thus, the phytic acid and trypsin inhibitor of transgenic rice are biologically equivalent to the non-GM rice and commercial rice.

2.2.2 Anti-Nutrients in Soybean

Zhang et al. analyzed phytic acid level and trypsin inhibitor activity in the RNAi-mediated anti-virus transgenic soybean and their nontransgenic counterparts (Zhang et al. 2014a, b). Lectins were measured by ing the soy sample extracts by rabbit red blood cell agglutinate characteristics according to a modification of reported methods (Klurfeld and Kritchevsky 1987; Strathmann et al. 2002). Trypsin inhibitor activity was evaluated by using a modified method (Hamerstrand 1981). The levels of two important anti-nutrition factors (lectins and trypsin inhibitor) are generally at a low level, no more than in the previous research range (ILSI 2004), and this may not change the nutrition of soybean seed. The concentration of lectin and trypsin inhibitor of the anti-virus transgenic soybean harvest seeds was observed, compared with the control harvest nontransgenic seeds, at a confidence level of 95%. These results indicate that the anti-nutrient of transgenic lines was substantially equal to nontransgenic counterparts.

Lepping et al. analyzed lectins and phytic acid level and trypsin inhibitor activity in DAS-444Ø6–6 (AAD-12 + 2m EPSPS + PAT) herbicide-resistant soybean and nontransgenic soybean (Lepping et al. 2013). These anti-nutrients were evaluated using methods formerly reported by Herman et al. (Herman et al. 2010). The study shows that the content of anti-nutrients in grain was similar between DAS-444Ø6-6 entries and the isoline for each of the analytes examined (lectin, phytic acid, trypsin inhibitor). High statistical values observed in two of five DAS-444Ø6-6 entries compared with the isoline and for trypsin inhibitor in two of five DAS-444Ø6-6 entries compared with the isoline. Lectin and trypsin inhibitor with higher levels in the result do not support the remaining three DAS-444Ø6-6 items in all circumstances, and mean values observed were the same with the commercial line within the scope of research and literature nontransgenic soy. In addition, the exogenous lectin and soybean trypsin inhibitors were inactivated in the standard process of food consumption (Padgette et al. 1996). Berman et al. analyzed lectin and phytic acid level and the activity of trypsin inhibitor insect-protected glyphosate-resistant soybean MON 87701 × MON 89788 and traditional soybean extending all over the world in different areas and different growing seasons (Berman et al. 2011). The method used for the anti-nutrients is derived from earlier references (Berman et al. 2009, 2010). Studies have shown that no consistent difference in product average anti-nutrients (agglutinin, phytic acid, trypsin inhibitor) was observed between MON 87701 × MON 89788 and the control.

2.2.3 Anti-Nutrients in Other Crops

Anti-nutrients include free gossypol, determination of total gossypol, and tannin cottonseed, after the introduction of the *Bacillus thuringiensis* gene, Bollgard II cotton event 15,985, DPH37B, and 5415 (Tang et al. 2006). Contents of free gossypol content, the total gossypol, and tannin were analyzed by high performance liquid chromatography (Waters, Milford, MA, USA). For gossypol, no significant statistical differences were observed between the 15,985, DPH37B, and the control 5415. Free gossypol and far range as previously reported levels of total gossypol cotton varieties (Co, 2000).

Anti-nutrients include phytic acid and trypsin inhibitor in transgenic RNAi-mediated anti-virus bean (event EMB-PV051-1) and non-GMO counterpart (Carvalho et al. 2015). Phytic acid detection is based on method 986.11 recognized by the AOAC. Trypsin inhibitor is analyzed based on AOCS method Ba 12–75. The Brazilian Agricultural Research Corporation determined anti-nutritional components in the grain from both the transgenic bean event Embrapa 5.1 and its non-GMO counterpart (cv. Olathe). In addition, the GM event and four other traditional Brazilian beans had been measured, varieties growing in various locations in 8 years (nutritional bean data bank). All the nutrition and anti-nutrition components measured in the modified soybean event were within established reference ranges observed in the bean databank.

2.3 Nutrient Use Efficiency

In addition to the issue of food safety of genetically modified crops, some other security issues, including crop nutrition, natural toxins, nutrient use efficiency, and unintended effects are also important aspects of food evaluation. However, so far, less attention has been paid to these problems. In recent years, more and more genetically modified crops with new traits such as improving the nutrition value, prolonging the shelf life of fruits and vegetables, improving the flavor of food, and so on have been developed successfully. Safety assessment for these new genetically modified crops should include nutritional value evaluation, especially nutrient use efficiency. In 1993, the OECD put forward the concept of "substantial equivalence." However, for most nutrition improvements in genetically modified crops with changed nutritional components, the principle of substantial equivalence could not account for the safety of these products. Further evaluation of nutrient use efficiency of the main nutrients (such as protein, amino acid, fat, fatty acids, calcium, phosphorus), absorption utilization ratio, and the apparent digestibility of genetically modified organisms and their products should be evaluated based on the "case-by-case" principle.

Nutrient use efficiency is an important part of the nutritional evaluation of GM products, which mainly includes protein efficiency ratio (PER) and apparent digestibility. The aim of measuring nutrient use efficiency is to evaluate nutritional value and unintended effects of GM crops on animals or humans.

In the case of substantial changes in plant composition (GM plants with output traits or GM plants of the second generation), studies measuring the digestibility/availability of some nutrients or nutrient precursors are necessary. Mostly such studies are done with model animals (mice, rats, rabbits) or small target animals (chicks, quails, piglets) because of the high costs and the limited feed amounts available in some cases. Similar studies are necessary to demonstrate the efficacy of enzymes expressed in plants or to show the higher phosphorus (P) availability in plants with lower phytate content. Phytate is one of the most important inhibitors of P-availability. In a study with pigs (Spencer et al. 2000b), low-phytate maize showed the same results as traditional maize supplemented with 2 or 1.5 g inorganic P/kg feed.

2.3.1 Present Studies on Nutrient Use Efficiency

The objective of efficiency trials is to measure the effect of food/feed from GM plants on the performance of food-producing animals, and to compare the results with an isogenic counterpart and at least four comparable commercial products.

Many feeding studies have been conducted in the past decade to show the substantial similarities of foods and feed derived from GM plants of the first generation. Most of the studies were done as efficiency trials, and GM food/feed was compared in adjusted diets with their isogenic counterparts and some conventional commercial varieties (one to ten in some cases). The experimental designs were done according to the recommendations by ILSI. Questions concerning the tolerance of some food/feed in animals (tolerance studies) may be also included in efficiency trials.

In recent years, several reviews on the nutrition and safety assessment of food/feed from GM plants (mostly plants from the first generation) have been published (Clark and Ipharraguerre 2001). Furthermore, the documents by ILSI (Hartnell 2007; ILSI 2003) and EFSA also summarize the present state in the feeding of GM plants derived food/feed to target animals. More than150 studies with target animals showed no biologically relevant differences between animals fed GM plants or non-GM plants as supplemental diets. There is wide agreement that GM plants (first generation) diets did not significantly affect animal health, performance, or composition and quality of animal products. Side effects might be expected in GM plants, particularly for GM plants with multiple modifications (multi-stacked events) (Zhang et al. 2004a), and these must be scientifically analyzed in detail. Furthermore the high biological range for many parameters should be considered. Presently, no other foods or feeds are as extensively analysed and tested in various studies as is the case for GM plant-derived products of the first generation. It can be concluded that the safety and nutritional assessment of GM and non-GM plants is not well balanced (Kok et al. 2008).

Based on the results just mentioned, the necessity of animal feeding studies with GM plants (first generation) diets is often questioned. According to various guidance documents (EFSA 2011; Sørensen, 2012), such studies are not urgently needed. No animal feeding studies are required if the differences in compositional analyses between isogenic and transgenic plants are small or negligible (first-generation GM plants) because of the costs of such studies and reduction of numbers of experimental animals.

On the other hand, feeding experiments with the first-generation GM plants to target animals may contribute to showing the nutritional equivalence and the safety of the feed to the public, and therefore improve the public acceptance of GM plants and derived products. Under the present regulations only animals fed with non-permitted GM feed cannot be used in the food chain: thus, if a GM feed with its isogenic counterpart and four commercial varieties are tested in a feeding study, more than 80% of the animals can be used for human nutrition. Therefore, efficiency feeding studies with first-generation GM plants could be useful in some cases.

Nutritional safety assessment of the second-generation GM plants should receive more attention. Composition variation in GM plants and their derived products may also influence the composition of food of animal origin, as has been exemplarily demonstrated for soybeans with a modified fatty acid pattern.

Stearidonic acid (SDA) may be transferred into the body fat of non-ruminants or may be used as precursor for longer fatty acid chains (e.g., C20 and C22 fatty acids) in non-ruminants and in ruminants. The proportion of SDA in stearidonic soybean oil ranges from 20% to 30%. Rymer et al. (2010) fed animals with GM soybean oil diet (240 g SDA/kg oil) and its conventional control diet for 13 weeks. The results indicated that SDA oil addition in the diet did not influence feed intake, weight gain, and feed conversion rate in the animals. However, higher concentrations of SDA as well as C20 and C22 polyunsaturated fatty acids in various body fats, were observed, which was in accordance with previous studies (Bernal-Santos et al. 2010; Kitessa and Young 2011).

Gibbs et al. (2010) considered the introduction of SDA oils in animal feed as a change to increase the intake of long-chain n-3 PUFA in men. However, for some polyunsaturated fatty acids there exist upper limits for human nutrition, and one should be careful with supplementing of such oils in animal nutrition. Therefore, animal body composition may be also an endpoint of animal feeding studies to measure the transfer of some ingredients of GM plants of the second generation into animal tissues, milk, or eggs. Animal body samples or products from animals such as milk or eggs should be considered and analyzed adequately.

2.3.2 *Protein Efficiency Ratio, PER*

A protein efficiency ratio (PER) test, which has been used widely since 1919, was conducted to assess the quality of a particular food protein in food/feed. PER can be calculated based upon the division of body weight gain of a test subject versus intake of the tested protein during the animal trial. In 1984, PER was recommended to evaluate nutrition value of GM products by the Association of Official Analytical Chemists (AOAC). As an independent, globally recognized, and non-profit organization, AOAC provides appropriate science-based solutions for the development of microbiological and chemical standards, which are used worldwide to promote trade and to improve public health and safety.

According to AOAC Official Method 960.48 Protein Efficiency Ratio, reference casein, balanced salt mixture, and vitamin mixture should be prepared before the PER test. Proximate analysis including N, fat, ash, moisture, and crude fiber will contribute to making animal diets having the same content of these ingredients. Therefore, proximate analysis is always conducted to balance the nutrition in animal diets. Male pathogen-free Wistar or Sprague-Dawley rats 3 to 4 weeks old are used in the AOAC test. After acclimatization for 3–7 days, the animals with body weight variation ≤5 g are divided into treatment groups (≥10 rats/group) randomly. One control group that will receive ANRC reference casein should be provide in assay of the target substance. During the assay period, each rat should be kept in an individual cage with adequate diet and sterile water ad libitum. All conditions of the environment should be maintained as uniformly as possible with respect to each of the groups being compared to ANRC reference casein during assay period. Body weight and food intake of each rat should be recorded at regular intervals, not more than 7 days. The animals are executed on the 28th day after the beginning of the assay period. Weight gain and intake of target protein per rat for each group should be calculated after the animal trial: PER = weight gain/protein intake. After calculating the PER of each group, determine ratio × 100 of PER for each assay group to PER for ANRC casein reference group. Protein quality of the sample is reported as ratio × 100 of sample PER to ANRC reference casein PER.

The AOAC assessment method of nutrient use efficiency has been accepted extensively worldwide. However, some especial standard methods are also established in some countries. For example, in China a standard method named "Food safety detection of genetically modified organisms and derived products—Detection of nutrients apparent digestibility by rat test" has been put forward and will be published before 2018. The main content of the method is as follows:

(a) Apparent digestibility is based on the differentials between intake and fecal excretion of a test a nutrient divided by its intake during the test period. Apparent digestibility is recommended to evaluate nutrition value of a particular nutrient of GM products in China. In the animal experiment, 3-week-old Wistar or Sprague-Dawley male and female rats (peak growth period) are treated with GM product. Intake and fecal excretion of particular nutrients (e.g., protein, fat, crude fiber, amino acids, fatty acids, vitamin A, vitamin E, calcium, phosphorus, iron, magnesium, manganese, selenium) of the GM product are detected to evaluate the absorption ratio of particular nutrients from the GM product.
(b) Assemble groups of 20 or more rats (half male and half female): in assay of particular nutrients, provide at least three groups that will receive AIN93G reference diet, GM product-added diet, and non-GM parental reference-added diet. Proximate analysis including N, fat, ash, moisture, and crude fiber will contribute to making animal diets having the same content of these ingredients. Therefore, proximate analysis is always conducted to balance the nutrition in animal diets. Each animal should be fed alone in a metabolism cage and adapted

to environment for 3 to 5 days before the official test. During the animal experiment, observations of clinical behaviors, growth condition, weight gain, food intake, and collection of fecal excretion twice every day are continued for a test period of 28 days. Particular nutrients of the diets and fecal excretion are tested by related reliable methods, and the apparent digestibility is calculated as already described.

2.4 Nutritional Evaluation for GMOs Using Animals

Nutritional evaluations of GMOs are based on the "case-by-case" principle and should not be considered as subacute toxicity evaluation. Nutritional evaluation is designed to determine nutritional variation of the GM products and can be regarded as a supplementary assessment to verify "substantial equivalence" between a GM crop and the non-GM parental crop. The animal trials are conducted on laboratory animals (e.g., rat) and large domestic animals (e.g., cattle, pig, sheep) for 28 days to evaluate the nutritional value of GMOs. Chicken and catfish are sensitive to the nutritional value of their diet in their growth rate, and can be selected as model animals as well. The parameters including clinical behavior, food intake, weight gain, gross necropsy, organ weight, and hematochemistry should be conducted in the animal study. Digestion efficiency of GM products in ruminants need to be monitored as well. Before the animal trial, a principal component test of GM and non-GM crops should be done, as animal diets with balanced essential nutrients are important to the animal experiment design. The limit of these animal trials is the tolerance of the test animal for particular food ingredients that are not expected.

Many studies have been performed with the first-generation GM crops on laboratory animals and/or large domestic animals to evaluate the nutritional value and food safety of these GMOs. GM crops (maize, sugar beet, soybean, etc.) and their non-GM parental controls were grown under identical conditions in the experimental fields. Component analysis of these crops including moisture, ash, crude fiber, fat, protein, amino acids, and fatty acids, minerals, undesirable substances were conducted by acknowledged methods. Nutritional evaluation of these GM crops including parameters such as food intake, digestibility, clinical behavior, gross necropsy, organ weight, and hematochemistry were conducted during the animal trial. Reproduction experiments carried out with quail and laying hens are also regarded as an important part of the nutritional evaluation of GMOs.

Presently, there is no systematic evaluation standard for investigating the nutritional value and food safety of the second generation of GM plants. Flachowsky and Böhme (Flachowsky and Böhme 2005) reported some approaches for assessing those GM foods. The proposal includes the following aspects. GM plants with improvement of specific nutrients such as, vitamins, unsaturated fatty acid, amino acids, and minerals may be beneficial to nutrient availability and conversion in animals or humans. In addition, the excretion of other nutrients such as nitrogen or phosphorus may be reduced. The details of the animal experiment procedure including the number of animals, experimental period, composition analysis, conditions of environment, clinical observation, and examinations are presented by EFSA (2004) and ILSI (2003, 2004). Considering the large variation of the second generation of GM plants, different experimental designs should be conducted based on the principle of "case by case."

However, transgenic operations may not only change the expression of intended desirable substances. Some research has indicated that unintended effects might occur and lead to unfavorable influences on the nutritional value or food safety of GMOs (Cellini et al. 2004). Unintended effects of GMOs should be paid more attention in the nutritional and safety evaluation in the future. Based on the principle of "case by case," specific animal experiments are necessary to address these issues.

2.4.1 Poultry

Hammond et al. first tested the nutritional safety of GM plants on poultry in 1996 (Hammond et al. 1996). The results indicated that the GM soybean did not significantly influence the fattening and slaughter performance of the broilers compared with isogenic control hybrids. Similar results were reported by Brake and Vlachos (1998) when GM maize diets were fed to broilers with a high percentages of incorporation. A test with large portions of GM canola meal (25%) fed to broilers was conducted by Taylor et al. (2004), and no significant influences on slaughtering data and body composition were observed. Even with a higher percentage incorporation of GM maize (50% of the diet) in a diet feeding study on layers and broilers, no significant differences in digestibility and in energy content were observed (Aulrich et al. 2001).

Furthermore, many feeding studies with GM plants such as insect-resistant (Bt) maize and glyphosate-resistant maize, soybean, wheat, canola, or potatoes were conducted on poultry. The incorporation percentage of GM products ranged from 27% to 78% to allow a high proportion of the test material. Component analysis of the GM diets was performed to identify the componential difference between GM and non-GM lines. Undoubtedly, these feeding studies did not show any significant differences in the various production parameters monitored in broilers and layers.

Some other studies indicated significant differences between GM and non-GM feeding groups; however, none of these were considered as a safety concern. A higher body weight gain was observed in a GM maize (Bt) diet group compared to the control group (Piva et al. 2001), which might have been ascribed to a lower mycotoxin content in the Bt maize compared to the conventional counterpart.

2.4.2 *Pig*

The pig is a model animal for both feed and food safety assessment of GM plants. Many nutritional assessment experiments have been conducted using pigs with transgenic maize and soybean, etc.

Component analysis and nutritional value assessment of BT-maize were conducted on pigs in previous studies (Reuter et al. 2002a, b). Component analysis included parameters such as proximates, starch, sugar, non-starch polysaccharides, fatty acid and amino acid profiles, minerals, and mycotoxins that were measured in both the GM and non-GM lines. The diets were incorporated with 70% (w/w) GM or non-GM maize seed powder in as high a ratio as possible to keep a balanced nutrition supply. The digestibility of the GM and non-GM maize seed was measured with six pigs per group, and the excrement was collected three times during the experimental period. The results indicated that the genetic modification in maize did not affect the composition or digestibility of nutrients and the energy level significantly. Interestingly, the concentration of mycotoxins in GM maize was lower than that in non-GM maize, which was in accordance with previous studies by Valenta et al. (2001). The introduction of the BT gene into maize improved the capacity of resistance against field infections by *Fusarium* spp. (Dowd 2000). In addition, the reduced level of *Fusarium* toxin in Bt maize was also reported in other studies (Bakan et al. 2002; Munkvold et al. 1999), indicating a beneficial effect on food safety.

The digestibility of the organic substance of GM maize and sugar beets with a herbicide-resistant trait was also studied by substantial equivalence with their non-GM lines. In a protection procedure, GM and non-GM crops were grown under different environments. The maize was dried gently and then made into diets. However, the sugar beet was freshly fed to pigs after chopping. The proportion of the GM and non-GM crops addition in diets was limited to 30% of dry matter in total food intake. Most of the variations between GM and non-GM groups were within the normal range, such as crude protein, crude fiber, dry matter content, and digestibility of the organic substance. The variation of digestibility of organic substance should be paid more attention.

Unintended effects of GM soybean (glyphosate tolerant) were evaluated by Armstrong et al. (2001) with 100 pigs. Other parameters such as sensory qualities, shear forces of pork, and loss in cooking were also determined in the study. The results indicated that the GM soybean was as safe as its non-GM control line. In another study on pigs with GM maize, Hyun et al. (2004) found that the GM maize did not significantly change food intake, growing situation, carcass yield, and body fat rate of animals compared with the conventional maize line. Component analysis and nutritional value of Bt maize were assessed in pigs by Piva et al. (2001). Both the Bt maize and conventional maize were cultivated under the same conditions; however the non-GM line was more easily contaminated with *Fusarium* toxins. In consequence, the body weight gain of pigs fed with Bt maize diet was significantly elevated compared with those fed the non-GM maize, indicating a potential safer level of Bt maize.

2.4.3 Sheep

Nutrient use efficiency of Bt corn (Event 176) and its conventional control line were tested by French scientists (Barrière et al. 2001) with Texel sheep. Twelve Texel sheep were acclimated in the environment for 1 week, then divided into two groups (six sheep per group). The animals were housed in digestibility crates independently and fed with the GM (Rh208Bt) and non-GM (Rh208) corn diets. The diet formula was adjusted to balance nutritional requirements of the laboratory animals. The results of component analysis of both lines showed no statistically significant differences. In addition, digestibility of organic substance, crude fiber, and neutral detergent fiber were not significantly different between Rh208Bt and Rh208.

Nutrient use efficiency of sugar beet roots and top silage were conducted on German blackface breed sheep. The experimental period was 14 days; the feces were collected over 10 days (Böhme et al. 2001). The animal diet was given as basal ration, and 50% or 60% the GM diets was replaced by sugar beets or top silage to make the GM sugar beet diets. The coefficient of digestibility of sugar beets was greater than 0.90. For most of the animal results, there were no significant differences between the GM and non-GM hybrids. Interestingly, the digestibility of GM hybrids was improved by 3.3% compared with the non-GM hybrids. However, a tendency toward a decline in NFE digestibility was found in the GM hybrids group. The digestibility and the energy content of sugar beet top silage for ruminants of the GM hybrids was slightly lower than that of the control group. Considering the differences were only noticed between the control and the conventionally treated transgenic hybrid, they could be regarded as having no biological significance, which was also indicated by the fact that the differences in digestibility and energy content were small.

2.4.4 Cattle

The nutritional value and food safety of GM plants products were also evaluated using dairy cows and growing cattle. In the first evaluation test in cattle, the authors reported that cows fed with glyphosate-tolerant soybeans achieved higher milk performance, which was probably a result of weaknesses in the experimental design. Subsequent studies with GM plants on cattle did not find any significant differences between GM and non-GM diet groups. Bt maize and its isogenic maize (variety Cesar) diets with 33% content incorporation were fed to bulls in a long-term feeding experiment. The results shown no significant differences between the two lines, either in the constituents or in the digestibility of the crude nutrients or energy content. The animals consumed an average of 18.7 kg and 18.8 kg fresh weight of the GM and non-GM silages, respectively, and the body weight gains were 1482 g and 1487 g, respectively. The results also indicated that meat and fat yield were accordance with the health of the animals.

GM maize (Bt) and non-GM (Cesar) line silages were fed to German Holstein bulls by Daenicke et al. (1999). Forty bulls with age of 165 days were divided into GM and non-GM maize silage groups (20 bulls per group). The animals with an initial average weight of 188 kg were fed both diets until they weighed about 550 kg. No significant differences in the nutrient composition of the corn silages were observed. Bulls fed with a GM maize silage diet showed less food and energy intake than that of the non-GM group. However, average daily gain (ADG), hot carcass weight, dressing percentage, and abdominal fat were not different for bulls fed the GM and non-GM corn silage.

Nutrient availability of GM maize (Event 176 and Bt 11) and non-GM hybrids were also investigated on 12 lactating Holstein cows at Iowa State University (Faust, 2000; Faust and Miller, 1997). Fresh silage from the two GM maize and a non-GM control hybrid were incorporated into the diets and fed to the cows for 14 days. The results showed no significant differences among treatment groups in the parameters such as food intake, milk production, and composition in milk. Another nutritional evaluation experiment with GM maize hybrids were conducted on 16 lactating Holstein cows at the University of Nebraska (Folmer et al. 2000) for 21 days. The trial was to assess the influence and unintended effects of early (N4242) and late (N7333) maturing maize with or without the Bt gene from Event Bt 11 on laboratory animals. To the animal diets was added a maximum of 40% maize silage, and the nutrition was balanced for cows. Indeed, no statistically significant differences in the parameters including milk production, milk composition, milk component yields, efficiency of FCM production, ruminal pH, concentration of VFA in rumen fluid, or in situ NDF digestion kinetics were observed among these treatment groups.

Nonlactating pregnant cows (about 2 years old) were used to evaluate nutritional value and food safety of GM maize (Bt) residues by Hendrix et al. (2000). Grazing preference was recorded in regulated intervals. The body weight gain and grazing preference of the experimental treatment groups did not differ from each other. Therefore, the Bt maize residue was as safe and nutrient effective as its non-GM conventional counterpart.

2.5 Perspective of Nutrient Assessment of GMOs

It is estimated that the world's population will increase from 7.3 billion people in 2016 to approximately 9.7 billion people by the year 2050. However, the global supply of food required to adequately meet human nutritional needs over the next 35 years is estimated to be quantitatively equal to the amount of food previously produced throughout the entire history of humankind. Modern methods of biotechnology, particularly transgenic technology, contribute to enhance crop yield, improve nutritional value, and increase resistance capacity for inimical growing conditions of crops. In spite of the many benefits brought by transgenic technology, great attention should be paid to the safety of these biotechnological products in the future. The food and feed safety of future products produced by transgenic technology should be determined as safe as their conventional counterparts. In this case, it was concluded that "Additionally there is a growing body of scientifically valid information available that indicates no significant risk associated with the consumption of DNA or the resulting proteins from GM crops that are registered in any of these countries. Based on the safety analyses required for each crop, consumption of milk, meat and eggs produced from animals fed GM crops should be considered to be as safe as traditional practices.", as reported by Beever and Kemp (Beever and Kemp 2000) in an review.

However, in the future biotechnology methods may be more flexible, powerful, and faster. We may expect crops with newly expressed proteins in crops and/or changed composition, but also crops with increased resistance against stresses such as drought and saline soils and more efficient in using limited natural resources. In addition, crops with more than one event (stacked events) are in development or already in cultivation. Such crops (feeds) should be more or less equivalent in composition and nutritive value to their isogenic counterparts. Changes in composition and possibly nutritive effects are not intended, but still have to be considered as possible unintended effects of the genetic modification during nutritional and safety assessment.

2.5.1 GM Crops with Additional Healthful Nutrients

It is reported that about 250 million preschool children worldwide are deficient in vitamin A, according to data from WHO. Unfortunately, 1% to 2% of vitamin A-deficient children go blind and half of them may die within 12 months of losing their sight each year. Most of the vitamin A-deficient children are located in Africa and restricted to poor sources of nutrients (Hefferon 2015). Golden rice is a variety of rice (*Oryza sativa*) produced through genetic engineering to biosynthesize beta-carotene, a precursor of vitamin A, in the edible parts of rice. It is one of the best recognized examples of the use of genetic engineering technology to improve a crop's nutritional value. The expression level of beta-carotene in the first generation of golden rice was 6 μg/g brown rice. In 2005, the second generation of golden rice, which produces up to 23 times more beta-carotene than the original golden rice, was developed successfully (Paine et al. 2005). Nutritional evaluation of the golden rice shown that bioavailability of the carotene from golden rice has been confirmed and found to be an effective source of vitamin A for humans. In addition, the golden rice yields well, has good taste and cooking qualities, and did not cause any significantly adverse effects on animal and human health (Demont and Stein 2013). It was reported that the golden rice would be released to the public over a decade (Hefferon 2015).

A field test was conducted to evaluate agronomic traits or any environmental influence of the first generation of golden rice (Datta et al. 2007). Violeta Villegas, coordinator of the International Rice Research Institute (IRRI) Golden Rice project, gave update on the status of the project at regular intervals. Dr. Villegas said, "The project is progressing with a new lead transgenic event, GR2-E, which is characterized by better agronomic traits and higher expression of beta-carotene." Several countries, such as the Philippines, Bangladesh, and Indonesia, would plant the new variety first. In 2016, small field tests of the Golden Rice GR2-E were performed in Philippines and Bangladesh, respectively. Before planting in confined field tests, the rice varieties underwent preliminary assessment inside the greenhouse. If any food regulatory approval is received in one of the participating countries, IRRI will supply the new Golden Rice GR2-E to an independent third party to further evaluate its efficacy at alleviating vitamin A deficiency.

Although many persons and organizations are opposed to the Golden Rice project, the coordinator of the project pointed out that the development of golden rice will meet the needs of farmers and consumers and contribute to economy and human health. IRRI declared that "Golden Rice will only be made available broadly to farmers and consumers if it is successfully developed into rice varieties suitable for Asia, approved by national regulators, and shown to improve vitamin A status in community conditions. If Golden Rice is found to be safe and efficacious, a sustainable delivery program will ensure that Golden Rice is acceptable and accessible to those most in need."

Golden rice increases concentrations of beta-carotene, which is only one goal of conventional crop breeding and genetic engineering. Projects for increasing nutrients such as iron and zinc in crops as different as wheat, pearl millet (*Pennisetum glaucum*), and lentil (*Lens culinaris*) are at varied stages of development. Currently, projects of improving nutrients in crops are at various stages of completion and testing.

2.5.2 Genetically Modified Crops with Lower Concentrations of Toxins

Acrylamide is a chemical compound that is classified as an extremely hazardous substance. The discovery of acrylamide in some cooked starchy foods in 2002 prompted concerns about the carcinogenicity of those foods. Acrylamide can be produced in starchy foods during processing under high temperatures. For example, acrylamide will generated in the processing of potatoes for French fries and potato chips. Baking bread also produces acrylamide. Acrylamide is considered a potential occupational carcinogen by U.S. government agencies and is classified as a Group 2A carcinogen by the IARC. However, as of 2014 it is still not clear whether dietary acrylamide consumption affects people's risk of developing cancer. Experimental results that are based on feeding acrylamide to animals might not be applicable to humans.

In food processing, acrylamide can be generated in a chemical reaction between a reducing sugar and asparagine. In this case, the concentration of acrylamide is expected to be decreased by decreasing the concentration of either reducing sugar or asparagine. A GM potato strain with significantly reduced expression of asparagine posed the advantageous cooking property that only 5% of the acrylamide was produced when cooked at high temperatures (Rommens et al. 2008) compared with the non-GM control line.

A low-acrylamide potato produced by Simplot Plant Sciences was deregulated by the USDA in 2014. Safety evaluation information of the potato was submitted to the FDA, and there were no problems with respect to the company's assessment of composition or safety (FDA 2015). It is expected that reduced acrylamide in potatoes should lower the overall acrylamide intake for many people. The average concentration of acrylamide in McDonald's French fries was 288 ppb as reported by the FDA. However, Wheatena Toasted Wheat Cereal had 1057 ppb acrylamide, which is much more than that from fast-food French fries. The level of acrylamide in any toasted bread is expected to be high. Consequently, the low-acrylamide potato may decrease total human exposure. In addition, the limits for exposure to acrylamide were established by the EPA and actually the current exposures are generally below the limits. The low-acrylamide potato is just one of the events with a lower food toxin concentration. Some other GM crops with lower natural toxin concentrations are in the stage of development.

Fungal toxin including aflatoxins and fumonisins can be produced by *Fusarium* and *Aspergillis* fungi in moldy maize (Bowers et al. 2014). Aflatoxins are regarded as human carcinogens based on sufficient evidence of carcinogenicity from studies in humans by the National Toxicology Program. Aflatoxins are regularly found in improperly stored staple commodities such as cassava, chili peppers, corn, cotton seed, millet, peanuts, rice, sorghum, sunflower seeds, tree nuts, wheat, and a variety of spices. When contaminated food is processed, aflatoxins enter the general food supply and cause a global health problem (Wild and Yun 2010). Fumonisins can also cause many physiological disorders and may be carcinogenic to humans. Genetic engineering is expected to lower the levels of natural toxins in foods and improve human health. However, more effort should be made to develop GM crops with new beneficial traits. In addition, nutritional and food safety evaluation of these GM events is required in the future.

2.5.3 *Industrial and Pharmaceutical Genetically Modified Organisms*

Industrial and pharmaceutical proteins can be produced in GM plants as well. There are many advantages of GM plants manufacturing over conventional fermentation systems such as low cost, safer, easier large-scale production, and better storage at room temperature (Daniell et al. 2009; Twyman et al. 2003).

Studies on pharmaceutical proteins or peptides produced in GM plants have progressed rapidly in the past decade. For instance, pharmaceutical proteins or peptides including vaccines, hormones, cytokines, bioactive peptides, and human serum albumin (HSA) have been developed (He et al. 2011; Wakasa and Takaiwa 2013; Yano and Takekoshi 2004). Many substances have already been produced in GM plants and are in different stages of clinical trials. Recently, HSA was expressed in rice with a productivity rate of 2.75 g/kg brown rice by Wuhan University, China (He et al. 2011). This production rate meets the needs of large-scale commercial production. Many experiments related with "substantial equivalence" between rice-extracted HSA (rHSA) and plasma-derived HSA (pHSA) were performed by He et al. (2011). Amino acid sequence and crystal structure of rHSA were exactly the same as with pHSA. Both rHSA and pHSA have equivalent biological characteristics, including efficiency of promoting cell growth and immunogenicity in vitro and in vivo (He et al. 2011).

Plants have great advantages and significance as bioreactors to produce pharmaceutical proteins, which can substantially meet the need of the rapid volume increases of therapeutic antibodies. Many plants are suited for use as bioreactors. Recombinant proteins with production greater than 20 kg per hectare have been reported utilizing alfalfa, soybean, maize, tobacco, and rice (Just et al. 1995; Khoudi et al. 1999). There are also some other pharmaceutical GM plant expression systems such as algae, moss, duckweed, and plant suspension-cultured cells, which offer the opportunity of fast turnover and high-yield molecular farming in a highly contained and completely controlled environment (Liénard et al. 2007).

Although the GM plants expressing industrial and pharmaceutical proteins will not be sold in the food market, some concerns for the food safety of these GM plants must be addressed before large-scale applications. This matter is of particular concern for the potential release of GM forms of staple crops such as rice, wheat, corn, and potatoes into conventional plant populations (Halford and Shewry 2000).

2.5.4 *Future Challenges of Nutrient Assessment for GMOs*

The overall prospects of the acceptance of GM plants and derived products will depend on positive public perception of this technology. Particularly, agricultural biotechnology companies should comply with laws and regulations and conduct studies on biosafety strictly. Similar to other newly developed technologies, genetically modified technology faces its own unique challenges and barriers, particularly public concerns on nutritional and food safety, environmental safety, and hurdles to global trade. Scientific and systematic regulations of GMOs for development, field testing, and marketing should be determined rigorously to clear ambiguity and potential lawsuits. To introduce specific traits into a plant, numerous technological challenges that must be overcome. In addition, enormous investment is required for developing and obtaining authorization for the commercialization of a GM event, which limits the development of GMOs for selecting traits of wide interest only. Agricultural biotechnology companies such as Monsanto, Syngenta, Dupont, Bayer Crop Science, and Dow Agro should pay more attention to the biosafety of GMOs and the farming community.

At present, numerous successes have been documented, particularly with regard to agronomic traits such as herbicide resistance, pest resistance, and drought tolerance. GM crops are already highly adopted in many developing and developed countries such as the USA, Canada, Brazil, and Argentina. GM lines with new traits including nutrient improvement, stress resistance, yield increases, and industrial and pharmaceutical product introduction are responsible for launching a new generation of improvements in human nutrition and health in the future. However, this is a challenge for developers, government monitoring regulation, and third-party detection institutions to determine the nutritional safety of the newly developed GM plants.

Chapter 3
Overviews of Food Allergy and Evaluating Methods Used in Allergenic Assessment of GMOs with Application Examples Conducted in Our Laboratory

3.1 The Background of Food Allergy, Allergenic Worries in GMOs, and the Decision-Tree Approach to Evaluate the Allergenicity of Genetically Modified Foods

3.1.1 Food Allergy

Allergenicity is a widespread adverse effect of foods, several plants, tree and grass pollens, industrial chemicals, cosmetics, and drugs. The first allergen to be sequenced was allergen M from codfish, and during the past 30 years a large number of native and recombinant allergen molecule sequences have become available. Up to the present, the number of known allergenic protein sequences probably amounts to more than 2500, and the figure is constantly growing (Poulsen 2004).

Food allergy, which has continued to be a serious health concern from ancient times, is defined as an adverse health effect raised from a specific immune response that reproducibly occurs on exposure to a given food (Boyce et al. 2010). Self-reporting of lifetime allergic responses to each of the most common food allergens (milk, eggs, wheat, soy, peanuts, tree nuts, fish, shellfish) ranges from 1% to 6% of the population (Nwaru et al. 2014). The most common type of food allergy is mediated by allergen-specific immunoglobulin E (IgE) antibodies. IgE-mediated reactions are known as immediate hypersensitivity reactions. The pathogenesis of immunoglobulin E (IgE)-mediated food allergies can be divided into an afferent sensitization phase wherein the immune system develops an IgE response to one or more constituents (allergens) of a food and an efferent eliciting phase in which a clinical allergic reaction occurs after ingestion of the food in question (Poulsen 2004). IgE-mediated reactions may occur to pollens, mold spores, animal dander, insect venoms, and other environmental stimuli as well as foods (FAO 2001). Peanuts, soybeans, milk, eggs, fish, crustaceans, wheat, and tree nuts are common allergenic foods that are associated with IgE-mediated reactions. In IgE-mediated food allergies, exposure to a specific food and the proteins contained therein elicits

K. Huang, *Safety Assessment of Genetically Modified Foods*,
DOI 10.1007/978-981-10-3488-6_3

the development of food allergen-specific IgE antibodies. These IgE antibodies attach to the surfaces of mast cells and basophils, thus sensitizing the individual to react upon subsequent exposure to the specific food (FAO 2001).

However, food allergies also encompass delayed hypersensitivity reactions. In cell-mediated reactions, the onset of symptoms occurs more than 8 h after ingestion of the offending food. The most common cell-mediated hypersensitivity reaction affecting all age groups of the population is celiac disease, also known as gluten-sensitive enteropathy. Gluten-sensitive enteropathy or celiac disease is a T-cell-mediated immunological response triggered by gluten (gliadin) that affects genetically disposed individuals. The active phase of the disease consists of an inflammatory process in the small intestine leading to malabsorption with body wasting, anemia, diarrhea, and bone pain as well as other symptoms (FAO 2001).

Both IgE-mediated food allergies and non-IgE-mediated reactions are treated with specific avoidance diets. Because the threshold dose is low and not precisely defined in both cases, affected individuals can experience difficulties in the adherence to the avoidance diets (FAO 2001).

3.1.2 Overview of GMOs

Genetically modified foods have become more and more available to people in our daily lives. GMOs are a result of genetic engineering that uses recombinant DNA technology to transfer genes from one organism to another (Spök et al. 2005). The introduction of novel proteins into plants may bear an increasing risk of eliciting allergic reactions in already sensitized allergic individuals or inducing de novo allergic sensitizations (Spök et al. 2005).

However, existing methodologies cannot meet the urgent requirements for rapid biosafety assessment of the increasing number of new and sophisticated GMOs. The allergenic potential of novel foods and food ingredients is mentioned in EU Regulation 258/97 ('Novel Food Regulation,' which was replaced by Regulation 1829/2003 in April 2004) and in the Recommendation 97/618/EC. There are, however, some problems in the assessment: these applications contained basically no experimental work to assess the allergenic potential of the GMPs (Good Manufacturing Practice). Only one of the applications included study of the IgE reactivity of the foreign protein. All other applications contained only arguments as to why the GMP does not present an allergenic risk. Mainly the oral route was considered as a possible route for allergic sensitization, whereas no consideration was given to other exposure or sensitization scenarios (Spök et al. 2005).

Currently, there is no single predictive test that can perform a hazard assessment in relationship to allergenic properties of newly expressed proteins in genetically modified organisms (Poulsen 2004). Therefore, a new, more efficient, and objective biosafety assessment methodology is needed. We suggest that the assessment procedures should be equally applied to GMOs as well as natural cultivars to establish effective measures for allergy prevention. A decision tree for evaluating allergenicity of genetically modified foods was developed by FAO/WHO in 2001.

The foreign DNA can be coated on particles that are inserted directly into the cells of the plant using a particle gun. Another possibility is to insert the foreign DNA into vectors such as plasmids or plant viruses. These vectors serve as molecular carriers to infect or transform plant cells and insert the desired DNA into the plant genome. The most commonly used bacterium is *Agrobacterium tumefaciens*, which has the ability to infect many types of plants. However, advances in research have shown that several antigen sources (e.g., maize, bacteria, yeast) that were thought not to act as as allergen sources do contain allergenic molecules. It should also be emphasized that even a protein that does not represent an allergen in its natural source may become a potent allergen through a new route of presentation.

Speakers and some members of the public have suggested that the prevalence of food allergies has increased because of genetically engineered (GE) crops. The Committee on Genetically Engineered Crops examined records on the prevalence of food allergy in the United States over time. As is clear from a survey by Jackson et al. (2013), the prevalence of food allergies in the United States is rising. For a rough comparator, the committee examined data on hospital admissions for food allergy in the United Kingdom over time. UK citizens eat far less food from GE crops. The data suggest that food allergies are increasing in the United Kingdom at about the same rate as in the United States (although the types of measurement are different) (Gupta et al. 2007).

Although there is almost no scientific evidence that genetically modified organisms (GMOs) exhibit increased allergenicity compared with the corresponding wild type, significant concerns have been raised regarding this matter. In principle, it is possible that the allergenic potential of GMOs may be increased by the introduction of potential foreign allergens, to potentially upregulate expression of allergenic components caused by the modification of the wild-type organism (Spök et al. 2005). The concerns about possible allergenicity of GMOs gained increasing support when it was found that a Brazil nut allergen had been engineered into transgenic soybeans and other transgenic plants (Spök et al. 2005).

3.1.3 Decision-Tree Approach to the Evaluation of the Allergenicity of Genetically Modified Foods

3.1.3.1 Introduction

A decision tree of evaluation of allergenicity of genetically modified foods was developed by FAO/WHO in 2001. An assessment approach similar to that developed by IFBC/ILSI was advocated that included the following criteria: source of the transferred genetic material, molecular weight, sequence homology, heat and processing stability, effect of pH and/or gastric juices (digestive stability), and prevalence in foods. Furthermore, the 1996 Consultation made several recommendations relative to allergenicity of genetically modified foods (FAO 2001):

- The transfer of genes from commonly allergenic foods should be discouraged unless it is documented that the gene transferred does not code for an allergen.
- Foods found to contain an allergen transferred from the organism that provided the DNA should not be considered for marketing approval unless such products can be clearly identified in the marketplace and this identity will not be lost through distribution and processing.
- Involved organizations should consider the appropriateness of actions to take in respect to foods containing new protein(s) that are determined to have the characteristics of an allergen, even though no patient population is known to exist with an allergy to this gene product.
- The identification of food allergens and the characteristics of these allergens that define their immunogenicity should be encouraged.
- If a genetically modified food contains the product of a gene from a source with known allergenic effects, the gene product should be assumed to be allergenic unless proven otherwise.
- If information about the genetic content does not positively determine it to be allergenic, additional data should be considered for addition to the decision-tree approach.

3.1.3.2 Food Containing a Gene Derived from a Source Known to Be Allergenic or Not Known to Be Allergenic

When the expressed protein stems from a origin known as allergenic, the analysis put forward in the FAO/WHO 2001 decision tree describes both sequence homology and subsequent evaluation of allergenic capacity of the expressed protein with sera of patients who are allergic to the original material (FAO 2001).

When the expressed protein stems from a origin known to be allergenic, the FAO/WHO 2001 decision tree requires (1) sequence homology to known allergens (food and environmental), (2) targeted serum screening for cross-reactivity with sera from patients allergic to materials that are obviously correlative to the origin material of the gene, (3) pepsin resistance, and (4) immunogenicity testing in animal models. Under these circumstances the search for homologous allergens is built on two steps (FAO 2001).

3.1.3.3 Post-marketing Surveillance

The FAO seminar considers the pre-market allergenicity evaluation methodology of the genetically modified food as a satisfying security guarantee. However, in consideration of the polytropism of genetics in human beings and food ingestion of different areas, if the product of a genetically modified food has appeared in the market, more assessment of the negative effects of the genetically modified food should be considered. (FAO 2001) Recorded data should be checked with respect to these points (FAO 2001):

- The results of clinical tests for allergenicity
- The relevance between the recorded negative interaction and the determined genetically modified food/food material

3.1.4 *Standardization of Methodologies*

3.1.4.1 Sequence Homology as Derived from Allergen Databases

The commonly used protein databases are currently not fully up to date. Now, people recommend procedures about finding amino acid status among the expressed protein and known allergens using a specialized allergen database that is under construction.

3.1.4.2 Specific Serum Screening

According to the FAO/WHO decision tree of 2001, a lack of sequence homology of the new protein with allergen sequences should lead to a serum screen.

With known allergies to related provenience materials, the activity of IgE antibodies in the sera of patients has been assessed; a proper in vitro method was supposed to be used. Some kinds of credible immunoassays are feasible for the objective (FAO 2001).

In case of an allergenic source organism, a 'specific serum screen' and otherwise a 'targeted serum screen' should be conducted. Although the former term means that sera from clinically well-characterized patients who are allergic to the provenience of the special protein are applied, the latter suggests the use of sera from patients allergic to related proteins of the newly inserted protein. However, some studies indicate that testing of the newly inserted protein alone is not sufficient. Instead, we suggest extending the serum screening from the target protein to the whole GMP to include effects that may indirectly increase the allergenicity of the GMP through pleiotropic effects (Spök et al. 2005).

3.1.4.3 Targeted Serum Screening

In expressed protein and known allergens, if no sequence homology was obtained, it is not certain whether a homologous allergen is involved because of a structural deletion on the related allergen. In allergic testing, choosing random serum samples has only a small probability of being helpful. However, in some situations, some directed method may be more proper (FAO 2001).

For this "targeted serum screen," six groups of source organisms are distinguished: yeast/molds, monocots, dicots, invertebrates, vertebrates, and "other." A panel of 50 serum samples with high levels of IgE of allergens in the relevant group is used to search for IgE antibodies that are cross-reactive with the expressed protein. If a positive reaction is obtained with one of these sera, the expressed protein is considered to be an allergenic risk and further evaluation for allergenicity would typically not be necessary. If a gene were obtained from a bacterial source, no targeted serum screening would be possible, because no normal population of individuals is known to be sensitized (IgE mediated) to bacterial proteins (Poulsen 2004).

3.1.4.4 Pepsin Resistance

Expressed protein when refined (non-heated and non-processed) is supposed to be placed into a situation of pepsin degradation. Moreover, the expressed protein is supposed to be evaluated in its edibility form under the same pepsin degradation circumstance to that applied to check the expressed protein. Both known nonallergenic (soybean lipoxygenase, potato acid phosphatase, or equivalent) and allergenic (milk beta lactoglobulin, soybean trypsin inhibitor, or equivalent) food proteins are supposed to be vested in comparators to confirm the relevant rank of the expressed protein pepsin resistance (FAO 2001).

However, it may be a misconception that allergens are always characterized by resistance to proteolytic digestion. Several studies have clearly demonstrated that stability against digestion does not represent a feature that is strictly associated with allergenicity. The focus on resistance to digestion also may be overestimated considering studies which indicate that sensitization and boosting of the allergic immune response via the respiratory tract may be more important in allergic sensitization than mechanisms operating via the gastrointestinal tract (Spök et al. 2005).

3.1.4.5 Animal Models

In the development, additional assessment of potential allergic expression protein can generate information data using animal models (FAO 2001). Animal models are being used to assess very different mechanisms: (1) a mechanistic approach to the understanding of IgE-mediated disease and (2) determining the allergenicity of novel proteins (Helm 2002). Several animal models have provided important information for understanding some of the mechanisms of allergenicity. A number of interesting animal models of food allergy have been developed, and mice, rats, guinea pigs, dogs, and swine have been employed as experimental animals (Spök et al. 2005).

3.2 In Vitro Evaluation Method Used in Allergenic Assessment of GMOs and Several Application Examples in Different Aspects Conducted in Our Laboratory

3.2.1 Bioinformatic Analysis

Bioinformatics analysis is widely used to assess the potential allergenicity of a protein. Assessment of the FAO/WHO recommends foreign protein allergen.

A key to the evaluation of the endpoint security novel protein and protein amino acid sequence similarity may be the potential safety hazard, that is, known toxins or allergens. For this reason, amino acid sequence comparison of the current food safety portion of the decision-tree strategy is recommended by FAO/WHO (2001) and Codex Alimentarius (2003). Immunoglobulin E (IgE) between new expression of the protein and a known allergen should be considered when there may be more than 35% of identity for a period of 80 or more amino acids (FAO 2001; Xu et al. 2009).

We have done some research with bioinformatic analysis, as listed in the following.

Research 1-a Safety Assessment of Cry1Ab/Ac Fusion Protein: this research was published in the journal *Food and Chemical Toxicology* (Xu et al. 2009).

1. Introduction

Insect pests cause the destruction of 20–30% of the world's crops annually (Estruch et al. 1997). GM crops are becoming more and more important characteristics of the agricultural landscape. As one of the world's most widely grown GM crops, *Bacillus thuringiensis* (Bt) corn was introduced in 1987, having a transgenic Bt gene against crop damage. Cry1ab/ac genes in this study are the result of the cry1ab gene (GenBank accession no. X54939) and the Cry1ac gene (GenBank accession no. Y09787), and its coding is highly toxic to Lepidoptera pests, such as *Chilo suppressalis*, *Scirpophaga incertulas*, and *Cnaphalocrocis medinalis*.

Vazquez and others have stated CrylAc is a powerful system, and mucosal immune and protoxin (pCry1Ac) is bound by immunohistochemical testing on the surface of the intestinal mucosa of rats (Vazquez-Padron RI et al. 1999, 2000). Moreover, this protein induced in situ temporal changes in the electrophysiological properties of the mouse jejunum. The foregoing data indicated a possible interaction in vivo of Cry proteins with the animal bowel, which could induce changes in the physiological status of the intestine (Vazquez-Padron et al. 2000). However, other studies concluded that GMO (Bt-Cry1Ac gene) cottonseed meal had no deleterious effect on growth performance, blood biochemicals, and various carcass characteristics of growing broiler chickens (Elangovan et al. 2006). The GM cottonseed expressing Cry1F, Cry1Ac, and PAT proteins had no adverse effects in 90 days of feeding test (Dryzga et al. 2007); in contrast to Cry1Ac protein, Cry1Ab was proved to be harmless by several studies (Bondzio et al. 2008; Kroghsbo et al. 2008; Onose et al. 2008; Schrøder et al. 2007). In this study, Cry1Ab/Ac protein is a new member of the family and safety assessment of the Cry1Ab/Ac protein is necessary for the wider use of Cry1Ab/Ac genes of rice. Although Cry1Ab and Cry 1Ac are harmless, changes in the protein primary structure will lead to change of secondary structure, and then protein digestion properties and thermal stability will also change. Two fusion proteins may also result in a new protein relative to the allergens or new glycosylation sites. Furthermore, there is much influence of one food on another, so we do not know if Cry1Ab and Cry 1Ac had any type of reaction before this study. Rice is a very important food for humans, so the safety assessment of Cry1Ab/Ac protein is quite necessary.

2. Methods

2.1 Amino acid sequence homology search

Identifying any known Cry1Ab/Ac protein in an online search shows potential allergic problems include 35% of the amino acid and protein, a window of 80 amino acids. To a certain extent, this assessment may infer that there are concerns of the potential toxicity. Complete Cry1Ab/Ac protein amino acid sequence and protein sequences are found in all six large public database references: NR, Refseq_Protein, SwissProt, Pat, PDB, and Env_nr. The algorithm used was BLASTP 2.2.18 + and the scoring matrix was BLOSUM62. The same bioinformatics program (see the earlier section), a study by comparing Cry1Ab/Ac protein, is divided into eight amino acids, a possible allergen of antigen epitope of allergen databases (Hérouet et al. 2005).

2.2 N-Glycosylation search

The entire length of the Cry1Ab/Ac protein is found in any N-glycosylation NetNGlyc 1.0 server web site (http://cn.expasy.org). The study is also supplemented by its determination in vitro.

3. Results

3.1 No homology with known toxins or allergens

The latest computer analysis results show that there is no evidence that there is any similarity between Cry1Ab/Ac protein and toxic or any known allergy proteins; that is, less than 35% of the identity and the known toxins or allergens in the length of 80 amino acids and 8 continuous amino acids with known allergen.

3.2 No N-glycosylation sites

Many protein allergen saccharifications increase the possibility of glycosyl groups that may cause allergic reactions (Jenkins et al. 1996). This concept may be considering a new protein glycosylation pattern associated with allergic reaction may differ with local counterparts. In this study, there are four such sites on the Cry1Ab/Ac protein, found by searching for potential N-glycosylation sites, but there was no N-glycosylation site on the Cry1Ab/Ac protein as actually proved by the in vitro assays.

Research 2-a A Modified Weight-of-Evidence Approach to Evaluate the Allergenic Potential of Food Proteins: this research was published in the journal *Toxicology Research* in 2015 (Sun et al. 2015a).

1. Introduction

Assessment of the potential allergenicity of food proteins, including novel proteins, is an important issue. Current methods for the study of the potential allergenicity of novel proteins are based on the effectiveness of the evidence put forward by the Codex Alimentarius commission. The recommended evaluation needs to consider the introduction of protein sources, the amino acid sequence homology comparisons to known allergens, and structural properties (e.g., susceptibility to acid and enzymatic digestion in vitro, heat stability), and, when appropriate, specific IgE-binding studies are considered. This strategy provides useful information to identify a novel food protein sequence homology or potential with known allergen (Sharma et al. 2011). However, considering the li0mitations of the open database and the scientific literature, we need to "rank" the known allergens in the interest of confirming whether the novel proteins have potential allergenicity. Furthermore, previous research has shown the correlation between serum IgE antibody levels and the severity of clinical symptoms (Blanco et al. 1998; Sun et al. 2013). In brief, these methods cannot separate or collectively provide direct evaluation of the inherent allergenic potential (Dearman et al. 2014).

To compensate for this shortage, there is considerable interest in the development of a more comprehensive strategy, including consideration of the antigen epitope of allergens, biochemical identification (i.e., resistance to digestion in simulated gastric fluid, heat stability), human serum analysis, and appropriate cell models and animal models, for assessing the allergenic potential of food proteins. Comparing homology in an eight-amino-acid sequence, which will represent the theory of a B-cell epitope and conservation of the T-cell epitope of minimum size, was originally based on this idea (Metcalfe et al. 1996). For those sequences that have no proteins homologous with known allergens, we need to consider direct analysis with the necessity of using the theory of bioinformatics B-cell antigen epitope. In human serum analysis, the specific IgE/IgG4 ratio may increase the measure of value of a specific IgE level to evaluate the clinical response of food allergens (Caubet et al. 2012), because the balance between the concentration of allergen-specific IgE and IgG4 may affect the clinical allergic sustainable development or tolerance (Ruiter et al. 2007). Furthermore, the rat basophilic leukemia (RBL) cell model is a useful tool for directly assessing potential sensitized food protein (Kaul et al. 2007; Sun et al. 2015b).

To this end, we report improved evidence of the effectiveness of the method evaluating potential allergy rather than novel proteins in food. The dominant allergens in hen's egg whites, ovalbumin (OVA, 45 kDa) and ovomucoid (OVM, 28 kDa), were selected as a model for antigen allergy research because its widespread use provides better comparison with previous reports.

2. Materials and Methods
2.1 Test materials

Lyophilized ovalbumin (OVA, A5503, grade VI, 98% purity), ovomucoid (OVM, T2011, type III-O, free of ovoinhibitor), and potato acid phosphatase (PAP, P3752) were purchased from Sigma-Aldrich (St. Louis, MO, USA) and dissolved in Milli-Q water to 5 mg ml^{-1}. The ovalbumin (OVA) and ovomucoid (OVM) solutions were put into a 100 °C water bath heating for 5 min, simulating the common boiled egg processing. The PAP was considered as a nonallergenic protein and selected as the negative control.

2.2 Bioinformatics

We analyzed the hydropathy (Hoop–Woods strategy), flexibility (Kparlus–Schuzl strategy), and antigenic index (Jameson–Wolf strategy) of the primary structure of OVA, OVM, and PAP by the DNAStar Protein system, based on their amino acid sequences acquired from the NCBI protein database (http://www.ncbi.nlm.nih.gov/protein). The overlaps of the prediction results obtained from the three parameters were considered as potential linear-epitope regions.

3. Results and Discussion

Actually, proteins do not cause an allergic reaction, but rather the conformation or linear antigen epitope IgE antibody. In young children whose gastric trace is not mature, allowing a greater increase the absorption of macromolecules, the conformational epitope initially may have an important role. After intestinal maturity, including its enzyme activity, more linear antigen epitopes are recognized and epitope transmission occurs (Beyer et al. 2003). It has been reported that linear and conformational epitopes in eggs and OVM technology act as primary and secondary IgE recognition, respectively. Egg-allergic patients who possess IgE specific for linear epitopes were reactive to foods in any form (even when extensively cooked) and tend to have a more persistent allergy, whereas those with IgE recognizing predominantly conformational epitopes exhibited partial clinical tolerance.

Bioinformatics research predicts higher OVM allergies may be compared with OVA, based on the analysis of the potential linear antigen epitope.

Research 3 The Analysis of Specific Allergenicity of Food Allergens Families: this research was published in the journal *Food and Agricultural Immunology* in 2013 (Wang et al. 2014c).

1. Introduction

Proteins can be divided into different families, according to their similarity in amino acid sequence and three-dimensional (3D) structure. Mills et al. have reported that most of the food allergens are contained in only a few protein families (Mills et al. 2004). Thus, bioinformatics can evaluate potential novel allergen proteins.

Of plant food allergens, 65% belong to four families: the Cupin superfamily, Prolamin superfamily, Bet v 1-related protein, and Profilin (Jenkins et al. 2005). It is reported that most plant food allergens display a relationship between the pathogenesis-related proteins (PRs) (Breiteneder and Ebner 2000). Accordingly, homologous PRs are divided into the plant defense systems of the family.

Most animal food allergens can be included in three allergens families: tropomyosins, EF hand proteins, and caseins (Jenkins et al. 2007). Differing from plant allergens, most animal food allergens share homology with human proteins, which may affect the human immune system identification method.

SDAP is the web server that provides a variety of computational tools to study allergic proteins. It can also provide some of the common allergens, the sequence of amino acids from the main proteins, and search the relative biological information server (PDB, SWISS-PROT, PIR). SDAP has listed 528 food allergens so far.

Although the food allergens can be classified as proteins from the SDAP family, not all the family proteins contain allergenic proteins, and consequently the classification based on protein families is overly inclusive. However, the AllFam database records only family allergen protein that contains allergenic proteins to form one; this is helpful for the study of allergens of special structure and function.

The AllFam Allergome-based database is currently the most comprehensive collection of allergen family data, providing information to determine the allergens that cause an IgE-mediated (allergic, atopic) disease (anaphylaxis, asthma, atopic dermatitis, conjunctivitis, rhinitis, urticaria). This database contains all the feature recognition and allergens, including those in terms of allergen in the official list. Allergens AllFam from the classification of the family database contains a protein family. Every allergen family is AllFam homologous with one or more family. AllFam allergens include 186 families and 1091 allergens. The allergens in the AllFam database are from a variety of creatures, such as fungi, insects, and tree pollen. Only food allergen detection and classification experiments are possible, because we focus on the study of food allergen characters.

Protean is one of the DNAStar modules that provides a variety of methods to analyze and predict the structure of proteins and nucleic acids. For analyzing the features of the antigenicity of major food allergens families, we adopted two methods: Rothbard–Taylor and Jameson–Wolf. With Rotherbard–Taylor the specific topics include an analysis method to predict the potential of T-lymphocyte antigen epitope; Jameson–Wolf is used to analyze the antigen index.

The Get SDAP food allergen test in our study is divided into different food allergens from AllFam families based on total allergens. Food allergen after the classification of the family involves the typical main allergen food allergen family selection and analysis in the DNA Star using the protean module.

2. Methods

2.1. Classification of food allergens

In this study, food allergens are divided into different allergens families in the AllFam list. This classification is based on the total AllFam database allergens in the family. The list of 528 food allergens (including the iso-allergens) is gained from SDAP (http://fermi.utmb.edu/SDAP/index.html). Pfam ID the allergens in the SDAP are obtained from the database. Using Shared Pfam ID as standard, the allergen in the same family can be from AllFam recognition and retrieval (http://www.

Table 3.1 Typical food allergens of five allergen families that contain most food allergens

Allergen family	Common allergen
Prolamin superfamily	Ara h 2 (peanut), Ara h 6 (peanut), Ber e 1 (Brazil nut)
Bet v 1-related protein	Api g 2 (celery), Dau c 1 (carrot), Pyr c 1 (pear)
EF hand domain	Gad c 1 (cod), Cyp c 1 (carp)
Cupin superfamily	Gly m 5 (soybean), Ara h 3 (peanut), Ses i 6 (sesame)
Profilin	Ara h 5 (peanut), Lyc e 1 (tomato), Mus a 1 (banana)

meduniwien.ac.at/allergens/allfam/). Then, all the allergens are ranked for every family on the basis of the number of families containing allergens.

Ara h 1 is an example. Ara h 1 is a major peanut allergens, and 7 s vicilins from it predict a superfamily. Pfam ID, obtained from SDAP, is a parameter to identify all other allergens in AllFam separated into family Ara h 1.

2.2. Analysis of antigenicity of food allergens

For the sequence of food allergens from NCBI, we download the FASTA files of typical major food allergens families and the ever-changing DNAStar analysis for the antigenicity of allergens.

3. Results

There are 466 to 528 food allergens listed in the SDAP allergens AllFam family accordingly; 466, or 60%, of food allergens belong to five allergens families: prolamin superfamily, EF hand domain, for predicting superfamily Profilin, and Bet v 1-related proteins. Typical food allergens families are shown in Table 3.1. The 28 allergens family contains fewer than three food allergens and accounts for 60% of all food allergens. It is shown that most of the food allergens can be divided into a food allergen family. This result is in keeping with the Mills research to some extent.

3.1 Prolamin superfamily

As one of the main plant food allergens, the prolamin superfamily contains 57 food allergens and 51 iso-allergens. Most allergens from the family have a large number of alpha-helix structures and similar three-dimensional structure, and the conservation of the skeleton is composed of eight cysteine residues in the protein sequence (Breiteneder and Radauer 2004; Kreis et al. 1985). The family members share a domain; also formed are four helicals wrapped in a right-handed superhelix folding topology (Ivanciuc et al. 2008). The domain, which is stabilized by disulfide bonds, has an internal cavity (Ivanciuc et al. 2008). In the analysis of the ever-changing prolamin superfamily of allergens, we know that most potential allergens have a T-lymphocyte antigen epitope that is an alpha-helix structure feature from the family. Several common allergens from soy, grains, vegetables, and fruits belong to the family; for example, 2 s albumin, nonspecific lipid transfer protein, grain of alpha amylase, and protease inhibitors. Well-defined peanut allergens, such as Ara h 2, Ara h 6, and Ara h 7, are also included in the prolamin superfamily.

3.2 EF hand domain

The EF hand domain includes 26 food allergens and 26 iso-allergens, most of which come from animals. Most allergic proteins binding protein and calcium sharing a topic with 12 by 12 circulation have residual alpha-helical domains (Ivanciuc et al. 2008). In the protean EF hand domain from the result of analysis of the typical allergens, we know the potential of the T-lymphocyte antigen epitope of the allergen alpha-helix structure. The biological function of these proteins can be divided into two categories: calcium signal and the buffer (or transport). Some reports suggest some EF IgE-binding ability, including allergens combined with calcium (Gao et al. 2006).

3.3 Cupin superfamily

Predicting the superfamily includes 24 food allergens and 26 iso-allergens, most allergens from beans, nuts, and seeds (Mills et al. 2004). The conservation of these proteins has two short consensus sequence patterns and a beta barrel structure. The Jameson–Wolf results show that a large proportion of the glutamine peptide antigen index is high. Also, some potential allergens sharing the same T-lymphocyte antigen epitope describe an alpha-helix structure. Members of this family mainly belong to the seed storage globulin and can be divided into two subfamilies: 7/8 s vicilins 11 s globulin and mature.

3.4 Profilin

Profilin, including 24 food allergens and 20 iso-allergens, most of which are 12- to 15-kDa proteins, mainly exists in eukaryotic cells. From profilin protean analysis of typical allergens, we know that most of the allergens in the family share three or four peptide antigens, so the index is high. In addition, Rothbard–Taylor analysis showed that the most potential allergens share three or four T-lymphocyte antigen epitope antigens, most of which are in beta-strand structure. From 10% to 20% of the patients allergic to allergens profilin also suffer from hay fever (Valenta et al. 1992). It is shown that these allergens have been regarded as risk factors of multiple sources of pollen allergy (Mari 2001).

3.5 Bet v 1-related protein

Bet v 1-related proteins include 18 food allergens and 44 iso-allergens, most of which are widely distributed in vascular plants, such as Rosaceae (apple, pear, drupe), Apiaceae (celery, carrot), and bean (soybean, peanut). Allergens from the family are made up of seven reverse beta-strands and three alpha-helices. They also share a C-terminal amphipathic helical motif, which consists of a helix. From protean analysis of some of these families' typical allergens, we found that a short peptide is found between the 120th to 140th amino acid and that all these 20 amino acids have high antigenic index, and the short peptide often forms an alpha (α)-helix structure. The result of Rothbard–Taylor analysis shows that many allergens of this family share two peptides, which may contain potential T-lymphocyte antigen epitope, at around the 50th–54th and 140th–148th amino acids. Those two peptides are also in α-helix structure. This family was classified by sequence similarity into three subfamilies: pathogenesis-related (PR)-10, Bet v 1 major allergen, and ripening-related (RRP) proteins (Liscombe et al. 2005). PR-10 includes a majority of allergens in those three subfamilies. Bet v 1 is the major birch pollen allergen. Many patients with pollen allergy are also allergic to vegetables and fruits, which suggests plant food allergens have IgE cross-reactivity with Bet v 1.

3.6 The cross-food allergen families include allergens from both plants and animals

Most allergens families include only plant or animal food allergens, which explains why most studies classify plant or animal food allergens separately. In our study, we found three angry food allergens in the family: EF hand domain, serpin serine protease inhibitors, and triosephosphate isomerase, including plant and animal food allergens. The EF hand domain includes only one plant food allergen, bras n 2 (from rapeseed); serpin serine protease inhibitors plus 2D (chicken) and 3D including 33 (wheat); triosephosphate isomerase, including arc s 8 (from) crustacean species, Cra c 8 (from North Sea shrimp), 31 (wheat), and 3 tpi (from wheat). Two of the three families from wheat contain allergens, suggesting that wheat allergens may share similar function or structure with animal food allergens.

Protean results on some typical proteins of this family of allergen analysis showed that most of these three food allergens show an alpha-helix structure. Allergens in the same family often share much specific antigen epitope peptide and other potential T-lymphocyte antigens. In addition, most of the structure of these peptides is an alpha-helix.

4. Discussion

We are limiting discussions of allergens to food allergens from plants and animal, respectively. However, we expand now to all food allergens, including plant food allergens and animal food allergens.

In this study, we use the SDAP food allergen list as a reference. According to the classification of SDAP, there are five main food allergen families: prolamin superfamily, EF hand domain, for predicting superfamily Profilin and Bet v 1-related proteins. Five main families comprise 60% of known food allergens, which suggests that most food allergens are included in a specific minority family of food allergens. Breiteneder and Jenkins also indicate that the majority of allergens belong to several protein families and some protein families even do not contain any allergenic protein, which is in line with the results of our study. Therefore, the use of total allergens family AllFam classification of food allergens rather than classification directly of allergic proteins to study the structure and function of the allergen food allergen family will allow us to understand most of the biological characteristics of food allergens.

In this classification, the prolamin superfamily, which includes only plant food allergens, is the biggest food allergen family. Some reports have shown that the cupin superfamily includes more plant food allergens than the prolamin superfamily. This difference may suggest that more plant food allergens in the prolamin superfamily have been found in the past 6 years.

Our classification is one of the highlights of the three angry food allergens in the family. We can adopt a kaleidoscopic classification and analysis of the three angry food allergens in the amino acid sequence. Therefore, we can find that these allergens share roles. Two of the three families from wheat contain allergens. Wheat allergens are included in the cause of two angry food allergies; the wheat allergen family may share similar functionality with animal food allergens, such as the metal-binding bra n 2 calcium-binding ability. However, the structure of the allergen is most important in evaluating potential allergen sex. Most angry food allergens have an alpha-helix, and the allergen cross-family often shares the same specific antigen epitope peptide containing the potential T-lymphocyte antigen. Because the spatial structure of most allergens from the three cross-families is still unknown, for further investigation of wheat allergen and structure, the relationship between animal food allergens is still necessary.

The study by Ovidiu reported that allergic proteins have specific motifs when compared with nonallergic proteins from the same protein family. In this study we also summarizes some specific topics of allergens in the same family. However, study of this subject is still limited. Although a particular theme of allergens assessment is a useful potential allergenicity of novel proteins, structure, and themes, for more information, it is necessary to use the allergen.

Protean analysis showed that the allergens from the same family may share some peptides and antigenicity. Some studies show that the structure is outstanding, which makes it easy to expose surface antigen. Because the beta-strand alpha-helix structure is stable, it is difficult to form an antigen. However, in this study, we found that the most potential T-lymphocyte antigen epitopes in alpha-helix or beta-strand structures, only some in antigen epitope structures, suggesting that it may not accurately predict antigen determined according to the amino acid sequence.

Food allergen study of the specific antigenic antigenicity of a family leads to a better understanding of food allergens. Scientific classification of food allergens is helpful to further study the biological characteristics of common food allergens and evaluate potential navel allergen protein.

Research 4-a Cell-Based Immunological Assay: Complementary Applications in Evaluating the Allergenicity of Foods with FAO/WHO Guidelines: this research was published in the journal *Food Research International* in 2014 (Sun et al. 2014**).**

1. Introduction

Novel protein as introduced into foods has caused more and more concern from the public about the risk of food allergies. The current approach for assessing the potential allergenicity of novel food proteins is based on a new decision-tree strategy proposed by the Food and Agriculture Organization (FAO) and the World Health Organization (WHO) expert consultation in 2001, which included a bioinformatical analysis, a pepsin digestion assay, serum screening, and animal model testing (FAO 2001). Bioinformatical analysis and pepsin digestion analysis can predict, but we cannot directly assess, potential protein food allergies. Based on serum IgE bound to the protein of a serum screening test can determine the potential risk of largeness for those known allergies (Goodman 2008). However, there was little correlation between the serum IgE antibody level and the severity of the clinical symptoms (Blanco et al. 1998; Mj et al. 1997; Sharman et al. 2001; Sun et al. 2013; Van and Aalberse 1999). Conversely, animal models can detect the potential of a protein to induce IgE-mediated allergic reactions in vivo, but the natural complexity makes it difficult to identify a validated animal model for evaluating the allergenicity to food proteins (Sharma et al. 2011).

To overcome these problems, a quite interesting cell immunoassay is the study of food allergies protein activity, a sensitive cell-screening model to create development of highly active cellular immunological methods. Human basophilic granulocyte and peritoneal mast cells of mice need purification; as these cannot be kept for a long time in the main culture, its widespread use is limited. The permanently growing rat cell line, rat basophilic leukemia (RBL), which was originally obtained by Eccleston and coworkers in 1973, was cloned by a limited dilution technique from leukemia cells isolated from Wistar rats (Eccleston et al. 1973). RBL cells are similar to mast cells and basophilic granulocyte reaction, and threshing after cross-linking IgE-bound Fc epsilon RI allergens and the release of the medium causes allergic reactions. However, RBL cells initiated at different times release a medium ability of tumor development (Passante et al. 2009). Among the RBL cells, RBL-2H3 and RBL-1 cells have been widely used in recent research on food allergies.

Thus, the allergenic potential of different food proteins, including glycinin (Holzhauser et al. 2009), ovalbumin (Jiménezsaiz et al. 2011), and potato acid phosphatase (Selgrade et al. 2009), were evaluated using an RBL cell-based immunological assay and decision-tree strategies. Based on these results, we further tested RBL and began RBL analysis of the characteristics of the complementarity between the application and the decision-tree approach.

2. Methods
Bioinformatical analysis

The amino acid sequences of Gly, OVA, and PAP were obtained from the protein database of the National Center of Biotechnology Information (NCBI). Based on these sequences, the hydropathy (Hoop–Woods), flexibility (Karplus–Schulz), and antigenic index (Jameson–Wolf) were analyzed by the DNAStar Protein system. The overlaps of the prediction results obtained from the three parameters were considered as potential linear-epitope regions.

3. Results and Discussion

Currently, potential linear bioinformatical analysis can only forecast the protein antigen epitope. However, conformational epitopes may be identified if bonded with a protein structure analysis, such as nuclear magnetic resonance spectroscopy (NMR) or X-ray diffraction methods. Because the hydropathy, flexibility, and antigenic index of Gly, OVA, and PAP analyzed by the DNAStar Protein system may be applied to comprehensively illustrate the variety of linear epitopes, we predicted that the ranking of the potential allergenicity was Gly > OVA > PAP. Only the contribution of linear antigen epitope predicts IgE-binding allergens, but it is not suitable for all allergens. However, cellular immune assay can determine the ability of linear and conformational epitope-inducing allergic reaction, which test and verify the predictions by bioinformatical analysis.

Research 5-a Food Proteins from Different Allergen Families Sensitize Balb/c Mice to Family-Specific Immune Responses: this research was published in the *Journal of Immunotoxicology* in 2014 (Wang et al. 2014b).

1. Introduction

Classification of food allergens shows that food allergens from the same family always have a similar structure or function (Jenkins et al. 2007; Jenkins et al. 2005). As such, differences in protein structure or function of the immune response can lead to different modes of exposure to the host.

2. Materials and Methods

2.1 Preparation of food proteins

OVA (ovalbumin, grade VI) and PAP (potato acid phosphatase) were purchased from Sigma-Aldrich (St. Louis, MO, USA). 11S globulin (Gly m 6) was extracted from de-fatted soybean meal that was extracted in 0.02 M phosphate-buffered saline [PBS (pH 8.5)]:soybean meal:PBS = 1:16 (w/v) for 45 min at 40 °C. After centrifugation (2000 *g*, 10 min), the supernatant was collected, adjusted to pH 6.2, and placed in an ice bath for 12 h. The extract was then centrifuged at 2000 *g* (10 min) and precipitates bearing 11S globulin collected. The recombinant HSA (human serum albumin), LRP (lysine-responsive storage protein), and Cry 1Ab/Ac (crystal protein) were produced by our laboratory (Cao SH, Xu WT, Luo YB, He XY, Guo X, Yuan YF, Yuan CM, Huang KL. Expression, purification and refolding of recombinant Cry1Ab/Ac obtained in Escherichia coli as inclusion bodies. J Sci Food Agric 2009), had been expressed in BL21 *Escherichia coli* (TianGen, Beijing, China), and purified from a His TrapTM FF gel column (Amersham Biosciences).

2.2 Classification of food allergens

AllFam (http://www.meduniwien.ac.at/allergens/allfam/) is based on an Allergome database that is the currently most comprehensive collection of allergen data. The list of 528 food allergens (including the iso-allergens) is gained from SDAP (http://fermi.utmb.edu/SDAP/index. html). Pfam ID of the allergen is also obtained from the database in SDAP. With the shared Pfam ID as a standard, allergens in the same family could be identified and retrieved from AllFam. The method of classification previously described was applied to obtain the allergen family to which OVA, 11S glycinin, HSA, LRP, and Cry 1Ab/Ac belong.

3. Results

Based on the analysis of bioinformatics, 11 s globulin, OVA, HAS, and LRP belongs to, respectively, for predicting superfamily, serpin serine protease inhibitors, serum albumin, or Bet v 1-related proteins. Cry 1Ab/Ac is not according to the analysis of potential food allergen.

Research 6-a Allergenicity Assessment of a Genetically Modified Protein-Recombinant Human Lactoferrin: this research was published in the *Journal of Allergy & Therapy* in 2013 (Zhou and Sun 2013).

1. Introduction

Human lactoferrin (hLF) is known to be a multifunctional protein of the transferrin family. About 80-kDa glycoprotein exists in all parts of the human body including liquid milk secretion, saliva, tears, nasal secretions, and numerous neutrophils (Janssen and Bijsterveld 1983; Lönnerdal and Iyer 1995; Tenovuo 2002; Tenovuo et al. 1986). Human lactoferrin has been demonstrated to have antibacterial, antiviral, antiinflammatory, and antioxidant properties. Researchers have considered using hLF in a wide variety of applications for reasons of its bioactivities. Potential applications include the prophylactic treatment of inflammatory disease (Håversen et al. 2003; Håversen et al. 2000) and iron deficiency. It has also been proposed for use as a food additive to reduce microbial spoilage.

Mass production of hLF is necessary to meet the needs of the potential for its many uses. The latest development of biotechnology allows production restructuring hLF (rhLF) using mammalian cells in culture and the transformation of bacterial fermentation (Chong and Langridge 2000; Liang and Richardson 1993). Investigators have genetically modified cattle at China Agricultural University that secrete rhLF aat 2.5 g/l and the second rhLF in milk at 3.4 g/l.Transformation is through the microscopic injection of bacterial artificial chromosome (BAC) containing a gene cloning LF fibroblasts (~150 kb), followed by somatic cell cloning and transfer to the receptor cow uterus to produce a small number of transgenic cattle (Penghua , 2008).

For genetically modified protein, it is necessary to evaluate rhLF expressing allergenicity in GM cattle according to the international standard. In the study of bioinformatics analysis, the stable rhLF pepsin and serum reactivity test evaluates rhLF allergenic potential. The code of standards underestimates important considerations for the safe use of all kinds of regulations of the state history. Because hLF is secreted by glands of the human body including breast, and in tears, salivary glands, and polymorphonuclear leukocytes, this protein is not only a common human exposure but constant. There were no reports of this protein allergy. Thus, the possibility of allergic reaction to a foreign protein source, if it has the same amino acid sequence, is negligible.

2. Methods
Bioinformatics analysis

On the basis of the recognized international guidelines for evaluating GMOs, the allergenicity of rhLF expressed in the milk of transgenic cows was evaluated simply based on bioinformatics and resistance to digestion by pepsin. The amino acid sequence comparison for rhLF was conducted with three databases, FARRP (http://www.allergenonline.com), SDAP (http://fermi.utmb.edu/SDAP/sdap_src.html), and ADFS (http://allergen.nihs.go.jp/ADFS/), which are widely used in the world at present. The primary methods of evaluation were searches for "80 amino acid alignments with greater than 35% identity by FASTA" and "8 amino acid exact matches." Because of the peptide match of 6 continuous amino acids to known allergens will result in many false-positives and produce many random, irrelevant matches (Hileman et al. 2002), this method was not used in this evaluation.

3. Results
Bioinformatics comparison

Overall FASTA aligned results showed that the milk ferritin rhLF identity (bLF) was 71.4% of the shares, 52.2% and 51.9% ovotransferrin ovotransferrin precursor, using FARRP allergen databases. In "80 amino acids arranged" results suggest rhLF slide contains 532 80-amino-acid sequence alignment hits, 35% > bLF ovotransferrin or ovotraneferrin precursor FARRP website search using sliding window.

Research 10-a Prokaryotic Expression and Allergenicity Assessment of Hygromycin B Phosphotransferase Protein Derived from Genetically Modified Plants: this research was published in the *Journal of Food Science* in 2007 (Lu et al. 2007**).**

1. Introduction

The B hygromycin phosphoric acid transferase gene (HPT), by 4-phospotransferase aminoglycoside encoded by aph (4), an assistant of *Streptomyces hygroscopicus*, is a useful screening transformant selection marker gene transfer in experiments with plants (Gritz and Davies 1983). Detoxification of the gene requires encoding an enzyme through phosphorylation hygromycin B. Hygromycin B is an aminocyclitol antibiotic, inhibiting protein synthesis in prokaryotes and eukaryotes by ribosome translocation and ammonia acyl recognition by interference (Waldron et al. 1985).

The gene is found in several transgenic plants such as rape, rice, and corn (Li et al. 1997; Sawahel 2003). However, before these plants can be commercialized, it is necessary to conduct safety assessments (FAO/WHO 2001).

Because the level of HPT in these plants is low, HPT was produced in *E. coli* to obtain sufficient HPT protein for in vitro digestion and animal studies.

2. Materials and Methods
2.1 Materials

The *E. coli* strains *DH5α* and *BL21 (DE3)* (stored at −80 °C) were used for cloning and protein expression, respectively. The plasmid pET30a was used for preparing the His6·Tag expression vector. Restriction enzymes and other enzymes were purchased, respectively, from Promega (Promega, Beijing, China) and TaKaRa (TaKaRa Biotechnology, Dalian, China). Simulated gastric fluid and intestinal fluid digestion were purchased from Sigma (St. Louis, MO, USA). Mouse anti-rat IgE serum and HRP-goat anti-mouse IgG serum were purchased from PharMingen (PharMingen Bioscience, San Jose, CA, USA). His_6·Tag monoclonal antibody and Ni-NTA metal-affinity column were purchased from Novagen (Merck, Darmstadt, Germany). *O*-Phenylene diamine was purchased fromSigma. A rat histamine kit was purchased from RapidBio (Calabasas, CA, USA). The biochemicals and kit were used according to the recommendations of the supplier. All other common reagents were of analytical grade.

2.2 Comparison of the HPT amino acid sequence with that of other proteins

The sequence of HPT was analyzed by the procedure formulated by the FAO/WHO Consultation on Assessment of the Allergenicity of Genetically Modified Food (FAO/WHO, 2002). The sequence of HPT protein was aligned with the complete entries in the Protein Information Resource (PIR database http://pir.georgetown.edu) using FASTA (Gendel 1998).

3. Results and Discussion
Sequence similarity analysis

At present, about 200 kinds of the main allergens, including their amino acid sequences, are known. In general, if at least eight continuous amino acids are identical to those in the sequences of known allergens, the similarity is considered significant. The reason is that the shortest peptide chains that bind to T cells to trigger an allergic reaction include eight or nine amino acids. Also, antigenic determinant binding to IgE requires much longer peptide chains. So, it is a reliable way to determine similarity with a standard of eight amino acid residues (Rothbard and Gefter 1991). The HPT protein sequence was compared with each member sequence in databases containing known allergen sequences (Gendel 1998). The protein had no significant similarity with known allergy proteins (PIR database, NCBI, PubMed, EBI, and CBI: www.cbi.pku.edu.cn). This negative result indicates that the protein is unlikely to be cross-reactive to antibodies of known allergens, although sequence homology searches have certain limitations. In particular, comparisons are limited to the sequences of known allergens in publicly available databases and the scientific literature. Also, they have limitations in the ability to detect noncontiguous epitopes capable of folding themselves in space and binding specifically to an IgE antibody.

3.2.2 Serum Screening

3.2.2.1 Specific Serum Screening

For IgE antibody reactivity assessment of serum of known allergy patients, a suitable method should be used in vitro. Various validation analyses can be used for this purpose. It is agreed anyone can use these tests.

In addition to the earlier prevention measures for the examination such as references, select the appropriate serum glycosylation, and the importance of polysaccharide antigen epitope must also be taken into consideration. Protein expression in host plants may be modified after the translation, which may affect the potential of their allergies. The effect of glycosylation particularly has relevant consideration, for these reasons:

1. The degree of glycosylation may affect the susceptibility of the protein to processing and proteolysis.
2. Glycosylation may alter the epitope structure, either by shielding part of the protein surface (particularly if the glycosylation is extensive), or by introducing glycan epitopes.

Glycan epitopes are known to be highly cross-reactive

Glycans may be attached via either an N-link or an O-link. N-linked sites can be predicted with some accuracy, but the prediction of sites for O-glycosylation is still unreliable.

Cross-reactivity of the IgE antibody polysaccharide antigen epitope is more important because of their potential contribution to allergy symptoms (which may be the smallest in many cases), but because the protein's structure part of glycoprotein is completely irrelevant in this case, all with the polysaccharide structure will be cross-reactive protein. In large-scale glycoprotein screening, it is important to make a clear distinction between the IgE antibody polysaccharide on the one hand and the IgE antibody protein part on the other hand. In general, we suggest using serum samples without IgE antibody that absorbed the IgE antibody-irrelevant glycoprotein to obtain the same host, or performing such tests with non-glycosylated variants, such as expressed in a bacterial host.

Information is mainly based on the plant polysaccharide antigen epitope and allergies to glycoprotein and invertebrate glycoprotein. Little is known of the glycoprotein eukaryotic microorganism. However, it is probably similar and may require taking defensive measures (FAO 2001).

3.2.2.2 Targeted Serum Screening

When no sequence homology is found between the expression of protein and allergens, this does not mean that there are no such cognate allergens: this may be the result of lack of information related to allergens. Stochastic screening for allergy in serum samples of the population is unlikely to be useful. However, some of the more targeted approaches may, in some cases, be more appropriate.

- If the recombinant protein is derived from a monocot, it is proposed to test serum samples from patients with high levels of IgE antibodies to monocot allergens such as grass and rice.

- If the recombinant protein is derived from a dicot, it is proposed to test serum samples from patients with high levels of IgE antibodies to dicot allergens such as tree pollen, weed pollen, celery, peanuts, tree nuts, and latex.
- If the allergen is derived from a mold, it is proposed to test serum samples from patients with high levels of IgE antibodies to molds, yeast, and fungi, such as *Alternaria* or *Cladosporium*, and of patients with aspergillosis or *Trichophyton* sensitivity.
- If the allergen is derived from an invertebrate, it is proposed to test serum samples from patients with high levels of IgE antibodies to invertebrates such as mites, cockroaches, shrimp, chironimids, or silkworms.
- If the allergens are from vertebrates, test the serum samples of patients with high levels of IgE antibody to pet mammals, experimental animals, milk, fish, chicken, chicken protein, and egg yolk/serum protein.
- If the allergen is derived from another source, such as a bacterium, no general screen using targeted sera is currently available.

Using a large pool (>5) of sera is discouraged because this will dilute any cross-reactive antibodies. Maximum sensitivity of individual sera should be tested.

Usually, to screen 25 patients, serum samples are selected from a group of high levels of IgE airborne allergens and IgE (if applicable); the choice of food allergen groups will be used (FAO 2001).

However, as a result of the serum levels of the serum category limit bank, this choice may not lead to high serum IgE responses. In this consideration, the test simulates digestion in the stomach and thermal stability or allergic test to evaluate in the body.

We have done some research on serum screening. I have listed these in the following sections.

Research 2-b A Modified Weight-of-Evidence Approach to Evaluate the Allergenic Potential of Food Proteins: this research was published in the journal *Toxicology Research* in 2015 (Sun et al. 2015a).

1. Introduction

The introduction is the same as the "Introduction" in Research 2-a.

2. Methods
Human serum studies
2.1 Patient sera

This study was approved by the Institutional Review Boards of China Agricultural University, and all donors gave informed consent. We studied sera from a total of eight egg-allergic patients (mean age, 40.6 ± 5.6 years; three men and five women). The donors were diagnosed as allergic to eggs on the basis of clinical histories and skin prick tests (SPT). The sera of five peanut-allergic patients without egg allergy were pooled and used as a negative control.

2.2 ELISA assay

Specific IgE and IgG4 levels against OVA or OVM were performed by enzyme-linked immunosorbent assay (ELISA) as described previously with some modifications. Briefly, the concentrations of OVA and OVM were determined by a Bicinchoninic Acid kit (Biomiga, San Diego, CA, USA): 96-well microtiter plates were coated with 100 μl 10 μg ml^{-1} OVA or OVM in carbonate buffer at pH 9.6. After overnight incubation at 4 °C, plates were washed and blocked with 150 μl 1% BSA–PBST for 1 h at 37 °C. Serum samples were diluted to 1:10 v/v in blocking buffer 1.5 h before addition to aspirated sample wells. Plates were incubated for 2 h at 37 °C. Wells were washed three times with PBST before the addition of 100 μl biotinylated goat anti-human IgE (1:1000 v/v in PBST, 16–10-04; KPL, MD, USA) or biotinylated mouse anti-human IgG4 (1:1000 v/v in PBST, ab99824; Abcam, Cambridge, UK). Plates were incubated for 1 h at 37 °C and then washed six times with PBST. Then, 100 μl NeutrAvidin–HRP (1:2000 v/v in PBST, 31,001; Thermo Scientific, MA, USA) was added to sample wells and incubated for 1 h at 37 °C. Plates were washed six times with PBST before addition of 100 μl TMB (KPL). Reactions were stopped after 10 min of incubation at 37 °C by addition of 50 μl 2 N sulfuric acid, and the ODs were read at 450 nm by Thermo Scientific Varioskan Flash.

2.3 Results and Discussion

Serum IgE binding, further performed to confirm the results of the bioinformatics, indicated that the level of OVM-specific IgE antibody was found to be significantly higher when compared to OVA in the eight egg-allergic patients tested. However, previous studies demonstrated there was little correlation between the serum IgE antibody level and the severity of clinical symptoms. One of the most important factors may be the blocking effect of specific IgG4. IgG4 antibodies, which act in the induction of tolerance, generally overlap with IgE antibodies in respect to linear epitope specificity in food allergic subjects. More importantly, a high IgE/IgG4 ratio, instead of a high IgE antibody level, correlates with allergy persistence. In our study, IgE/IgG4 ratios to OVM was found to be significantly greater compared with OVA in the eight egg-allergic patients tested ($p < 0.001$), demonstrating that OVM was more likely to induce persistent egg allergy compared with OVA.

Research 6-b Allergenicity Assessment of a Genetically Modified Protein-Recombinant Human Lactoferrin: this research was published in the *Journal of Allergy & Therapy* in 2013 (Zhou and Sun 2013).

1. Introduction

It is the same as the "Introduction" in Research 2-a.

2. Methods

Human serum test.

Some guidelines suggest the human serum test rhLF for further evaluation of the potential allergenicity of rhLF using serum from donors allergic to milk or eggs.

Voluntary blood donor sera were collected under the institutional review board, which approved 12 egg-allergic theme and 21 milk-allergic individuals for the study. The donors were diagnosed as allergic to egg or milk based on clinical history, skin prick test (SPT), and specific IgE tests (allergic-specific Pharmacia UniCAP tests). The level of egg- or milk-specific IgE in all the sera samples was > 3.5 KUA/l. Additionally, serum samples from individuals were mixed as a serum pool to be used as a negative control.

Immunoblots were conducted according to the previous protocol with some modifications. In brief, SDS-PAGE was conducted using 400 ng purified rhLF, bLF, and hLF (kindly provided by Professor Li Ning; purity, 98%) and 10 μg protein extract prepared from a standard egg powder (Lot. 1452807v; USA) or from milk powder (No. 1549; USA) with samples loaded in adjacent wells of the gel. Following separation of proteins by electrophoresis, proteins were either stained with Coomassie blue or transferred to nitrocellulose membranes (NC) for immunoblot. Membranes were then rinsed with distilled water and blocked by submersion in 3% bovine serum albumin (BSA) in phosphate-buffered saline (PBS) buffer for 2 h at room temperature. After washing, the blocked membranes were incubated for 2 h at room temperature in individual allergic sera or control sera, which had been diluted 1:20 (v/v) in blocking buffer 30 min before adding to the membrane. The membranes were washed six times in TBST (0.02 M, 0.05% Tween-20), then incubated for 1 h in monoclonal mouse anti-human IgE conjugated with horseradish peroxidase (HRP) (kindly provided by Professor Li Ning) that was diluted 1:8000 (v/v) in blocking solution. After six washes in TBST solution, detection was achieved using enhanced chemiluminescence (ECL; Amersham BioSciences, Piscataway, NJ, USA; no. RPN2106), with exposure for 3 min on the X-ray films.

3. Results

Bioinformatics comparison results indicated cross-reactive rhLF as a possible allergen for some consumers allergic to cow milk or eggs and based on the guidelines suggested that the human serum test should be performed. However, because hLF is widely expressed in every one of us, if not allergic to human lactoferrin, the data support, without risk, that there should not be a need to perform the serum test. To meet regulatory requirements and expectations of consumers to ask questions or regulators, we decided to test the serum IgE binding.

After SDS-PAGE and staining with Coomassie brilliant blue, there are mainly four protein bands, which are about 65 kDa (BSA), 34 kDa (casein, CAS), 26 kDa (caseins, CAS), and 18 kDa (β-lactoglobulin, LG) in the extraction of standard milk powder. Ten protein bands were evident in the egg powder, with molecular weights approximately 230 kDa, 150 kDa, 130 kDa, 83 kDa (ovotransferrin, OVT), 72 kDa, 60 kDa, 45 kDa (ovalbumin, OVA), 40 kDa, 34 kDa, and 28 kDa (ovomucoid, OVM). As the standard egg powder used in this study is whole egg powder, it is assumed that the high molecular weight is from egg yolk. Three obvious bands were rhLF, hLF, and bLF, respectively, about 80 kDa.

Western blot results show that the binding is not visible in the incubation of membrane control serum from healthy donors. A notable IgE-binding reaction was detected in the protein bands of 150 kDa, 83 kDa, 45 kDa, and 34 kDa in egg powder and of 65 kDa, 26 kDa, and 18 kDa in milk powder. These proteins may be the important allergens, which are named ovotransferrin (OVT), ovabumin (OVA), ovomucoid (OVM), bovine serum albumin (BSA), casein (CAS), and lactoglobulin (LG) in egg and milk, as in accord with the previous study. And finally, there is no visible IgE-binding reaction and purification of rhLF, hLF, and bLF in all the serum samples tested. Because of the sensitivity of the Western blot detection method, using ECL, IgE should be detectable at a picogram (pg) level, and the preliminary experiment of C-reactive protein and antibody testing used loading doses of 40 ng. Thus, the lack of obvious hLF and rhLF shows that rhLF would not have any IgE binding in said milk. Based on the results of the evaluation process, following the legal criterion and considering the history of safe use, there is no evidence to suggest that rhLF expression of genetically modified cattle can cause the risk of allergic reactions for consumers. So, rhLF may be added to the food or formula, improving the nutritional status.

3.2.3 Simulated Gastric Digestion and Heat Stability

Simulated digestion in the stomach (SGF) is recommended by FAO/WHO as another method in the guidelines for evaluating the potential of extrinsic allergic protein (FAO 2001). Previous research has shown that the protein, which can cause anaphylaxis, has the characteristics of digestion stability.

In addition to pollen protein, all allergic protein enzymes express high stability. So, the stability can be used as a potential indicator of the allergen protein (Astwood et al. 1996). According to the report, peptide and the single amino acid that is less than 3.5 kDa protein hydrolysis by pepsin can be considered unlikely to cause allergies (Tong Jen Fu et al. 2002). And if allergen protein causes allergic reactions through the intestinal mucosa, they must be able to resist peptic and trypsin digestion, the acidic conditions of the digestive system.

We have done some research simulating digestion in the stomach and thermal stability, as shown next.

Research 1-b Safety Assessment of Cry1Ab/Ac Fusion Protein: this research was published in the journal *Food and Chemical Toxicology* (Xu et al. 2009).

1. Introduction

It is the same as the "Introduction" for Research 1-a.

2. Methods

2.1 Heat stability study

The Cry1Ab/Ac protein was dissolved in 20 mM Tris–HCl and 5 mM EDTA buffer at a concentration of 0.1 mg ml^{-1} in 1.5-ml microcentrifuge tubes, which were tested for stability at temperature of 100 °C for periods of 10, 30, and 60 min.

The assay was terminated by placing the sample tubes on ice, and adding SDS-sample buffer (50 mM Tris–Cl, 8% sucrose, 2% SDS, with 5% 2-mercaptoethanol, and 0.02% bromophenol blue). Control samples of a 0-min incubation of the protein (kept at 4 °C) were also prepared. The resulting proteins were analyzed by SDS–PAGE and Western blot.

2.2 In vitro digestion of Cry1Ab/Ac protein

The concentrations of the Cry1Ab/Ac protein in simulated gastric fluid (SGF) or simulated intestinal fluid (SIF) (Sigma, USA) were 0.25 mg ml^{-1} or 0.1 mg ml^{-1}, respectively. The mixture was incubated at 37 °C and the samples were removed at intervals of 0 s, 15 s, 30 s, 60 s, 2 min, 5 min, 10 min, 20 min, and 30 min. According to the results of SDS-PAGE and Western blot, the stability of the protein to be digested in SGF/SIF was observed. The experiments also used the positive control of STI (soybean trypsin inhibitor) and negative control BSA (bovine serum albumin) or alpha casein for SGF or SIF, respectively.

3. Results
3.1 Heat stability

On SDS-PAGE, the Cry1Ab/Ac protein was clearly visible at about 67 kDa at time zero. When treated at 100 °C for 60 min, Cry1Ab/Ac protein still remained detectable by SDS–PAGE, that is, there was no protein degradation. Therefore, the Cry1Ab/Ac protein was not degraded or modified in a way that would affect their migration in SDS-PAGE at 100 °C for 60 min. In addition, their immunoreactivity was still detectable. The same heat treatment and the native protein identification of anti-Cry1Ab/Ac antibodies shows that conformational changes associated with epitope degeneration do not affect accessibility. It is interesting to note that this means that the epitope homology search, display, and allergic antigen epitope show no similarities; there is considerable weight in the safety assessment of the evidence.

3.2 Rapid degradation in digestive fluids

The digestibility of Cry1Ab/Ac protein was confirmed by using SGF and SIF. CBBG-stained gel and Western blot of the protein sample indicated that high molecular weight proteins were completely digested within 15 s, which is consistent with the silico analysis (PeptideMass). The reference proteins of the SGF/SIF assay, α-casein, or BSA and STI, were rapidly or slowly digested, respectively, which is in agreement with published data (data not shown).

The results show that but do not prove that the activities of the proteins have no allergy status, because the relationship between the protein hydrolysis resistance and allergic activity is not absolute (Tong Jen Fu et al. 2002).

Research 2-c A Modified Weight-of-Evidence Approach to Evaluate the Allergenic Potential of Food Proteins: this research was published in the journal *Toxicology Research* in 2015 (Sun et al. 2015a**).**

1. Introduction

It is the same as the "Introduction" for Research 2-a.

2. Methods
Pepsin digestion assay

OVA, OVM, and PAP were dissolved in simulated gastric fluid (SGF, 35 mM NaCl) at pH 2.0, preheated for 5 min at 37 °C, and subjected to an in vitro gastric digestion with porcine pepsin (4220 U/mg protein; P6887, Sigma-Aldrich) at an enzyme/substrate ratio of 10:1. At each time point (0 s, 15 s, 2 min, 30 min, and 60 min), 200 μl of reaction mixture was transferred to a sampling tube that contained 70 μl 5 × Laemmli buffer (40% glycerol, 5% 2-mercaptoethanol, 10% SDS, 0.33 M Tris, 0.05% bromophenol blue, pH 6.8) and 70 μl 800 mM $NaHCO_3$. All neutralized samples were then boiled at 100 °C for 5 min and subjected to tricine SDS-PAGE. Briefly, the samples (35 μl per lane) were loaded on a 10% acrylamide mini-gel with tricine buffers. The gels were stained with Brilliant Blue R Staining Solution (B6529; Sigma-Aldrich), destained, and photographed. Prestained Protein Marker (LC5925; Invitrogen, Carlsbad, CA, USA) was used for estimating the molecular weights of the electrop horesed proteins.

3. Results and Discussion

One of the important characteristics responsible for the allergenicity of food proteins is their stability against pepsin digestion. Thus, OVA, OVM, and PAP were evaluated individually for digestibility in the simulated gastric fluid (SGF). The nonallergenic protein PAP was completely digested by the first time point of 15 s, whereas native OVA and OVM were stable in SGF for at least 2 min. Heated OVA was digested within 15 s, although native OVA was stable in SGF for at least 2 min. These data demonstrated that the heat treatment made OVA more susceptible to digestion by pepsin. As for OVM, it is worth noting that a broad band representing intact OVM appeared at about 34–49 kDa in the SDS-PAGE gel because of the presence of five N-linked sugar chains, although the molecular weight of OVM is about 20 kDa based on its amino acid sequence. Both native and heated OVM were digested within 2 min, suggesting that heat treatment of OVM did not influence the digestion pattern of pepsin.

Resistance to digestion in the gastrointestinal tract allows the protein or its large peptides to have sufficient opportunity to interact with the intestinal immune system, causing allergic responses. In an in vivo study, a pepsin-resistant protein increased the intestinal permeability by 15 fold, thereby enhancing the chance of the sensitization of animals to the exposed protein. Rapid digestion was observed for both native OVA and OVM, and OVM had higher thermostability than OVA. Although the pepsin digestion assay is not effective for ranking the food allergens, it may predict the effect of heat treatment on the allergenicity of food allergens.

Further, we investigated the effect of heat treatment on the biological mediator-releasing activity of OVA and OVM. Our results showed that heat treatment significantly reduced the biological activity of OVA but not OVM, suspecting that heating, particularly at 90 °C, which exceeds the denaturation temperature of the protein of 78.3 °C, made the OVA form irreversible high molecular mass aggregates (>66 kDa) leading to the loss of some epitopes. Conversely, the conformational change of OVM is reversible during heat treatment, which makes it thermally stable. Nevertheless, this does not mean that OVM is native even after heating. Newly formed epitopes might occur by heating. Furthermore, heat treatment of digested OVA and OVM significantly reduced the biological mediator-releasing activity. Similarly, a previous report indicated that all mice challenged with native OVA and OVM presented the symptoms of anaphylaxis whereas the mice challenged with the corresponding heated allergens did not develop anaphylaxis, because heat treatment of OVA and OVM completely abrogated the delivery of immunologically intact forms of allergen across the intestinal mucosa.

Research 4-b Cell-Based Immunological Assay: Complementary Applications in Evaluating the Allergenicity of Foods with FAO/WHO guidelines: this research was published in the journal *Food Research International* in 2014 (Sun et al. 2014**).**

1. Introduction

It is the same as the "Introduction" in Research 4-a.

2. Methods

Gly, OVA, and PAP were added into an in vitro gastric digestion containing porcine pepsin (4220 U/mg protein). The ratio of enzyme/substrate was 10:1. Equal samples were taken at 0 s, 15 s, 2 min, 30 min, and 60 min of gastric digestion for the following SDS-PAGE. The method is shown in detail in the Supplementary Material.

3. Results and Discussion

In 2001, the FAO/WHO recommended the decision-tree strategies, of which the pepsin digestion test is an important section. Some protein or its large peptides that resist digestion in the gastrointestinal tract can survive in an immunogenic form for sufficient interaction with the intestinal immune system and stimulation, and then cause allergic responses. Obviously, the ability of food proteins to undergo digestion by pepsin is rated as a potential risk factor for allergenicity. Therefore, we conducted the pepsin digestion test to explore the allergenic potential of Gly, OVA, and PAP. The results showed that PAP had no stability in the SGF and OVA was less stable compared with Gly, which predicted that the ranking of the potential allergenicity was PAP < OVA < Gly. Nevertheless, only one pepsin digestion test in vitro cannot completely reflect which state the food proteins actually maintain when digested in the gastrointestinal tract. Although the epitopes of food proteins will be damaged in the digestion by pepsin, the hidden linear IgE epitopes might also be exposed. Therefore, it is known that some relatively unstable proteins, for instance, vegetable food allergens, can cause an allergy, whereas some very stable proteins, such as thaumatin-like proteins from grapes and apples, seldom cause food allergy or possibly only slight reactions. Fortunately, an immunological assay based on cells can also estimate the potential allergenic activity of food proteins in digestion, which compensates for the lack of the pepsin digestion test.

Research 5-b Food Proteins from Different Allergen Families Sensitize Balb/c Mice to Family-Specific Immune Responses: this research was published in the *Journal of Immunotoxicology* in 2014 (Wang et al. 2014b**).**

1. Introduction

It is the same as the "Introduction" to Research 5-a.

2. Methods

Simulated gastric fluid (SGF) was prepared as follows: 62.4 mg pepsin (4220 U/mg) and 0.2 g NaCl are added to 70 ml distilled water and the pH adjusted to 1.2 with HCl. Then, mixture volume is brought to 100 ml with distilled water. For digestive analysis, the mixing ratio of SGF and test protein was 19:1 (v/v). The reaction was carried out at 37 °C for 15 s, 2 min, 30 min, and 60 min after the introduction of test protein (at 5 mg/ml). At each time point, the reaction was stopped by the addition of 70 μl 0.2 M $NaHCO_3$. Next, SDS-PAGE was conducted on the resulting products in the samples

3. Results

The results of the pepsin digestion assay indicated that with 11S globulin [composed of A (acid) and B (basic) subunits], the A subunit was completely digested during the period between 2 min and 30 min, whereas the B subunit resisted digestion up to 60 min. In contrast, Cry 1Ab/Ac and OVA were completely digested between 2 min and 30 min, indicating that both these had weak resistance to digestion. The results of LRP and HSA digestion assays suggested that both could be rapidly digested by pepsin; thus, they also were not digestion stable.

Research 6-c Allergenicity Assessment of a Genetically Modified Protein-Recombinant Human Lactoferrin: this research was published in the journal *Allergy & Therapy* in 2013 (Zhou and Sun 2013**).**

1. Introduction

It is the same as the "Introduction" in Research 2-a.

2. Methods
Stability to pepsin digestion

The pepsin digestion resistance of rhLF was tested based on previous methods. Stimulated gastric fluid (SGF) was prepared by adding pepsin (4220 U/mg) obtained from Sigma Chemical (Sigma-Aldrich China, Shanghai). The ratio of pepsin and test protein was about 5:1 (w/w, 20 U/μg) throughout the study. Bovine serum albumin (BSA) and bovine β-lactoglobulin (BLG), which were used as the control proteins, were also purchased from Sigma-Aldrich. It has been reported that the BLG is stable to SFG whereas the BSA is labile to SGF under standard conditions. The resulting products in samples after digestion were then evaluated in SDS-PAGE analysis under standard conditions. The images of stained gels were captured and analyzed with gel imaging equipment (GelDoc-It Imaging System, P/N 95–0441-02; USA).

3. Results
Stability to pepsin digestion

The samples of BSA were fully digested in SGF at 15 s, whereas the samples of BLG were still clearly visible at 60 min. The rhLF was completely digested within 15 s by pepsin, which was the same as the report in a previous study.

Research 10-b Prokaryotic Expression and Allergenicity Assessment of Hygromycin B Phosphotransferase Protein Derived from Genetically Modified Plants: this research was published in the *Journal of Food Science* in 2007 (Lu et al. 2007**).**

1. Introduction

It is the same as the "Introduction" in Research 10-a.

2. Methods
In vitro digestion of HPT protein

HPT protein was added to simulated gastric fluid with porcine pepsin in 0.03 M NaCl and the pH was adjusted to 1.2. The final concentration of HPT protein was 2 μg/ml. Next, the reaction mixture was adjusted to pH 8.0 with 100 mM Tris-HCl (pH 9.5) and 2 mM $CaCl_2$ for intestinal fluid digestion (IFD; porcine trypsin and bovine chymotrypsin) (Sigma Chemical). The ratio of protein and digestive enzyme was 1:100. The reaction was conducted at 37 °C for 40 s. The reaction was stopped with a buffer solution (pH 9.5; 100 mM NaCl, 50 mM Tris-HCl). The samples were taken out at different time points, mixed with protein-denaturing 2× sample buffer, boiled, and subjected to SDS-PAGE and Western blotting as described previously. The positive control (OVA) and negative control (BSA) were also processed as in the method given earlier.

3. Results and Discussion
3.1 Antigen-specific IgG response

The immunoreaction was significantly different between the positive group, which was treated by gavage with OVA, and the negative group. Also, the sera of BN rats in the positive group, which were diluted 32 fold, had a much higher level of IgC-specific antibody than those of the negative group ($P/N > 2$). In contrast, the immune response of the test group was basically matched to that of the negative group. This result suggested that OVA, not HPT, triggered an obvious immunoreaction.

3.2 Evaluation of IgE response

Food allergy is an abnormal response of the immune system to some specific proteins in an otherwise harmless food or food component. Several types of immunological responses are caused by true food allergies, mainly including cell-mediated reactions and IgE-mediated reactions. The most common one is the latter, which is also known as immediate hypersensitivity reaction, because the symptoms occur within only a few minutes to hours upon the ingestion of disagreeable foods.

Therefore, specific IgE antibodies were produced by the immune system when the test animals were fed with allergic proteins, which could be tested by ELISA. The optical density (OD) values of the sera in the OVA group were significantly higher than those of the water group (P/N < 2). In contrast, the OD values related to HPT protein were similar to those of the negative group. This result indicated that OVA, not the HPT protein, evoked specific IgE antibodies.

3.3 Results of histamine test

There was no significant difference in the level of histamine between the test group and the negative group, although the level of histamine of the positive group was twofold higher than that of the test group ($p < 0.05$).

FAO/WHO has recommended the decision tree to evaluate the potential allergenicity of novel genetically modified food products. If the new gene encodes an already known allergen, the recombinant protein could be assessed by in vitro tests using sera to determine its allergenicity. If the source of the introduced gene has no history of allergenicity, alternative methods are needed for the stability and structure of the foreign protein. These methods must be strictly carried out and based on a case-by-case principle. Because the HPT protein has no history of allergenicity, it could be evaluated mainly by gastrointestinal digestion in vitro and in an animal model. At present, gastrointestinal digestion in vitro has been accepted as a method to assess the allergenicity of food proteins; however, animal models are not yet accepted. Although several strains of rodent have been developed, they are still controversial. Presently, BN rats are widely used in studies related to allergen. In these studies, the potential allergen is usually provided via gavage without adjuvant.

The 3R International Principle proposes that the number of test animals should be controlled to six at most and the animals should live under a healthy environment. We demonstrated that the BN strain acts as a good model for allergy study under these conditions.

3.2.4 Cell Models

A cell-based immunological assay is more appropriate for assessing the allergenicity because of fewer influential factors and the simplicity of the testing system (Sun et al. 2015b).

3.2.4.1 Current Cell Models (Sun et al. 2015b)

Allergic reactions caused by food are usually mediated by the release of mediators, such as histamine and leukotrienes, which is in response to the allergens crosslinking of specific IgE bonding to FcεRI on basophils or mast cells. Thus, it is promising to use the in vitro cell-based immunological assay, which is in view of the activation and mediator release of effector cells to learn the bioactivity of food proteins. In recent years, to establish a cell model for allergen identification and clinical diagnosis, varieties of cell models have been put forward. An ideal cell model should have the following characteristics:

- Release mediators after antigen/IgE sensitization
- Be specific for distinguishing food allergens from allergenic to nonallergenic proteins
- Be easily acquired and reproducible in laboratories
- Be sensitive for recognizing threshold beyond predicting which allergic response is significant.

Mast Cell Model

Mast cells, which were discovered in 1877, have a crucial function in promoting immediate allergic reactions. The mature cells exist exclusively in tissues and particularly in regions. They are usually located in interfaces with the external environment, for example, lungs, skin, and mucosal surfaces. Mast cells are pivotal in evoking immediate allergic reactions. Crosslinking of their IgE-bound FcεRI caused by allergens leads to a rapid release of newly synthesized and preformed mediators. Newly synthesized mediators such as leukotrienes C4 and D4, prostaglandin D2, and platelet-activating factor are produced and secreted after cell stimulation. Preformed mediators, such as histamine, chemotactic factors, heparin, chymases, and tryptases, are secreted via exocytosis and stored in granules. These mediators cause allergic symptoms ranging from simple urticaria to anaphylactic shock and even death.

Isolation of mast cells from small animals, such as rats and mice, via peritoneal lavage is easier than those in other tissues. It is proven that histamine release assay of peritoneal mast cells (PMC) is specific and sensitive for allergen identification. PMC of the sensitized mice were incubated with the corresponding allergen and presented high mediator release. Also, the degranulation of sensitized mast cells was 520% generally when they contacted the noncorresponding allergens. This level was much lower than specific degranulation. However, PMC are limited by some defects to act as an applicable cell model for assessment of the allergenicity of food proteins. On the one hand, the cells should be purified to reduce their ability to react to stimuli. On the other hand, it is out of the question to keep the cells in primary cultures over long periods of time.

Human Basophil Model

Basophils are granulocytes that develop from CD34+ pluripotent progenitor stem cells which differentiate and mature in bone marrow. They then circulate in the periphery in which they account for 51% of the white blood cell population. Basophils, similar to mast cells, are also key effector cells in allergic disorders.

They secrete several kinds of granule-stored mediators responsible for allergies and other inflammatory diseases. In fact, to respond to specific allergens, basophils not only secrete quantifiable bioactive mediators but also upregulate the expression of various markers that can be efficiently tested using flow cytometry with specific monoclonal antibodies. The related approaches are basophil mediator release assay and basophil activation test (BAT).

In general, the high-profile functional in vitro tests are basophil mediator release assays, such as basophil histamine release (BHR). The diagnostic specificity (82%) observed in BHT was superior compared to SPT (32%) and specific IgE (24%). However, the assay sensitivity was less impressive; for example, BHR (53%), SPT (82%), and specific IgE (82%). Basophils release histamine in a few minutes after the reaction levels reach the maximum blood concentration. Then, the histamine is soon metabolized to N-methyl histamine, which can be eliminated via the urine. Therefore, it is difficult to measure histamine in peripheral blood, and this method was restrictedly employed in clinical applications.

When the basophil mediator is released, CD63 is upregulated simultaneously. This discovery promoted the development of a flow cytometric technique that can quantify and analyze allergen-specific in vitro activation of peripheral blood basophils. Subsequently, a basophil activation test (BAT) based on anti-CD63 was developed to assess the activation of these cells using flow cytometry. Other cellular markers, such as CD203, were applied in subsequent research to optimize the technique. Since the 1990s, it has been proven that BAT based on CD63 and CD203c is reliable for the diagnosis of IgE-mediated allergies. Although BAT was less sensitive, especially when containing nonresponders, compared with conventional diagnostic tests for SPT or specific IgE, it was more specific, and could reach at least 95% specificity. Moreover, BAT could discriminate irrelevant IgE responses from clinically relevant food-specific IgE. However, the use of BAT might be adjusted when it is not feasible to determinate specific IgE or SPT or equivocal results were given according to the clinical history. Nevertheless, some technical limitations limited the popularity of this method in clinical laboratories. First, blood samples must be immediately processed after collection. Second, in terms of the sample type, although the whole blood may better reflect physiological status in vivo, compared with separated cells, other factors still exist to interfere with the assay. Cell separation may avoid such interference. However, it could lead to a loss of basophils and nonspecific activation.

Rat Basophil Leukemia (RBL) Cell Model

Mast cells and basophils were respectively obtained by peritoneal lavage and from peripheral blood. However, there are a few drawbacks that limit their extensive use. One is that both must be purified, and another is that it is out of the question to keep the cells in primary cultures over prolonged periods of time. Considering these difficulties, presenting some characteristics of both mast cells and basophils and the availability of a permanently growing rat basophilic leukemia (RBL) cell line bring hope to researchers.

Designated RBL-2H3, the most common cell line was cloned from leukemia cells isolated from rats treated with β-chlorethylamine using a limited dilution technique. RBL-2H3 cells, similar to mast cells and basophils, respond by degranulation and release a range of newly synthesized and preformed mediators. These mediators trigger a potent allergic response followed by crosslinking with allergens of the IgE-bound FcεRI. Two mediators, β-hexosaminidase and histamine, can monitor RBL-2H3 cell degranulation. The release of these two mediators is closely parallel in time. β-Hexosaminidase is more suitable to act as a biomarker of RBL-2H3 degranulation compared with histamine because of its simplicity and the efficiency of the detection.

RBL-2H3 cells sensitized with allergen-specific murine IgE have been widely applied for determination of the bioactivity of allergens or their extracts, because some functional characteristics of them are similar to those of mast cells. The species restriction of FcεRI in RBL-2H3 cells requires murine IgE to be used in this assay. There is no doubt that the most important aspect of this assay is the murine IgE used for passive sensitization of RBL-2H3 cells. Generally, it seems that the immunization schedule based on repeated applications of low doses of allergens is suitable to induce allergen-specific IgE in Balb/c mice. Although most of the major allergens can be detected by both human and murine sera, there are still some differences with the type of the detected allergens. For instance, the major allergens Ara h 1, Ara h 3/4 complex, Gal d 1, Gal d 2, Gal d 3, and Bos d 5, and the minor allergens Bos d 8 and Gal d 4, were recognized by both murine and human sera. However, Ara h 2 was not recognized by murine sera. Thus, the RBL cell-based immunological assay based on murine IgE may not be completely equivalent to the one based on human IgE.

To solve this problem, RBL-2H3 cells were transfected with cDNA, which codes for human high-affinity IgE receptor (FcεRI) chains. With human FcεRI, these cells can bind IgE from sera of allergic subjects and be activated in following allergen-specific manners. Mediator release by humanized RBL cells provided a suitable cell-based assay to evaluate the bioactivity of allergens in vitro and has been proven to be specific, sensitive, and reproducible over time. Humanized RBL cells that were in combination with a human serum pool might then stand for a preferable test system. Further, murine sera may be profitable in cases of "rare" allergens, such as mutated recombinant allergens, because it is difficult to obtain human sera. Although the allergen-binding patterns of human and murine are not completely identical, the potency data of their systems was similar and they complemented each other. Therefore, this pattern could analyze varieties of allergens.

It was suggested that many serum factors (i.e., IgE-related) and allergen factors were important for their ability of inducing mediator release with RBL cell lines. Several studies have been conducted on the relationship between immune-chemical parameters of IgE and functional degranulation response in the transfectants. Dibbern et al. (Dibbern et al. 2003) indicated that the absolute amount of peanut-specific IgE is an important determining factor of the ability of a serum to effectively sensitize cells for allergen-dependent degranulation (r = 0.95, $p < 0.001$). However, the amount of total IgE was a relatively poor predictor. In contrast, Marchand et al. (Marchand et al. 2003) showed that there was no correlation ($r = 0.27$) between the proportion of bound allergen-specific IgE and the strength of degranulation response following sensitization with serum IgE. However, a significant correlation ($r = 0.97$) was observed on purified monoclonal IgE. Thus, RBL cell activation mediated through serum IgE is a consequence of complicated relationships including not only the dependence on allergen-specific IgE content but also the ability to efficiently sensitize cells and trigger signaling responses leading to degranulation. Moreover, some serum factors may hinder the binding of IgE with FcεRI expressed on the cell membrane and result in adverse effects on cells. These factors and their mode of action on cell have not yet been identified. However, the IgE-immunopurification of sera was a rapid, simple, and efficient method to remove interference factors caused by background release, sensitivity, and reproducibility of the assay. Additionally, it may be necessary to determine the optimum dilution scope of the serum of each allergic subject. It was found that the incubation with 420% serum or undiluted serum could cause cytotoxic effects in the RBL cells, such as decrease viability, a result of natural occurrence of enoreactive antibodies. Moreover, the optimal concentration for allergens or allergen extracts may also be of importance because of inducing IgE-mediated responses with such cell lines. However, high concentrations of allergens or allergen extracts might induce a non-IgE-mediated release on stimulation of non-sensitized RBL cells without added human serum, which might be a result of the exsistance of endotoxin, protease, endotoxin, or (1,3)-β-D-glucans. Therefore, in any RBL cell-based immunological assay, the optimal configuration of experimental parameters before the assay was required for allergen identification, clinical diagnosis and/or immunotherapy.

A summary of useful cell models for allergenicity evaluation and clinical diagnosis is listed in Table 3.2.

We have done some research on cell models. I listed them in the following.

Research 2-d A Modified Weight-of-Evidence Approach to Evaluate the Allergenic Potential of Food Proteins: this research was published in the journal *Toxicology Research* in 2015 (Sun et al. 2015a).

1. Introduction

It is the same as the "Introduction" in Research 2-a.

2. Methods
RBL cell-based immunoassay

The rat basophilic leukemia cell line RBL-2H3 (National Platform of Experimental Cell resources, China) was grown with 15% defined fetal bovine serum (FBS) in Eagle's minimal essential medium (EMEM). RBL cells were seeded at 2 × 105 cells ml^{-1} on a 96-well cell culture plate (200 μl per well) and cultured at 37 °C for 24 h. The cells were sensitized by Balb/c mice serum at 37 °C for 24 h. Then, cells were washed with sodium PIPES (NaPIPES) buffer (100 mM NaCl,

Table 3.2 Cell models for allergenicity evaluation and clinical diagnosis

Cell models	Protocol	Strengths	Weaknesses
Murine peritoneal mast cells	Histamine release	Sensitive, specific	Requires purification, semipermanent cultivation
Human basophils	Histamine release	Specific	Measure of histamine is difficult
Human basophils	Basophil activation	Specific	Consecutive blood sampling to obtain the basophils
RBL cells	β-Hexosaminidase release assay	Easily available; permanent cultivation; sensitive, specific, and reproducible	Allergen-binding patterns of mouse and human sera might not be fully identical
Humanized RBL cells	β-Hexosaminidase release assay	Easily available; permanent cultivation; sensitive, specific, and reproducible	Rare allergens for which human sera are difficult to obtain

25 mM disodium PIPES, 5.6 mM glucose, 5 mM KCl, 1 mM $CaCl_2$, 0.4 mM $MgCl_2$, 0.1% BSA; pH 7.2). Degranulation was initiated by incubating the sensitized cells with gradient concentrations of different allergens in NaPIPES buffer. Controls were conducted with no allergens to detect the spontaneous release.

The total release was quantified by lysing cells with 1% Triton X-100; 30 μl of supernatant was transferred into a 96-well plate to measure the released β-hexosaminidase. To detect the enzymatic activity of β-hexosaminidase, 50 μl 4-nitrophenyl N-acetyl-β-D-glucosaminide was added as a substrate (1.3 mg ml^{-1} in 0.1 mol l^{-1} citric buffer, pH 4.5). After incubation at 37 °C for 1 h, 100 μl glycine solution (0.2 M, pH 10.7) was added and the absorbance was measured at 405 nm. The consequence of the allergen-specific release is provided as a portion of the total release after the correction of spontaneous release. The bioactivity was shown as the reciprocal value of the ED_{50}, which is defined as the allergen dilution at half the maximal release.

3. Results and Discussion

RBL cell assays based on the activation and mediator release of effector cells are of great interest to assess the allergenicity of food proteins. It is considered that there is close correlation between the extent of mediator release and the severity and persistence of anaphylactic reactions. Moreover, our previous study showed that the endpoint of RBL cell assay could distinguish allergenic proteins from non-allergenic proteins. In this study, the allergenic activity of native OVM and OVA was evaluated by determining their capacity of inducing degranulation of RBL cells. Although there was no significant difference in MaxR induced by native OVM and OVA ($p > 0.05$), OVA needed sevenfold more molecules than OVM to obtain the same release, which implied that OVA had lower allergenic activity compared with OVM. Our results findings are in accordance with the findings of skin prick tests reported in previous studies.

Research 4-c Cell-Based Immunological Assay: Complementary Applications in Evaluating the Allergenicity of Foods with FAO/WHO Guidelines: this research was published in the journal *Food Research International* in 2014 (Sun et al. 2014**).**

1. Introduction

It is the same as the "Introduction" in Research 4-a.

2. Methods

Cell-based immunological assay

RBL-1 cells and RBL-2H3 (National Platform of Experimental Cell Resources, China) were sensitized by Balb/c mice serum or purified mouse IgE at 37 °C for 24 h. Degranulation was induced by incubating the sensitized cells with gradient concentrations of different purified allergens or anti-mouse IgE (eBioscience, San Diego, CA, USA) in NaPIPES buffer (100 mM NaCl, 25 mM disodium PIPES, 5.6 mM glucose, 5 mM KCl, 1 mM CaCl2, 0.4 mM $MgCl_2$, and 0.1% BSA; pH 7.2). The specific release was detected by the determination of β-hexosaminidase in the supernatant. The bioactivity was shown as the reciprocal value of the ED_{50}, which is defined as the allergen concentration that induced the half-maximal release.

More details are provided in the Methods section of the Supplementary Material, including descriptions of RBL cells and culturing, analysis of binding mouse IgE to RBL cells, passive sensitization, and activation of RBL cells.

3. Results and Discussion

It was proven that the RBL-2H3 cell-based immunological assay was sensitive. The cells that were sensitized with highly active mouse sera could be induced to show optimal degranulation at a concentration of 10–100 ng/ml of allergens. This assay was also very specific. The cells sensitized with mouse sera containing anti-OVA IgE did not degranulate after exposure to Gly. Also, this assay could distinguish the allergenic activity of OVA, PAP, and Gly. It is also demonstrated that the RBL-2H3 cell was a reliable model to evaluate the ability of a protein to induce IgE-mediated degranulation. A bell-shaped dose–response curve of increasing concentrations of antigens was exhibited, which was verified in our study. This phenomenon has been previously reported involved the degranulation of not only mast cells but also basophils. However, there is still some restrictions of the RBL-2H3 cell assay. The species restriction of the Fcε receptor requires murine IgE in this assay. Because the allergen-binding patterns of human and mouse sera are not completely identical, the RBL-2H3 cell-based assay cannot fully equate to humans. Such as it is, some studies have indicated that both the mouse and human systems revealed quite similar potency data to most of the major allergens. In contrast, in the case of 'rare' allergens, mouse sera may be more beneficial, for it is difficult to obtain human sera owing to its low prevalence of the respective sensitization.

Therefore, it is possible for the RBL-2H3 cell assay to serve as a substitute for the human IgE-based assay in cases where the mouse IgE can recognize the allergens relevant for humans.

3.3 In Vivo Evaluation Method Used in Allergenic Assessment of GMOs and Several Application Examples in Different Aspects Conducted in Our Lab

3.3.1 Animal Models

Because predictive animal testing for allergens in GE and non-GE foods is not sufficient for allergy assessment (Wal et al. 2015), researchers are working to discover or develop an animal model that can predict sensitization to allergy (Ladics and Selgrade 2009). However, none has been proved predictive so far.

In 2001, the FAO/WHO expert consultation recommended developing an in vivo animal model as part of the decision-tree strategies. An ideal animal model should react similarly to proteins on an "allergenicity" scale that is comparable with what is known in the matter of their prevalence and the severity of responses in humans (Ladics et al. 2010; Selgrade et al. 2009).

The animal models have one thing in common that specific IgE antibodies are produced to the specific proteins. Some models, including different strains of mice and rats and mice, comprise intraperitoneal injection and the analysis of specific IgE and IgG responses. It is claimed that the proteins that cause food allergy in humans with ease can produce more pronounced IgE responses relative to IgG responses, whereas the proteins that produce food allergy with difficulty are opposite in effect. Adjuvants are often applied to induce the immune response (Trials 2008).

An animal model has been developed to determine the potential allergenicity of food components. Brown Norway (BN) rats, which have high IgE responders, are sensitized with or without an adjuvant before oral and intraperitoneal exposure to the test compound. To avoid the induction of tolerance, these rats are reared for at least two generations on an allergen-free or test protein diet before challenge with the test compound. The outcome of such experiments should be carefully evaluated. It should be recalled, for example, that a rat experiment has failed to demonstrate the allergenicity of the 2S albumin from Brazil nut transferred into soybean, whereas individuals allergic to Brazil nuts reacted positively to the novel product (Melo et al. 1994).

Other models use mice or rats orally exposed to proteins, in which the IgE response and mast cell mediator release after a period of sensitization to the protein is analyzed. These latter models have the advantage of higher relevance of exposure and a clinical outcome (Jonge et al. 2007).

Many researchers realize that the results of animal models vary greatly depending on the protocol. Particularly, the strains, exposure modes, dosage and frequency of administration, and the endpoint of evaluation are of great importance (Ahuja et al. 2010). Therefore, the operating complexity makes it hard to identify a validated animal model for assessing the allergenicity of food proteins. Nevertheless, animal models are more suitable for exploring the mechanisms of allergic responses.

However, what should be mentioned is that induction of specific IgE in these animal models is not always linked with clear clinical signs of food allergy occurring in food-allergic patients. In fact, although other phenomena related to allergy have been noted, such as eosinophilia, delayed-type hypersensitivity, and mucous secretion, the majority of the animal models developed until now can only indicate sensitization such as the induction of IgE. It needs to be confirmed that whether there is correlation between induction of specific IgE as well as related immune responses in these models and the ability of the food proteins to cause food allergy in humans (Trials 2008).

We have done some research about animal models. I have listed these in the following.

Research 2-e A Modified Weight-of-Evidence Approach to Evaluate the Allergenic Potential of Food Proteins: this research was published in the journal *Toxicology Research* in 2015 (Sun et al. 2015a**).**

1. Introduction

It is the same as the "Introduction" in Research 2-a.

2. Methods

2.1 Animals and maintenance

Female Balb/c mice 3–4 weeks old were obtained from Vital River Company (Beijing, China) and fed on an egg-free diet under specific pathogen-free conditions with a light:dark cycle of 12 h. Temperature and relative humidity were maintained at 23(±3) °C and 40–70%, respectively.

2.2 Immunization protocol

The mice were sensitized by gavage with 1 mg OVM, OVM, or PAP plus 10 μg cholera toxin (CT) (C8052; Sigma) on days 0, 7, 14, 21, and 28. The control mice were given 100 μl saline alone or 10 μg cholera toxin in 100 μl saline each time. Two weeks after the last sensitization, the sensitized mice were orally challenged with 10 mg OVM, OVA, or PAP. All animal studies described herein were approved by the Beijing Municipal Science and Technology Commission of China (no. SYXK 2010–0036).

2.3 Systemic anaphylaxis assessment

For quantitative determination of vascular permeability, three mice from each group were administrated 100 μl 0.5% Evan's blue dye (Acros Organics, Morris, NJ, USA) via tail vein injection closely followed by intragastric challenge with OVM, OVA, or PAP. At 30 to 40 min after dye/antigen administration, the intestines and footpads of the mice were inspected for signs of vascular leakage. Also, 40 min after administration, the vascular permeability was quantified by measuring albumin levels in the peritoneal lavage (PL) fluid based on previous reports. The albumin content was also quantitatively determined in the supernatants with a Bicinchoninic Acid kit (Biomiga).

2.4 Measurement of plasma histamine levels

Plasma was obtained with tubes containing EDTA-K_2 20 min after administration. Plasma histamine levels were determined by a commercial enzyme immunoassay kit (Cusabio Biotech, Wuhan, China) according to the directions.

2.5 Histology and mast cell staining

At 24 h after challenge, the animals were euthanized by cervical dislocation. Spleen, lungs, and small intestine were fixed in 10% (v/v) formaldehyde, embedded in paraffin wax, and stained with hematoxylin and eosin (Sigma) for microscopic examination (Leica DM1000; Heerbrugg, Switzerland). Moreover, the mast cells of intestine tissue and spleen tissue from three independent cross sections per mouse were stained by toluidine blue and counted under a microscope. A mast cell with degranulation was defined as a toluidine-positive cell with five or more distinct stained granules entirely outside of the cell.

3. Results and Discussion

Even though there are still no validated animal models for assessing the allergenicity of food proteins, these may be applied to explore mechanisms of allergic responses. In this study, the Balb/c mice model was put into use to dissect the specificity of allergic reactions triggered by OVM and OVA. The results showed that both OVM and OVA could evoke systemic allergic reactions of the Balb/c mice and that OVM might also induce local allergic reactions in intestinal and pulmonary tissues. The data further indicated the allergenicity of OVM was higher than that of OVA. To a great degree, the mechanisms of OVA and OVM allergy are ascribed to their structure and the stability of their epitopes in the gastrointestinal tract. The structure of ovomucoid is composed of nine disulfide bonds and the carbohydrate content accounts for one fourth. Three well-separated domains compose the ovomucoid with one trypsin-binding site residing in the second domain. OVM digested by pepsin remains its trypsin (protease) inhibitor activities. Although the result showed that OVM was digested rapidly, some pepsin-digested fragments retained immunogenic epitopes. The display of immunogenic epitopes to T lymphocytes by antigen-presenting cells prompts T lymphocytes to differentiate into Th2 cells, which secret Th2 cytokines, inducing the generation of antigen-specific IgE antibody. Therefore, the binding of antigen with specific IgE on effector cells, similar to mast cells and basophils, results in mediator release including histamine and leukotrienes, leading to increasing vascular permeability. Moreover, mast cells also release proteases and cytokines that promote the intestinal permeability and damage the physiology of intestinal mucosa. Remarkable damage of intestinal mucosa was presented in the OVM-sensitized mice, such as necrosis and villous atrophy. An increase of intestinal permeability after degranulation triggers a massive passage of allergen molecules, resulting in more severe local and even systemic reactions. Thus, immunogenic epitopes of OVM that retained pepsin-digested fragments lead to its allergenicity. However, OVA did not retain trypsin inhibitor activities, which induced only 30% of the IgE-binding activity at the end of duodenal digestion. Therefore, digested OVA provoked a significant decrease of degranulation.

Research 4-d Cell-Based Immunological Assay: Complementary Applications in Evaluating the Allergenicity of Foods with FAO/WHO Guidelines: this research was published in the journal *Food Research International* in 2014 (Sun et al. 2014**).**

1. Introduction

It is the same as the "Introduction" in Research 4-a.

2. Methods

Animal model assay

Three to four-week-old female Balb/c mice were obtained from Vital River Company (Beijing, China). In the preliminary experiment, to determine whether the mice were immunologically naïve before exposure to the proteins, they were sensitized with 1 mg Gly, OVA, or PAP with 10 μg cholera toxin (CT) (Sigma-Aldrich) one time by gavage. After 30 min of exposure, the mice were exsanguinated. In the final experiment, to determine the allergenic responses to the studied proteins, the mice were sensitized orally with 1 mg Gly, OVA, or PAP with or without 10 μg CT on 0, 7, 14, 21, and 28 days. The controls were administrated 100 μl saline alone or 10 μg CT in 100 μl saline at the corresponding time point. After the aforesaid induction, the mice were not exposed to any allergen for 1 week. Two weeks after the last sensitization, the sensitized mice were administrated orally with a higher dose of allergens (10 mg/mouse). All animal studies described here were approved by the Beijing Municipal Science and Technology Commission of China (no. SYXK 2010–0036).

See the Methods section in the Supplementary Material for detailed descriptions of animal maintenance, antibody (IgG1 and IgE) titer determination, plasma histamine levels, systemic anaphylaxis assessment, histology, and mast cell staining.

3. Results and Discussion

In this study, Balb/c mice were sensitized with several allergenic food proteins, including OVA, Gly, and PAP to determine whether the Balb/c mouse model had the ability to discern different allergenic food proteins. The results showed that OVA, Gly, and PAP induced significantly different IgG1 and IgE antibody responses, histamine release, and systemic anaphylactic reactions in Balb/c mice and that PAP, OVA, and Gly represented nonallergenic, moderately allergenic, and strongly allergenic proteins, respectively, which was similar to the allergic responses in humans.

4. Conclusion

In conclusion, the RBL cell assay might be a practical tool for assessing allergenic activity of food proteins based on its high specificity, high sensitivity, and the ability to discern different allergenic food proteins. In view of our results and other investigations, we put forward a framework for evaluating the allergenic potential of food proteins as follows. First, a bioinformatical analysis and a pepsin digestion test are used to predict the allergenicity. Second, the RBL cell assay acts as a "referee" or "verification" to determine allergenicity. Finally, we explore the mechanisms of the allergic responses and the potential changes based on the modification of the allergenic proteins with the animal model assay. Therefore, complementary application of decision-tree strategies and the RBL cell assay tree is an advisable choice for overall assessment of allergenic potential.

Research 5-c Food Proteins from Different Allergen Families Sensitize Balb/c Mice to Family-Specific Immune Responses: this research was published in the *Journal of Immunotoxicology* in 2014 (Wang et al. 2014b**).**

1. Introduction

It is the same as the "Introduction" in Research 5-a.

2. Methods
2.1 Animals

Female Balb/c mice 3–4 weeks old were purchased from Vital River (Beijing, China) and acclimated to their new housing for 1 week before the experimental protocols. The animals were fed specific pathogen-free conditions with a light:dark cycle of 12 h. Temperature and relative humidity were maintained at 23–26 °C and 40–70%, respectively. Mice could access filtered water and standard rodent diet ad libitum during the course of the experiment.

3. Sensitization and Challenge Protocols

The mice (eight per group) were administrated with 1 mg test protein in 100 μl saline by gavage once per week for 4 consecutive weeks. Retro-orbital plexus blood samples were collected at day 0 and the last day of each week and processed for sera/plasma. After the induction period, the mice were not exposed to any protein for 2 weeks. On day 42 of the study, all sensitized animals were orally administrated 10 mg protein. Mice in the negative control group were administrated the same volume of saline. At 20 min after challenge, a blood sample was drawn from the retro-orbital plexus of each mouse. Part of the blood was collected into tubes containing EDTA-K2 for plasma. The rest was collected into other tubes and kept standing to clot for 1 h at room temperature for serum. In all cases, the blood samples were centrifuged at 3000 *g* (10 min, 4 °C) for plasma used in histamine measurement or sera used in antibody analyses. The serum and plasma samples were preserved at −20 °C until use.

At 40 min after the challenge, the mice were euthanized with chloral hydrate and intraperitoneally injected with 3 ml PBS containing 10 mM EDTA. Peritoneal lavage (PL) fluid was slowly aspirated after gentle massage of the abdomen. Then, aspirate was centrifuged at 600 *g* for 6 min at 4 °C. The supernatant was collected and stored at −70 °C until analysis.

4. Determination of Protein-Specific IgE and IgG1 Levels

Protein-specific IgE and IgG1 antibodies were tested with an indirect ELISA. The titers of protein-specific antibodies were then detected as described previously.

5. Determination of Histamine Levels

Plasma histamine levels were detected with a mouse Histamine (HIS) ELISA Kit (Cusabio Biotech, Wuhan, China) according to the instructions.

6. Vascular Leakage

Vascular leakage was detected by determining albumin level in the PL fluid after 40 min of the challenge as described earlier. Albumin content was determined in cell-free supernatants with a Bicinchoninic Acid kit (Biomiga), and its concentration in each sample was calculated by extrapolation from a BSA standard curve.

7. Results

Balb/c mice were sensitized by gavage over 4 weeks. Because OVA is a model of food allergen in rodent models, the experimental results of OVA were comparable to those of negative groups in the assessment of the utility of the Balb/c model in this study. All the indicators of mice in OVA group were significantly higher than those in the negative groups, which indicated that a T-helper (TH)-2 cellular response has been induced in the OVA-sensitized mice.

Other mice were sensitized with one of three proteins with different allergenicity (PAP, OVA, or 11S globulin) to determine whether this model could distinguish allergenicities of different food proteins. OVA and 11S globulin are already known as food allergens, whereas there is little evidence that PAP exerts allergenicity. It was also found here that it was obviously easier to develop oral tolerance against OVA than 11S globulin, which had a strong allergenicity in the mouse model. The results of protein-specific IgE antibody levels, histamine levels, and vascular leakage were all in agreement with the known allergenicities of these three proteins (PAP < OVA < 11S globulin). However, there was no significant difference between the histamine levels of the OVA mice and that of 11S globulin mice; rather, an evident increase in the level of PAP-specific IgG1 antibodies was observed.

It was shown that 11S globulin had a strong allergenicity. High levels of histamine and 11S globulin-specific IgE were observed. Up to day 35, the vascular leakage obviously increased; however, it decreased thereafter. In contrast, the level of 11S globulin-specific IgG1 antibodies was relatively low, and there were no significant weekly differences.

As an allergen that can induce oral tolerance with ease, OVA could induce a strong specific IgG1 response, the level of which declined at day 21 and day 28. The level of specific IgE of the OVA group was relatively lower than that of the 11S globulin group, which decreased from day 28 on. Also, the level of vascular leakage and histamine in the OVA-sensitized mice significantly increased in this period.

The three exogenous proteins HSA, LRP, and Cry 1Ab/Ac induced lower levels of histamine and vascular leakage compared with OVA and 11S globulin. The levels of LRP-specific IgG1 at day 42 were higher than before, although the levels of LRP-specific IgE were quite low. However, there was no significant weekly difference. HSA induced a relatively weak specific IgE and IgG1 response. The levels of HSA-specific IgE increased from day 28, whereas that of HSA-specific IgG1 reached a maximum at day 21. The results showed that there was no significant difference of specific IgE and IgG1 antibodies with the nonallergenic exogenous protein, Cry 1Ab/Ac, from week to week.

Research 7 Allergenicity of Recombinant Human Lactoferrin to an Animal Model, Brown Norway rats: this research was published in the journal *Food and Agricultural Immunology* in 2014 (Zhou et al. 2014).

1. Introduction

Food allergy is a prevalent disease all over the world. The development of approaches for the allergenicity assessment of novel foods has been increasing rapidly in the recent years. The evaluation of potential allergenicity of novel transgenic foods mainly includes the source of the gene, bioinformatics (sequence homology to known allergens), stability in pepsin, etc., according to the regulatory guidelines (Goodman et al. 2008). It is demonstrated that animal models can be used in studies on the mechanism of allergenic response and provide useful information for evaluation of the allergenicity of food proteins (Griffiths et al. 2005; Hausding et al. 2008; Knippels and Penninks 2003). An indispensable character of an ideal animal model is to mimic the mechanism and clinical symptoms of humans with respect to the same route of exposure. Most important of all, it should distinguish a high percentage of known allergic proteins from nonallergenic proteins and predict their allergenicity (Aldemir et al. 2009). Several studies have been carried out toward this end (Bøgh et al. 2009; Gaudry et al. 2004; Vinje et al. 2011). Although no animal model has been validated as completely ideal so far, many studies have been shown that the Brown Norway (BN) rat is a high responder strain (Knippels et al. 1999a; b), whose recognition of induced antibodies to specific protein is comparable to that noticed in sera from allergic patients (Knippels et al. 2000). Recombinant human lactoferrin (rhLF) is a novel protein that was derived from cloned cows developed with transgenic technology. It was indicated that the properties of iron binding and releasing as well as the antibacterial effect of the rhLF were identical to the native hLF. Nevertheless, rhLF has the advantage of better iron absorption and higher resistance to decay in performance, which makes it an important novel protein for the improvement of health conditions of the human body. Also, rhLF can be added to milk powder as a nutritional supplement to increase the iron absorption of infants. However, it should take into consideration that the incidence of milk allergy in children is about 1% to 5%, or even 7.5%. Thus, it is important to assess the allergenicity of rhLF before its addition. Although hLF is not considered to be an allergen according to the long history of human with exposure to it, the allergenicity of rhLF is suspended. Therefore, we comprehensively researched the potential allergenicity of recombinant human lactoferrin (rhLF) with BN rats in this study.

The result of a previous bioinformatics study indicated that rhLF has strong potential allergenicity, so an in vitro stimulated gastrointestinal digestion test was conducted to determine its stability to gastric and intestinal fluid. BN rats were also used in the animal experiment to assess the allergenicity of rhLF directly. The results showed that the allergenicity of rhLF is weaker than that of bovine lactoferrin (bLF).

2. Methods

Female specific-pathogen free (SPF) Brown Norway (BN) rats in separate groups were gavaged with saline, ovalbumin, rhLF, and bLF for 42 days without adjuvant. The content of LF in milk and OVA in egg is 0.3 g/l and 3 g/100 g, respectively. The recommended intakes of milk and egg are 400 ml/day and 60 g/day, respectively, in China. Chinese average body weight (BW) is generally about 60 kg. Thus, the recommended intakes of LF and OVA for humans are 0.2 mg/100 g·BW/day and 3 mg/100 g·BW/day, respectively, via calculation. The treatment dose to the experimental animal is set to five times human recommended intake. Therefore, the doses of rhLF and bLF were 1 mg/ml and 1 ml/day per animal and the dose of OVA was 15 mg/ml and 1 ml/day per animal, respectively. All these proteins were dissolved in saline. On day 41, the blood pressure of all animals was measured as a baseline. On day 42, all animals were challenged by intragastric intubation with given protein orally (1 ml 10 mg rhLF or bLF/ml, 1 ml 0.15 g OVA/ml, or 1 ml saline). During the experiment, two rats of the same group were maintained in one stainless steel wire cage with ad libitum access to fresh water and diet without milk and egg. The cages were placed under conditions with 15 air changes/h and a light:dark cycle of 12 h at 23 ± 2 °C and 50 ± 10% relative humidity.

On day 0, all animals were bled, and their plasma and sera were pooled as a negative control. On days 14, 28, and 42 of the experiment, blood samples were collected from the orbital venous plexus of rat left eyes for detection of cell counts. Then, blood samples for detection of IgE, IgG, and IgG2a were centrifuged at 4 °C and 3000 *g* for 10 min following coagulation at room temperature for 2 h for sera. Within 30 min after treatment, animal blood samples for plasma histamine were collected into centrifugal tubes with heparin. All the sera and plasma above were stored at −20 °C until analysis of antigen-specific IgE, IgG, IgG2a, and histamine level using ELISA. On day 42, blood pressure of all rats was measured 7 h after the treatment.

At the end of the experiment, all animals were anaesthetized and killed by cervical dislocation for gross and histopathological examination. A thorough necropsy of major organs was conducted by visual inspection. Then, organs including spleen, thymus, heart, liver, lung, and kidney were excised, weighed, and examined. The relative weight of each organ was standardized based on final body weight.

Samples of the foregoing organs were fixed in 4% buffered formaldehyde and embedded in paraffin. Then, sections 5 μm thick were affixed to slides and stained with hematoxylin and eosin for histopathology examination.

3. Results and Discussion

The results of antigen-specific IgG2a level in both rhLF and bLF groups indicated that both bLF and rhLF induced a similar immune response in animals. In terms of the level of antigen-specific IgE, although that in the rhLF group increased at a low level on day 14, it could not be detected on days 28 and 42. However, the animals in the bLF group produced a certain amount of antigen-specific IgE. Thus, it suggested that the ability to evoke humoral-mediated allergic reaction of rhLF was weaker than that of bLF. No significant difference was observed in the variation trend of count size of basophils among different groups, which was consistent with the previous study. As to eosinophils, a rising trend was noticed in animals of the OVA group and the positive control group, whereas the count size of eosinophils in the rhLF group and bLF group was not. This finding may predict that the count size of basophils in BN rats blood is not affected by food allergy or the relationship between them is not obvious except for eosinophils. Also, neither of these two lactoferrins caused an apparent cellar immune response.

Oral challenge of OVA did not induce an evident effect on blood pressure in most animals of OVA group, although systolic blood pressure (SBP) in some animals slightly decreased temporarily, which was in agreement with previous studies. However, SBP of animals in rhLF, bLF, and saline groups was kept at a normal level without large fluctuations. Also, histopathological changes of the respiratory organ lung in different groups were observed in this study. A dramatic increase in relative weight of lung and eosinophil infiltration lesions was noticed in more than half of the rats in OVA group compared with those of other groups, which indicated that local and systemic immune responses were absent in BN rats challenged with rhLF and bLF. Neither rhLF nor bLF can evoke an evident allergy clinical manifestation.

4. Conclusion

In theory, rhLF is a protein existing in human milk and should not be allergic to human body. Some modification has occurred in the recombinant expression of rhLF. Thus, it is necessary to evaluate its allergenicity with a series of methods. The results in this study showed that the potential allergenicity of rhLF is not conspicuous. However, the allergenic potential of rhLF is still not explicit and needs to be identified further with other methods. In future, the structure analysis of the protein and human serological experiment are of great importance and should be developed to explore its allergenicity.

Research 8 Allergic Reactions Compared Between BN and Wistar Rats After Oral Exposure to Ovalbumin: this research was published in the *Journal of Immunotoxicology* in 2013 (Sun et al. 2013**).**

1. Introduction

The morbidity of food allergy, an adverse immune response to foods, has been escalating in recent years. Food allergy occurred in about 5% in young children and in about 3% to 4% of adults in Westernized countries (Sicherer and Sampson, 2010). Currently, food allergens mainly include eggs, milk, shellfish, fish, peanuts, wheat, soy, and tree nuts. However, with the rapid development of novel foods, including foods and food products derived from genetically modified crops, more attention has been paid to available approaches to evaluate the potential allergenicity of novel gene products (Miller 2003; Party et al. 2002).

Through the investigation of animal models, including rats, mice, and guinea pigs, used in assessments of allergenic potential (Piacentini et al. 2003; Vinje et al. 2009), it was found that some weaknesses limit the use of guinea pigs as an applicable animal model, such as the significant differences in immunophysiology compared with other species and the lack of approaches to study its immune system (Ahuja et al. 2010a). In terms of mice, the natural complexity of the allergic reactions makes it hard to look for a single reliable marker to quantify the sensitization potential of a protein (Aldemir et al. 2009). Overall, the rat, which is one of the most commonly used species in toxicological tests, has a number of advantages compared with other animal models. Many studies have been conducted on the rat immune system and the available tools for immune-related studies (Penninks and Knippels 2001). In this study, we evaluated the extent of allergic reactions in both BN and Wistar rats with oral exposure to ovalbumin.

2. Materials and Methods

Four-week-old female BN and Wistar rats were orally exposed to OVA on days 1 and 14, and thereafter daily from day 15 to day 42. The dose of OVA was 150 μg/g·BW, 1 ml per animal. The rats in control groups received 1 ml saline at the same time. After this induction period, none of the animals was exposed to OVA for 1 week. On day 49, all OVA-sensitized animals were orally challenged with 750 μg OVA/g·BW, 2 ml per animal. The rats in control groups received 2 ml saline at the same time. Sera and plasma were collected and screened for OVA-specific antibodies and histamine. The levels of OVA-specific IgG and IgG2a in serum were detected by an indirect ELISA with OVA-coated microtiter plates as described previously. A capture ELISA was employed for the determination of serum OVA-specific IgE. Then, levels of histamine in plasma were determined with a Rat Histamine (HIS) ELISA Kit. On day 49, all the OVA-sensitized animals were orally treated with OVA. Then, blood pressure was detected by noninvasive blood pressure cuff at intervals during a period of 7 h. On day 50, differential cell counts and histopathological examination were conducted.

3. Results and Discussion

The IgE-dependent allergic reaction consists of two phases. One step is the induction phase, in which the immune system of the host is sensitized by the allergen and then specific IgE antibodies are produced to bind to specific cell-surface receptors on mast cells in target organs. The other step is the triggering phase, mediated by the allergen, which binds to the IgE and stimulates mediator (histamine) release from mast cells. In evaluation of the allergenicity of food antigens, both phases should be determined by suitable tests (Fritsché 2009).

OVA-specific antibody responses acted as the markers of the inducing phase. From day 28 to day 42, OVA-specific antibodies titers in both OVA-sensitized BN and Wistar rats were significantly higher than that of the negative control group. However, OVA-specific IgE titers in BN rats significantly increased and reached the maximal level on day 28 of oral exposure, whereas the OVA-specific IgE responses in the Wistar rats could only be detectable after day 14. The results indicated that in terms of the IgE-dependent allergic reaction, BN rats were more sensitive to OVA compared with the Wistar rats. The levels of OVA-specific IgE antibody were also decreased in BN rats with prolonged oral exposure to the antigen. This finding is likely to contribute to the observation that constant oral exposure may be related to development of oral tolerance, a process driven by T-regulatory cells and other associated mechanisms. Also, an almost continuous increase of OVA-specific IgE titers in Wistar rats was observed.

It has been proven that induction of specific IgE antibody is always accompanied by induction of the antigen-specific IgG (Vaz et al. 1970, 1971; Yamanishi et al. 2003), especially for subclasses of IgG that are regulated in a manner similar to IgE (as surrogates for IgE). In rats, IgG2a antibody is subject to cytokine regulation similar to IgE (Gracie and Bradley 1996; Kimber and Helm 2003; Saoudi et al. 1993). After oral exposure of OVA, IgG and IgG2a antibody responses stimulated in BN rats were generally expressed as higher titers than those in Wistar rat counterparts, with the exception of IgG on day 14.

The level of plasma histamine was recognized as a marker of the triggering phase. The levels of histamine in plasma of both BN and Wistar rats significantly increased on day 28. As another parameter of the triggering phase we examined the amount of basophils, eosinophils, and neutrophils that increased in the blood. Basophils are considered to be key cells in allergic inflammatory reactions because of their ability to release the histamine mediator. In contrast, eosinophils may inhibit allergic reactions owing to their ability to release deactivating histaminase († et al. 2010). The numbers of basophils in sensitized BN and Wistar rats significantly increased, and eosinophil levels in the BN rats were also significantly elevated in their lungs and blood. Therefore, it is possible that although basophil numbers in both BN and Wistar rats were elevated. and possibly the levels of histamine release by those cells as well, the added presence of histaminase-releasing eosinophils might lead to inactivation/degradation of some histamines whose release was triggered by the OVA in the BN rats. If so, it can be assumed that the BN rats may release more histamine than Wistar rats, but the ultimate influence of this could be mitigated in situ with the appearance of more and more eosinophils.

For further explanation of the clinical results related to any changes in antibody formation, histamine release, or immune cell populations, systolic blood pressure was determined on day 49 after the OVA exposures. The systolic blood pressure data would help to better define mechanisms related to the pathology of IgE-mediated allergic reaction in turn. The systolic blood pressure in some BN rats dropped temporarily and returned to normal values within 1 h, which is in accordance with previous studies. However, an almost continuous decrease of blood pressure in some Wistar rats was noticed after challenge. Because these clinical manifestations were only found in some animals, these results could reflect the situation seen in food-allergic patients as well.

It has been demonstrated that eosinophilia is the first inflammatory cell linked firmly to asthma pathogenesis (Dunn et al. 1988; Marsh et al. 1985), followed by the T cell soon after (Mosmann and Cherwinski 1986). Histopathological examination of the lungs of OVA-challenged BN rats revealed characters associated with elevated IgE levels and allergic asthma, namely, eosinophilic aggregates. Similar characteristics were noted in the blood analysis; for example, the number of eosinophils in sensitized BN rats significantly increased compared to that of control groups. In a strain-specific outcome, a slight interstitial pneumonia was also observed in OVA-challenged Wistar rats. These results were in accordance with previous studies. Furthermore, increased eosinophils were noticed in patients who will develop late asthmatic reactions 24 h following allergen provocation. As it is considered that T-helper (Th)-2 cytokines, particularly IL-5 and to a certain extent IL-4, are also key factors in allergic asthma because of their ability to promote eosinophil proliferation in the lungs (Bates 2009), measurement of systemic and local IL-5 levels may also be of value in future studies on the inflammation that occurs as a part of the pathology related to allergic asthma.

4. Conclusion

In summarize, immune responses and clinical manifestations were different in OVA-sensitized BN and Wistar rats. These results indicated that these two rat strains may differ in the immunological mechanisms of allergy. Moreover, no correlation was found between immune responses and the severity of clinical symptoms. Clearly, the data from this study should be recognized as 'preliminary' because only a single protein allergen was examined. Hence, further studies are needed to compare the allergic reactions in BN rats with those in Wistar rats with purified strongly, weakly, and nonallergenic proteins based on our study here. Additionally, as a certain correlation exists between the presence of selected inflammatory mediators and clinical symptoms, evaluation of these mediators might be of diagnostic value and possibly be helpful in monitoring the severity of clinical symptoms in the future.

Research 9 Potential Allergenicity Research of Cry1C Protein from Genetically Modified Rice: this research was published in the journal *Regulatory Toxicology and Pharmacology* in 2012 (Cao et al. 2012b).

1. Introduction

Bt proteins, which are insecticidal and derived from *Bacillus thuringiensis*, have been incorporated into GM crops for pest resistance. However, it was shown recently that some insects have developed resistance to Bt proteins under laboratory and greenhouse conditions (Cao et al. 2009; Hama et al. 1992; Hiratsuka et al. 1995; Mcgaughey 1985). The Cry1C protein was developed in answer to insect resistance to Bt proteins. Because the Cry1C protein binds a different midgut brush border membrane site in insects compared with Cry1Aa, Cry1Ab, and Cry1Ac toxins (Alcantara et al. 2004), it can combine with Cry1A proteins or other groups of Bt proteins to block or delay the occurrence of pest resistance. The Cry1C toxin is efficient against varieties of lepidopteran pests including rice stem borers. Thus, Cry1C toxin can act as a potential alternative to Cry1A toxins and combine with other Cry1A genes for development of two-toxin Bt crops.

As a new member of the BT insecticidal protein family, the safety of Cry1C protein must be evaluated before Cry1C genes are widely used in rice or other GM crops. In this study, the potential allergenicity of Cry1C protein was tested in BN rats. In the animal studies, the Cry1C protein was produced in *Escherichia coli* (*E. coli*). As the exposure of human to GM rice was oral, an oral sensitization protocol without adjuvant to mimic physiological conditions was used. The levels of IgE, IgG2a, cytokine, and histamine, and numbers of blood eosinophils and mast cells, were tested following protein exposure.

2. Materials and Methods

We use the *E. coli* strain PET-30a(+)-Cry1C-rcp-BL21 (DE3) (Laboratory of Food Safety, China Agricultural University, Beijing, P.R. China) to highly express Cry1C protein. The Cry1C fusion protein was expressed as in the previous study. The purified Cry1C protein was identified by SDS-PAGE, Western blot, LC-MS/MS analyses, and biological activity tests. The equivalence of *E. coli*- and plant-produced proteins was presented in our previous study.

Four-week-old female BN rats were randomly divided into five groups ($n = 6$). Animals were challenged with 0.1 mg polynitroxylated albumin (PNA), 1 mg PAP, 1 mg OVA, or 5 mg purified Cry1C protein dissolved in 1 ml water by gavage once a day for 42 days without adjuvants. Blood samples were obtained in one of two ways: from the orbital plexus under light CO_2 anesthesia before killing or by exsanguination from the abdominal aorta at the time of death.

To detect the protein-specific IgG2a, 96-well microtiter plates were coated at 4 °C overnight with 100 μl/well of a 10 μg/ml solution of PNA, PAP, or OVA or with a 100 μg/ml solution of Cry1C protein in 0.05 M carbonate buffer, pH 9.6. The control serum pool acted as the negative control, measured at a 1:4 dilution. Three times the standard deviation was added to the average extinction in negative control wells, which provided the reference value to detect the antibody titer. At the start, each test serum was titrated at a dilution of 1:4. The reciprocal of the furthest serum dilution providing an extinction value higher than the reference value was read as the titer. Every sample was conducted in duplicate. Negative and positive control samples were incorporated for each 96-well plate.

Antigen-specific serum IgE antibodies were measured with ELISA techniques. The titer of the test proteins in this study was detected as follows: a water control group sera pool was chosen as the negative control. The pooled control group sera were measured at a dilution of 1:4. Three times the standard deviation was added to the average extinction in negative control wells, which provided the reference value to determine the titer in the test sera. At the start, each test serum was titrated at a dilution of 1:4, and the reciprocal of the furthest serum dilution providing an extinction value higher than the reference value was read as the titer. All analyses were conducted in triplicate.

IL-4, interferon (IFN)-γ, and histamine serum concentrations were detected with RapidBio ELISA Kits (RapidBio.org, USA). All tests were conducted following the manufacturer's instructions.

cDNA was transcribed with an iScript cDNA Synthesis Kit (BioRad Laboratories, Hercules, CA, USA), and total RNA was isolated with the Qiagen RNeasy Kit (Qiagen, Hilden, Germany). Gene expression was measured by polymerase chain reaction (PCR) with an ABI 7500 thermal cycler (Applied Biosystems, Foster City, CA, USA) and specific Taqman probes (Applied Biosystems) for each gene of interest. We used β-actin as a housekeeping gene for analysis of changes in cycle threshold values. The fold induction above SEB alone was tested based on changes in the Δ-cycle threshold values.

Animals that had been sensitized previously with PNA, PAP, OVA, or Cry1C were orally challenged 10 days after the final oral gavage of the sensitization period. Animals sensitized with certain protein and those in the control group were orally challenged with 2 ml of a protein solution (5 mg ml^{-1}) suspended in tap water. Blood samples were gathered from the orbital plexus under light CO_2 anesthesia at 0, 0.5, 1, and 1.5 h after protein administration. The sera were isolated and used for histamine quantification with ELISA.

Blood was gathered into tubes coated with EDTA, and absolute eosinophil numbers were measured by Hemavet 950 FS (Drew Scientific, USA).

Jejunum and ear tissues were collected, fixed in formalin, and embedded in paraffin. Then, the tissue sections were stained with toluidine blue or hematoxylin and eosin for the determination of mast cells. Mast cell numbers and activation status were measured through counting cells with dense metachromatic granules and compact shape versus those with dispersed granules that clearly extended outside the cell body.

3. Results and Discussion

In this study, we used the BN rat as an animal model to study food allergy. The strong allergen OVA or PNA induced antigen-specific IgE and IgG2a. The weak allergen PAP induced lower antigen-specific IgG2a and IgE, and later. There was no antigen-specific IgG2a or IgE in Cry1C protein-sensitized rats. In PNA-treated rats, the IgE-mediated allergy was indicated by increased histamine levels immediately following PNA stimulation and increased numbers of mast cells. In addition, eosinophils were found in the intestinal villi clearance and intestinal glands clearance of PNA- or OVA-sensitized rats. Differing from other protein groups, the rats in Cry1C group showed no allergy symptoms. In conclusion, there is no evidence that Cry1C is more allergenic than PAP, and Cry1C is safe for use in food or animal feed, which is consistent with our previous conclusions. In this study, several methods were employed to determine the allergenicity of a new protein. These approaches could be verified with each other. In addition, various proteins with different allergenic properties were used to judge the strength of Cry1C protein sensitization. The foregoing technique provides a new methodology to evaluate GM food allergenicity.

Research 10-c Prokaryotic Expression and Allergenicity Assessment of Hygromycin B Phosphotransferase Protein Derived from Genetically Modified Plants: this research was published in the *Journal of Food Science* in 2007 (Lu et al. 2007).

1. Introduction

The same as the "Introduction" in Research 10-a.

2. Methods

2.1 Allergenicity evaluation of HPT protein in Brown Norway (BN) rats

Four-week-old male Brown Norway (BN) rats were randomly divided into three groups ($n = 6$). Ovalbumin (OVA), a potent respiratory and food allergen, was applied as the positive control. Animals were challenged with 1 mg OVA or 5 mg pure HPT protein, both dissolved in 1 ml water, by daily gavage without adjuvants for 42 days. The third group received only 1 ml water as the negative control.

2.2 Measurement of IgE, IgG, and histamine

Total and specifically induced serum IgE antibodies for OVA and recombinant HPT protein were measured with ELISA; 96-well microliter plates (Dynex Technologies, Chantilly, VA, USA) were coated at 4 °C overnight with a 10 μg/ml solution of OVA (Sigma A-5503) or with a 100 μg/ml solution of HPT protein in carbonate buffer (50 mM carbonate/bicarbonate, pH 9.6). Plates were washed three times with 200 μl/well of PBST (containing 0.1% Tween-20), followed by addition of 150 μl/well of blocking buffer (PBS containing 5% skim milk). After incubation at 37 °C for 1 h, the plates were washed and 1:10 dilutions of rat serum in blocking buffer were added into the well. Plates were then incubated at 37 °C for 1 h. After washing, a 1:200 dilution of 100 μl/well of mouse antirat IgE serum (PharMigen Bioscience) with blocking buffer was added. After incubation at 37 °C for 1 h and washing, 100 μl/well of HRP-goat antimouse IgG serum (PharMigen Bioscience) diluted 1:5000 with blocking buffer were added and incubated at 37 °C for 1 h. Then, the plates were washed six times and 100 μl/well of fresh substrate *O*-phenylene diamine (Sigma) was added. Next, the plates were incubated at 37 °C for 30 min. Finally, the reaction was ended with 50 μl/well of 2 M sulfuric acid.

Optical density was read spectrophotometrically at 492 nm with an ELISA plate reader (Finland LABSYSTEMS). IgG was analyzed similarly, and histamine was determined with an arathistamine kit (RapidBio).

3. Results and Discussion

3.1 Antigen-specific IgG response

The BN rats in the positive group treated with OVA by gavage showed significant immunoreaction, and the level of IgC-specific antibody in their sera, which were diluted by 32 fold, was much higher than that of the negative group (P/N > 2). In contrast, the immune response of the test group was basically equivalent to that of the negative group. This result showed that OVA triggered an obvious immunology reaction but that HPT protein did not.

3.2 Evaluation of IgE response

Food allergies are adverse reactions to an otherwise harmless food or food component that involves an abnormal response of the body's immune system to specific proteins in foods. True food allergies may induce several types of immunological responses, mainly including cell-mediated reactions and IgE-mediated reactions. The most common type is the latter, also known as immediate hypersensitivity reactions, for the reason that the symptoms occur within minutes to a few hours after ingestion of the offending food. Thus, when animals are fed allergic proteins, their immune system could produce specific IgE antibodies that can be tested with ELISA. The OD values of sera in the OVA-group were significantly higher than those of the water group (P/N < 2). In constrast, the OD values corresponding to HPT protein were similar to those of the negative group. This result indicated that HPT protein did not evoke specific IgE antibodies, whereas high levels of specific anti-OVA IgE levels were observed.

3.3 Results of histaminase test

There was no significant difference between the level of histamine of the test group and that of the negative group, although that of the positive group was two times higher than that of the test group ($p < 0.05$).

FAO/WHO has recommended a decision tree to evaluate the potential allergenicity of novel gene products. If the transgene encodes a known allergen, the recombinant protein may be determined for allergenicity using sera by the classical in vitro tests. If the source of the introduced gene has no history of allergenicity, alternative approaches are needed based on the structure and stability of the foreign protein. These methods must be critically appraised and conducted on a case-by-case basis. Because the HPT protein belonged to the latter group, it was evaluated mainly by gastrointestinal digestion in vitro and by an animal model. So far, in vitro digestion is accepted as an model system. However, good (oral) animal models to study food allergies and the allergenicity of food proteins have not yet been available. Several strains of rodent have been developed but are still under consideration. Currently, most studies have been conducted with BN rats receiving the potential allergen by gavage without adjuvant. In this study, we adopted the 3R International Principle, which recommends that the number of test animals should not exceed six and that healthy conditions should be maintained. It is demonstrated that the BN strains could be a good allergy model under these conditions.

3.3.2 Human Trials

Ethical issues and feasibility prevent including human trials in the transgenic food safety evaluation system, but trials are often conducted in diagnosis of food allergies. Diagnosis is traditionally based on the clinical history, sensitization [skin prick test (SPT) or specific IgE (sIgE)] and, optionally, a double-blind placebo-controlled food challenge (DBPCFC) test (Soares-Weiser et al. 2014).

The skin prick test (SPT) is a method to rapidly determine sensitization for food-induced allergic disorders. Several different methods of interpreting SPT have been employed; however, it remains uncertain whether these can accurately predict food allergy reactions (Knight et al. 2006; Serup and Staberg 1985; Sussman et al. 1982). Among the foregoing methods, the most commonly evaluated endpoint is wheal size. Generally, when the mean wheal diameter is more than 3 mm compared with the negative control, the results of the food allergy SPT are recorded as positive. However, this method may not sufficiently reflect the actual size of those wheals with irregular shapes (Eigenmann and Sampson 1998). Furthermore, SPT cannot be applied to predict symptomatic IgE-mediated food allergy (Ocmant et al. 2009).

It has been considered that double-blind placebo-controlled food challenges (DBPCFC) are the gold standard for evaluation of food allergy. However, they are costly and potentially dangerous because they can trigger severe reactions in some subjects (Fabienne Rancé 2002). Hence, reliable, safe screening tests are still required to evaluate reactivity to food allergens.

3.4 The Prospect of Allergenic Assessment of GMOs and the Future Development Trend of Evaluating Methods Used in Allergenic Assessment of GMOs

It should be mentioned that no single one of these foregoing criteria can provide evidence of the absence of allergenic potential of the protein in question. For example, the amino acid sequences are not known for all allergens. Also, tertiary structures of proteins, which are significant determinants of allergenicity, are not well predicted from amino acid sequences. Failure to determine potential allergenicity in tests may not eliminate that the food in fact is allergenic. For instance, there exists a correlation between protein stability and allergenic potential, but this correlation is not absolute (Tong Jen Fuet et al. 2002). Similarly, in vitro tests using human sera depend on the allergen specificities of the sera of the allergic individuals used. In vivo human studies, especially those involving provocation, are often regarded as unethical and cannot be conducted at will. Therefore, animal testing is involved as the alternative testing, although it should be stressed that no validated animal tests to determine potential allergenicity of foods for humans are available so far. Full assessment of allergenicity of foods should be based on a case-by-case, weight-of-evidence approach, taking all available information into consideration. Further, most proteins, including those already present in plants, have the potential to induce food allergenicity (IgE antibodies) in one or more persons, indicating the near impossibility of ascertaining an absolute lack of allergenic potential for most proteins.

At present, there are no internationally harmonized guidelines to test potential allergenicity of food proteins in laboratory animals, but it can be done on an experimental basis. A review of the application of animal models in assessment potential food allergenic activity has been published by Prescott and Hogan (2006).

In the future, cell models may be widely used in research studies to explore allergic pathways and mechanisms. When the typical markers in an allergy pathway are identified, we could select some of them to conduct subsequent experiments to understand the occurrence of allergenicity in depth. Thus, the food allergic reactions may be eliminated or reduced.

Chapter 4
Toxicology Assessment

4.1 Toxicology Assessment Used in Safety Assessment of GMOs and the Methods Used in Food Safety Assessment, with the Advantage and Disadvantage of Methods In Vivo and In Vitro

4.1.1 Toxicology Assessment Used in the Safety Assessment of GMOs

In recent years, more genetically modified organisms (GMOs) have entered the market. The issue of GMO safety is an issue of intense public concern, and in the case of foods, products containing GMOs or products thereof carry the risk of consumer rejection. The World Health Organization (WHO) defines GMOs as those organisms in which the genetic material has been altered by an in vitro artificial operation, although the WHO has declared that the GM products are as safe as the non-GM products. The GM products that are currently on sale in the international market have all gone through risk assessment by national authorities. The risk assessment of GM foods in general, and crops in particular for human nutrition and health, has not been systematically performed as indicated in the scientific literature (Domingo 2007). Although safety assessment had been conducted for every GM crop, evaluations for each GM crop or trait have been accomplished using different feeding periods, animal models, and parameters. The most common result is that GM and conventional crops are substantial equivalence in nutritional performance and animal health. Adverse microscopic and molecular effects of some GM foods in different organs or tissues have been reported, but these phenomena are common in animal feeding studies (Magana-Gomez and de la Barca 2009). Diversity among the methods and results of the risk assessments reflects the complexity of the subject. Probably one of the most important principles used in the safety assessment of GM foods/plants is the "substantial equivalence" concept. This notion is based on the

K. Huang, *Safety Assessment of Genetically Modified Foods*,
DOI 10.1007/978-981-10-3488-6_4

principle: "If a new food is found to be substantially equivalent in composition and nutritional characteristics to an existing food (the non-GM crops), it can be regarded as being as safe as the conventional food."

4.1.2 The Methods Used in Food Safety Assessment, and the Advantages or Disadvantages of Methods In Vivo and In Vitro

Strategies for assessing the food safety often including three approaches: the in silico method, investigating the potential toxicity and allergenicity by comparing the structure–function relationship and bioinformatics analysis; in vitro assays with proteins, enzymes, or cultured cell lines; and in vivo animal studies. Animal studies are conducted following the bioinformatics and vitro assays. The selection of animal studies is often based on the results of a molecule's structure, function, and in vitro toxicity results; at the same time, ethical criteria should be considered. Various methods including in silico and in vitro methods may contribute to the safety assessment of GM food and feed and components. The following methods are used in the safety assessments.

1. In silico searches for sequence homology or structural similarity compare the novel proteins or their degradation products with known toxic or allergenic proteins.
2. Simulated gastric and intestinal fluids research studies the digestive stability of newly expressed proteins and the heat stability test in in vitro systems for analysis of the stability of the novel protein under heat or other processing conditions.
3. In vitro genotoxicity test methods screen for point mutations, chromosomal aberrations, and DNA damage and repair. For example, the micronucleus test is a mammalian in vivo test that detects damage to the chromosomes or the mitotic apparatus induced by chemicals. The in vivo bone marrow cytogenic test is a mutagenicity test for the detection of structural chromosomal aberrations. The *Salmonella* typhimurium histidine(his) reversion system is a microbial assay that measures his^-–his^+ reversion induced by chemicals that cause base changes or frameshift mutations in the genome of this organism.
4. The acute oral toxicity methods can detected the averse effects occurring within a short time of oral administration of a single dose of a substance or multiple doses given within 24 h. It is an initial step in establishing a dosage regimen in sub-chronic and other studies and may provide initial information on the mode of toxic action of a substance.
5. Sub-chronic oral toxicity is the adverse effect occurring as a result of the repeated daily oral dosing of a chemical to experimental animals for a part of the lifespan not exceeding 10%. It provides information on possible health hazards likely to arise from repeated exposures over a limited period of time. It will provide

information on target organs, the possibilities of cumulation, and can provide an estimate of no-effect level of exposure that can be of use in selecting dose levels for chronic studies and for establishing safety criteria for human exposure.

Many in vitro methods are used in the evaluation of food safety for the advantages of relatively low cost and high throughput. The results observed in vitro test may be indicative of toxic effects; but these systems also have some disadvantages: some results rely on cell components or purified protein, or laboratory cultures of cell lines that do not correctly reflect the real circumstances of such cell components or cells in living animals or a human body. However, the observations on toxicity from in vitro tests can provided some clues for the in vivo tests. The application of in vitro methods has greater importance in the area of genomic research and microarray systems to monitor changes in gene expression. Results acquired from in vivo tests usually indicate potential toxic effects of one test substance in humans. However, the extrapolation of results from animal tests to humans is uncertain: unpredictable differences can include interspecies and interindividual differences in metabolism, physiological processes, and lifestyle. These uncertainties are usually addressed through the use of uncertainty factors.

4.2 Animal Tests Used in the Safety Assessment of GMOs and the Application of Animal Tests in Toxicology Research

4.2.1 Animal Tests Used in the Safety Assessment of GMOs

Animals have been widely utilized in the safety assessment of GMOs. The acute toxicity test usually involves administration of a large single dose followed by 7–14 days of observation. Sub-chronic toxicity, chronic toxicity, and carcinogenicity tests are maintained for 1 month or several months in an animal. A variety of tests with laboratory animals have been developed to identify and characterize health hazards by exposure to food or food additives. The use of these tests has recently been reviewed in detail by the European Concerted Action Food Safety in Europe (FOSIE) (Barlow et al. 2002). Most of these tests have been standardized, and OECD guidelines are available for their conduct and interpretation (OECD 1993b) as these tests were established for chemicals or other toxic substance. Some changes are presented when these tests are applied in the safety assessment of GMOs. First, it is easy to acquire the LD_{50} of one chemical and calculate the dose relationship; according to the results, it is easy to extrapolate the human acceptable daily intake (ADI). A safety factor of 100 fold is commonly applied to animal data to derive the acceptable daily intake (ADI) of food additives. The highest administered dose is expected to produce some toxicity, but it is often not possible to incorporate too much into the diet for its nutrient. In the sub-chronic study, 70% incorporation usually is the highest dose in rice; the incorporation is very high compared with the chemical, but this incorporation of rice can just top dozens of human consumptions, far below the chemicals. In an animal study conducted previously, feeding trials with whole foods at the highest incorporation dose, no adverse effect was observed. In this animal study, with no adverse effect observed, and the administered doses lower than expected human exposure levels, it is not precise to extrapolate the safety of one substance. Furthermore, it may be not easy to keep the nutritional balance when much novel foods or food extracts are incorporated into the diet, so the food incorporation needs careful consideration.

4.2.2 Species Selection

The use of animal model for hazard identification in the safety assessment is far from ideal, but is still widely considered to be the best approach available for its advantages.

4.2.3 Dose Selection

Three doses are usually selected in routine toxicological studies: the highest doses often cause some expected toxicity to humans or animals, and the lowest dose is set so no toxicity effect is observed. The purpose of these studies is to point out the nature of any adverse effect and both the effect and the No Observed Adverse Effect Level (NOAEL). As an alternative, the calculation of benchmark doses (BMD) also has been advocated (Crump 1984). Animal-based toxicological methods for hazard identification and characterization have recently been reviewed by Barlow et al. (2002).

4.2.4 Route of Administration

The test substance is often administered into the diet to acquire the pathway most closely resembling that in humans. Otherwise, administration in water is another choice that may be reasonable for beverages. Oral gavage may be one good method for developmental toxicity studies, but is less convenient in a study over a long time period. The test substance is often given for 6 days or 1 week.

In addition to this administration route, other methods also can be used; for example, i.p. (intraperitoneal) or i.v. (intravenous) administration. Thus, the mode of drug administration needs consideration of many factors, not only in relationship to identifying hazard potential but also in relationship to the mode and dosage of human exposure.

4.3 The Expression of Foreign Protein (the Exogenous Gene in GMOs) and Equality in In Vitro Analysis

4.3.1 The Expression of Foreign Protein

Many hosts can be used for the production of recombinant protein, including bacteria and yeasts. *Escherichia coli* is the preferred choice for its advantages: it is easy to culture, had a short life cycle, and its genetics are well known. *E. coli* was the first host used in biopharmaceuticals to produce recombinant DNA (rDNA). In 1982, Eli Lilly's rDNA human insulin produced by *E. coli* was approved to be used in the medical field. For its high production, the *E. coli* host is often used to produce therapeutic proteins to reduce the cost. In *E. coli*, we have two main technical problems to overcome: (1) little or no expression of some foreign gene and (2) recombinant proteins are insoluble when overexpressed. Structural, functional, and safety assessment of protein requires a large amount of protein with good quality. Many methods are used to improve the level of protein expression or non-expression in the *E. coli* host. We can regulate and control many important factors such as the secondary structure of mRNAs, solubility, codon optimization, or bacterial toxins; these factors influence the expression of foreign genes in *E. coli* [27]. When the expression and solubility of the foreign protein are less, four strategies may be effective in increasing the expression and solubility of the foreign protein: changing to another host strain, changing one vector, adding some chemicals, and optimizing the codon.

We have done some research about the expression of foreign proteins (foreign protein from GMO), as follows.

Research 1 Prokaryotic expression and allergenicity assessment of hygromycin B phosphotransferase protein derived from genetically modified plants. This research was published in the *Journal of Food Science* in 2007.

The aminoglycoside 4-phospotransferases encoded by the aph (4)-hygromycin B phosphotransferase gene (hpt), derived from Ta of *Streptomyces hygroscopicus*, is a useful selectable marker to screen for transformants in gene transfer experiments involving plants (Linda 1983). The enzyme coded by this gene can detoxify hygromycin B through phosphorylation. Hygromycin B is an aminocyclitol antibiotic that inhibits protein synthesis in both prokaryotes and eukaryotes by interfering with both ribosomal translocation and aminoacyl-tRNA recognition (Waldron et al. 1985). The gene is used in several GMOs such as rice, corn, and rape (Sawahel 2003). For its low expression in these transgenic plants, histidine phosphotransfer protein (HPT) was produced in *E. coli*. The exogenous recombinant protein HPT can be used for research in vitro digestion and animal studies.

4.3.1.1 Construction of Protein Expression System in Prokaryotic Cell

The hpt gene (GenBank accession number E00287) was cloned from the genome of genetically modified plants by polymerase chain reaction (PCR). Two primers were designed with 5c/CGC GGA TCC GCG ATG AAA AAG CCT GAA3c / and 5c / CCC AAG CTT GGG TCT ATT TCT TTG CCC TC3c. We defined the sequence of the hpt gene and introduced *Bam*HI and *Hin*dIII at the 5c- and 3c-ends of the PCR product, respectively. PCR conditions were set as follows: preincubation 5 min at 95 °C, 35 cycles consisting of denaturing 30 s at 94 °C, annealing 60 s at 58 °C, and extension 90 s at 72 °C, followed by a final extension of 10 min at 72 °C. The products of PCR were digested with the enzymes *Bam*HI and *Hin*dIII. The products were then ligated at 15 °C overnight through a standard T4 DNA ligase procedure. The T4 was shifted into into *E. coli* strain DH5α cells. When the expression system was built, the recombinant plasmids were screened by digestion with *Bam*HI and *Hin*dIII and confirmed by DNA sequencing. The constructed plasmid was designated as pET30a-hpt. The sequence of the hpt gene in pET30a-hpt was sequenced to ensure its exact construction. The sequencing result was compared with the sequence reported by GenBank (GenBank accession number E00287).

4.3.1.2 Expression and Purification of HPT Protein

Plasmid pET30a-hpt was introduced into *E. coli* BL21. The expression conditions of the HPT fusion protein were 200-ml cultures set at 37 °C in Luria-Bertani (LB) liquid culture. The cells were grown until the OD (600 nm) within 0.6 to 1.0, 0.6 mM isopropyl-β-D-thiogalactopyranoside (IPTG) was added, and the culture was incubated at 37 °C for 3 h. Cells were collected by centrifugation and resuspended in binding buffer (20 mM Tris-HCl, 0.2 M NaCl, 5 mM iminazole pH 8.0). The lysate was subjected to centrifugation and the supernatant was analyzed for HPT content by sodium dodecyl sulfate–polyacrylamide gel electrophoresis (SDS-PAGE). The purification of HPT was performed by Ni-NTA metal-affinity chromatography. Purified protein was obtained first by binding to a Ni-NTA-agarose column followed by removal of the His6·Tag with an enterokinase (EK) for 18 h at 37 °C.

4.3.1.3 Identification of the Recombinant HPT Protein

The purified HPT protein was identified by SDS-PAGE and Western blotting. Proteins were transferred to a nitrocellulose (NC) membrane (Millipore, Bedford, MA, USA), and the membrane was blocked with 1% bovine serum albumin (BSA) for 2 h. The membrane was incubated in His6·Tag monoclonal antibody (Novagen) liquid for 1 h. The membrane was washed with TBST six times, exposed to peroxidase-conjugated goat anti-mouse IgG secondary antibody (Sigma), washed six times, and the substrate, 4 chloro-1-naphthol, was added. After color appeared, the membrane was washed twice with distilled water.

4.3.1.4 Results

Construction of Protein Expression System in Prokaryotic Cell

The hpt gene (GenBank accession number E00287) was amplified by PCR and correctly inserted into the pET30 (+) vectors at the *Bam*HI site and *Hin*dIII site. DNA sequencing results (data not shown) further confirmed that there was no unwanted error or unwanted mutation in the hpt gene, and the hpt gene was inserted into the expression vector pET30 (+) correctly.

Expression, Purification, and Identification of HPT Protein

Escherichia coli BL21 (DE3) with the hpt gene was used to produce HPT protein, for it is difficult to acquire enough protein from genetically modified plants directly. Equivalence evaluation between the bacteria-derived and plant-derived protein was conducted. When all the results demonstrated the protein produced by *E. coli* was in equivalence with the protein from the GMOs, it is an appropriate substitute for that extracted directly from GMOs. The fusion protein was highly expressed in soluble form and the protein weight was about 45 kDa with His6·Tag. Results from SDS-PAGE and Western blot indicated that HPT fusion proteins were expressed with the right size of 45 kDa. The antiserum reached a very high degree of sensitivity to recombinant HPT protein. The cross-reactivity determined experimentally showed a high degree of specificity for the antiserum used, because values were all less than 0.1%

Research 2 Cloning, expression and characterization of recombinant elastase from *Pseudomonas aeruginosa* in *Picha pastoris*. This research was published in the journal *Protein Expression and Purification* in 2009.

The gene *lasB* from *Pseudomonas aeruginosa*, which encoded elastase, was cloned and first successfully expressed in *Pichia pastoris* strain KM71 under the control of the AOX promoter. The effects on the recombinant elastase activities of different pH, different temperatures, and different metal ions were assayed. The full-length gene (1497 bp) encodes a preproenzyme including an N-terminal signal peptide (23 amino acids, aa), a propeptide (197 aa), and mature elastase (301 aa). The recombinant elastase was secreted into culture supernatants using a signal sequence from *lasB* and showed a single band at about 34 kDa by SDS-PAGE. The recombinant elastase expression reached the highest level of approximately 450 mg/l, and the specific elastolytic activity of the recombinant elastase was 130 U/ml, which was approximately 26 fold higher than that of elastase obtained from *Pseudomonas aeruginosa*. The optimal temperature and pH of the recombinant elastase were 28 °C and 7.4, respectively. The enzyme possessed high resistance to heat and can be activated by Ca^{2+}. These enzyme properties suggested that it could be produced on an industrial scale and has the potential to be a commercial enzyme.

4.3.1.5 Materials and Methods

Microorganisms, Vectors, and Materials

The strain of *Pseudomonas aeruginosa* (no. 10647) was obtained from the Agricultural Culture Collection of China (Beijing, China). *E. coli DH5α* was used as the host for genetic cloning. *Pichia pastoris KM71* (arg4 aox1::ARG4) and vector pPIC3.5 K (Invitrogen, San Diego, CA, USA) were used for protein expression. Pyrobest Taq DNA polymerase and endonucleases were obtained from Takara. Elastin and Elastin-Congo Red were purchased from Sigma (St. Louis, MO, USA).

Cloning of Full-Length CDS of lasB and Sequence Analysis

The genomic DNA was extracted from *Pseudomonas aeruginosa*. The cloning primer sequences were designed according to GenBank (accession no. M19472) as follows: 5′ > CGGGATCCATGAAGAAGG TTTCTACGCTT < 3′ and 5′ > GCGAATTCTTACAACGCGCTCGG < 3′. The *Bam*HI and *Eco*RI restriction sites were inserted into the sense primer and anti-sense primer, respectively. After initial denaturation at 94 °C for 5 min, the PCR cycles were carried out using Pyrobest Taq polymerase. The conditions for each cycle were as follows: denaturation for 30 s at 94 °C, annealing for 30 s at 63 °C, and extension for 1.5 min at 72 °C. This cycle was processed for 35 cycles, and a final extension step for 10 min at 72 °C was added. Amplified PCR products were ligated into pGEM-T Easy vector and the recombinant plasmids were sequenced. Then, the correct plasmid was digested with *Bam*HI and *Eco*RI and ligated into pPIC3.5 K with the same enzymic sites, generating pPIC3.5 K/PAE. The insertion was identified by restriction analysis and sequencing.

Transformation of *Pichia pastoris* with pPIC3.5 K/PAE

Recombinant plasmid-pPIC3.5 K/PAE was linearized with *Sac*I, and electroporated into *P. pastoris* KM71 under the following conditions: 1.5 kV, 25 μF, and 200 Ω, using a GenePulser. The transformants were cultivated in MD agar plates at 28 °C (1.34% YNB, 2% glucose, 4 × 10– 5% biotin) for 2–4 days. The integration of the lasB into the genome of *P. pastoris* was confirmed by PCR using primers.

Screening of High Expression Transformants

According to Mei et al., the His^+ transformants were grown in microtiter plates until all clones were at the same density. Then, the His^+ transformants were spotted on YPD plates containing G418 at a final concentration of 0, 1.0, 2.0, or 4.0 mg/ml. The plates were incubated at 28 °C for 4 days.

Expression of Recombinant *Pichia pastoris* KM71 Strains in Shaking Flask

The colonies were inoculated into 10 ml BMGY medium [100 mM potassium phosphate (pH 6.0), 1.34% YNB, 4 × 10%–5% biotin, 2% peptone, 1% yeast extract, and 1% glycerol) and shaken (250 rpm) at 30 °C. The culture was shaken for 5 days at 30 °C. To maintain induction, methanol without water was supplemented every 24 h to a final concentration of 1% throughout the induction phase. The concentration of protein in supernatant was measured by the method of Brandford.

SDS–Polyacrylamide Gel Electrophoresis (SDS-PAGE)

The molecular mass of the recombinant enzyme was determined by SDS-PAGE. After electrophoresis, the gel was stained with Coomassie Brilliant Blue R-250, and destained by washing with a mixture of acetic acid–methanol–water (10:25:65, v/v/v).

Elastase Activity Assay

An agar plate containing 5% skim milk powder and 50 mM Tris–HCl buffer (pH 7.0) was used to identify elastase proteolytical activity. A cleared zone caused by proteolysis appeared within 1–2 h at room temperature.

The elastolytic activity was quantitatively determined by a modified method described by Morihara. The reaction mixture contained 1 ml of appropriately diluted enzyme and 2 ml 0.2 M boric acid buffer containing 10 mg Elastin-Congo Red. Digestion was carried out at proper temperature for 2 h, and terminated by 2 ml 0.7 M phosphate buffer (pH 6.0), followed by centrifugation at 3000 *g* for 10 min. Supernatant was measured by spectrophotometer at 495 nm. One unit of elastase activity was defined as the digestion of 1 mg Elastin-Congo Red at 28 °C in 2 h at pH 7.4.

Recombinant Enzyme Characterization

Estimations of the recombinant elastase activities at different pH and temperatures were conducted using the crude enzyme. To determine the optimal pH, buffers over the range pH 5.5–9.0 were used for diluting the enzyme and the substrate. The buffers used included 0.05 M MES buffer (pH 5.5–6.5), 0.02 M phosphate buffer (pH 6.5–7.5), and 0.2 M boric acid buffer (pH 7.4–9.0). The reactions were run in different buffers for 2 h at 28 °C. To estimate pH stability, the enzyme was pre-incubated in the different pH buffers for 4 h at 4 °C; later, the substrate was added and the standard assay was as described in the section on elastase activity assay.

The optimal temperature was determined by the standard activity assay at various temperatures from 4 ° to 65 °C in 0.2 M boric acid buffer (pH 7.4). To estimate thermal stability, the enzyme was pre-incubated for 15–120 min at the different temperatures. The standard assay as described in Sect. 3.7 was performed after pre-incubation. To investigate the effects of different metal ions and EDTA on the recombinant elastase activities, 10 mM $CuSO_4$, 10 mM $ZnSO_4$, 10 mM $CaCl_2$, 10 mM $FeSO_4$, 10 mM $MgSO_4$, 10 mM $MnSO_4$, 10 mM $CoCl_2$, and 10 mM EDTA were added separately to the reaction solution, and elastolytic activities were then measured under the standard assay as described in the section on elastase activity assay.

4.3.1.6 Results

Cloning and Sequence Analysis of Elastase Gene (lasB) from *Pseudomonas aeruginosa*

The lasB gene encoding elastase was amplified from the genomic DNA of *Pseudomonas aeruginosa* using the primers designed on the basis of the nucleotide sequences of elastase from *P. aeruginosa* reported in GenBank, and a 1497-bp PCR product was obtained. Compared with other elastase gene sequences reported in GenBank, the nucleotide sequence of the cloned DNA shared 99% homology with lasB from *P. aeruginosa* PA01 (GenBank accession no. M19472). Only three bases changed (1039 C → T, 1110 T → A, 1122 T → C), but there was no difference between the deduced amino acid sequence and native elastase from *P. aeruginosa* (GI: 151210). The obtained 1497-bp nucleotide sequence has been submitted to GenBank, and the obtained GenBank accession no. is EU265777. Comparison of the gene sequences suggests that the difference of nucleotide sequence may be attributed to strain diversity.

Expression of Elastase in *Pichia pastoris*

The constructed vector denoted as pPIC3.5 K/PAE was *Sac*I linearized and electroporated into *P. pastoris* KM71. Transformants were selected by Geneticin 418 (4.0 mg/ml) and confirmed by the genomic PCR assay and DNA sequencing. The elastase activity of supernatant was monitored after 1, 2, 3, 4, and 5 days of methanol induction by agar plate containing skim milk powder. Hydrolysis areas, appearing as transparent zones, were measured semiquantitatively. No elastase activity was detected in the supernatant of yeast transformant containing only pPIC3.5 K. The size of recombinant elastase was approximately 34 kDa by SDS-PAGE analysis, similar to that of the native elastase from *Pseudomonas aeruginosa*. The optimal induction time was also determined, and the highest level of the recombinant enzyme occurred after 4 days of induction. Elastolytic activity did not increase after 4 days of induction.

The recombinant elastase was secreted into culture supernatants using a signal sequence from lasB and showed a single band at about 34 kDa by SDS-PAGE. As the recombinant enzyme accounts for more than 93% of the total protein in supernatant (detected by densitometer), the purity of the recombinant elastase can meet further enzyme analysis.

Characterization of the Recombinant Elastase

One *Pichia pastoris* clone expressing the highest elastase activity was selected for characterization. The optimal pH and temperature of the recombinant elastase were 7.4 in boric acid buffer and 28 °C, respectively. The pH optimum changed because we used different buffers. We found it is obvious that the recombinant enzyme in boric acid buffer at pH 7.4 performed much better than in other buffers, so boric acid buffer (pH 7.4) was chosen as the buffer for the later experiments (data not shown).

The assays of enzyme resistance to different pH indicated that the recombinant elastase can preserve its activity in acidic solution (pH 3–7) at about 4 °C, differing from the native elastase from *Pseudomonas aeruginosa*, which was stable in pH 6–10. The hydronium in the reaction solution might play a role in the recombinant elastase activity, but the mechanism is still unknown.

The elastolytic activity of the recombinant enzyme was significantly affected by many metal ions (Zn^{2+}, Mn^{2+}, Fe^{2+}, Co^{2+}, Cu^{2+}, and Cd^{2+}), especially by Zn^{2+} and Cd^{2+}, and the chelating agent EDTA (ethylenediaminetetraacetic acid) inhibits half its activity. However, Mg^{2+} has little effect on its activity, whereas Ca^{2+}, which is required for elastase activity, could slightly increase its activity. These results were quite the same as that of the native elastase from *P. aeruginosa*, which also performed in the same conditions, except that EDTA inhibited nearly 95% of native *P. aeruginosa* elastase activity.

Research 3 Expression, purification, and refolding of recombinant Cry Ab/Ac obtained in *E. coli* as inclusion bodies. This research was published in the *Journal of the Science of Food and Agriculture* in 2009.

Since the first cloning of *cry* gene from strain *HD21Dipel* by Schnepf in 1981, 433 kinds of such protein genes have been isolated and cloned. These genes, according to the similarities of their amino acid residues, can be classified into 49 groups, 93 subgroups, 147 classes, and 328 subclasses. Among them, *cry1* is a group of relatively highly homologous genes that exhibit a highly specific toxin to *Lepidoptera*, and Cry protein encoded by the *Cry1ab/ac* gene, which fuses the *cry1ab* (GenBank accession no. X54939) and *cry1ac* (GenBank accession no. Y09787) into one gene, is highly toxic to *Chilo suppressalis*, *Scirpophaga incertulas*, and *Cnaphalocrocis medinalis,* three important lepidopteran pests of rice.

GM rice containing the *Cry1ab/ac* gene was provided by Huazhong Agricultural University (Wuhan, China). The genomic DNA samples used for PCR amplification were extracted from rice leaves (provided by Huazhong Agricultural University) according to the method of Xu et al. The amplified PCR product of modified *Cry1ab/ac* gene was purified and ligated into a pGEM-T Easy vector, transformed to *E. coli* DH5α, and identified as just described. We named the modified *Cry1ab/ac* gene as *Cry1ab/ac-rcp* (GenBank accession no. EU816953).

The *Cry1ab/ac-rcp* gene was amplified from the plasmid pGEM-T Easy vector using the sense primer (Bt-3F) of 5′-CGC GGA TCC atg gac aac tgc cgt cca tac a-3′ and the antisense primer (Bt-3R) of 5′-CCC AAG CTT tga cgt tgt gac ctc cga ctt att-3′ to introduce the *Bam*HI and *Hin*dIII restriction sites (underlined). The amplified PCR product was gel purified, digested with *Bam*HI and *Hin*dIII, then ligated into *Bam*HI and *Hin*dIII sites of PET-30a(+) fused to the N-terminal His·Tag sequence in frame. The plasmids were then transformed into *E. coli* DH5a cells. The recombinant expression plasmid was checked by restriction digestion, and then the positive ones were transformed into BL21 (DE3) to express Cry1Ab/Ac protein.

The protein was expressed in Luria-Bertani (LB) liquid medium containing kanamycin (100 μg/ml). To obtain the maximum protein, several cell growth conditions generated by varying temperature, induction time, and IPTG concentration were tested.

The recombinant fusion protein was purified and refolded from inclusion bodies. All mass spectra of MALDI-TOF-MS were obtained on an AXIMA-CFR plus MALDI-TOF-MS (Shimadzu, Japan). PMF (peptide mass fingerprinting) obtained was compared with theoretical Cry1Ab/Ac protein searched through the Peptide Mass database (http://www.expasy.ch/tools/peptide-mass.html).

Toxicity of the inclusion bodies protein, purified Cry1Ab/Ac fusion protein, and the Cry1Ab/Ac protein without His tag were checked by feeding assays on *Chilo suppressalis* (provided by China National Rice Research Institute, China). The quantified toxin was diluted in lysis buffer (50 mM Na_2CO_3, pH 10) in a way that the amount of the protein to be fed to an insect larva was present in a total volume of 100 μl. Seven such dilutions containing 0–17.6 μg of the protein were tested by mixing the 100-μl dilution with 1 g solidified insect diet in glass tubes. *Chilo suppressalis* larvae, reared on an artificial diet and maintained under controlled environmental conditions, were released, ten per tube. Five assays for each kind of Cry1Ab/Ac protein preparations were performed. On day 5 at completion, all the insects were checked for viability. The data obtained were subjected to probit analysis, and lethal concentrations were calculated according to Finney.

The genomic DNA samples used for PCR amplification were efficiently extracted from GM rice leaves with the CTAB method. The purified PCR product with Bt-1F/R primer was ligated into a pGEM-T Easy vector and then the ligation product was successfully transformed into *E. coli* DH5α. The positive transformants, selected by blue-white colony screening, colony PCR, and endonuclease digestion, were verified by sequencing. This analysis indicated that the *Cry1ab/ac* gene was successfully amplified and its amino acid sequence was not changed. The *Cry1ab/ac* gene was modified from the plasmid pGEM-T Easy vector with Bt-2F/R primer, then purified and ligated into a pGEM-T Easy vector. The modified *Cry1ab/ac* gene was amplified with Bt-3F/R; then, the purified amplified PCR product with Bt-3F/R primer was digested with *Bam*HI and *Hin*dIII, and ligated into the PET30a(+) vector that was digested with the same enzymes, fusing with the N-terminal His·tag sequence in frame. The recombinant expression plasmid was also selected by colony PCR, confirmed by endonuclease digestion. DNA sequencing results further confirmed that there was no error or unwanted mutation in the inserted sequence and that the Cry1ab/ac-rcp gene was correctly inserted into the expression vector PET-30a(+). The expected fusion protein was 659 amino acids long and contained a His tag at the N-terminus.

The expression of *Cry1ab/ac* gene was seriously affected by the presence of rare codons. No Cry1Ab/Ac protein was obtained with the PET-30a(+)-*Cry1ab/ac*-BL21(DE3) or PET-30a(+)-*Cry1ab/ac*-BL21(DE3)-CodonPlus combinations. The codon usage pattern in *E. coli* obtained from the graphical codon usage analyser (http://gcua.de) shows that AGA, AGG, and CGA (arginine); AUA (isoleucine); CUA and CUC (leucine); CCC, CCU, and CCA (proline); GGA and GGG (glycine); CUU, UUG (leucine); AAG (lysine); and ACA (threonine) are rarely used codons in *E. coli*. However, these triplets are common in *Cry1Ab/Ac* gene. Several studies have reported that the position and number of rare codons, especially continuous AGA and AGG codons within the first 25 codons of a mRNA or consecutive CUA codons when near the 5′-end of a messenger, will significantly decrease the gene expression, particularly in the later growth phase of *E. coli*. By amplification of the gene with the Bt-2F/R primers, the first 20 codons were modified in to improve the expression of Cry1ab/ac gene in *E. coli*.

Vector PET-30a(+)-*Cry1ab/ac-rcp* was transformed into *E. coli* BL21(DE3) to express the fusion protein in *E. coli*. The Cry1 Ab/Ac recombinant protein was partially found in inclusion bodies but also in the soluble fraction (data not shown). The Cry1Ab/Ac protein was expressed under several growth conditions: temperatures in the 15–37 °C range with variable use of IPTG concentration up to 2 mM. The soluble fraction could be obtained at 15 °C or 20 °C, but the protein was localized in the cytoplasmic inclusion bodies fraction invariably at 30 °C or 37 °C. The greatest expression of Cry1Ab/Ac protein was obtained at 37 °C after induction with 0.1 mM IPTG for 4 h, and recombinant Cry1Ab/Ac protein accounted for approximately

57.8% of the total inclusion bodies protein. This result is in accordance with published information.

The insoluble aggregates of the inclusion bodies fraction were extensively washed in buffer A, which could wash away cell debris that showed dark brown on the surface of the proteins. Cry1Ab/Ac protein from the washed inclusion bodies was completely solubilized in buffer B containing 8.0 M urea, successfully purified by a His TrapTM FF crude column under denaturing conditions, then refolded by 0.4 M L-arginine in a linear 8-0 M urea gradient refolding buffer. At this time the Cry1Ab/Ac protein was obtained with 96.4% purity in the pooled fractions. The His Tag was cut out from the Cry1Ab/Ac fusion protein by enterokinase, and removed by His TrapTM FF crude column. The Cry1Ab/Ac protein recovered after gel filtration on a fast performance liquid chromatography (FPLC) Superdex 75 HR 10/30 column; it was 99.2% pure as demonstrated by the single band obtained in SDS-PAGE. The final yield of the protein isolation is 0.206 g for 1 g cells.

As there was no clear distinction between the digested Cry1Ab/Ac protein and the protein with His tag on SDS-PAGE, Western blot with His tag monoclonal antibody was employed to test whether or not the His Tag was cut out. Both the inclusion bodies protein and purified Cry1Ab/Ac fusion protein can be detected with His tag by Western blot. The same protein purified in gel filtration was detected only in SDS-PAGE but by immunoblotting using Cry1Ab/Ac antibodies, confirming the successful cleavage of the His Tag.

The identity of the Cry1Ab/Ac protein was determined by Western blot. The same amount of protein was used to detect the identity of Cry1Ab/Ac protein of the inclusion bodies protein, the purified Cry1Ab/Ac fusion protein, and the Cry1Ab/Ac protein without His Tag. With the band became darker and darker, a substantial increase in purity was obtained by refolding and enterokinase digestion.

The primary structure of Cry1Ab/Ac protein was tested by MALDI-TOF MS. Eight peptides (range, 1000–3000 Da) are identified to have 20.2% homology with Cry1Ab/Ac protein sequence by searching the Peptide Mass database (http://www.expasy.ch/tools/peptide-mass.html) and and they are as follows: WYNTGLER, APMFSWIHR, WGFDAATINSR, QGFSHRLSHVSMFR, LEGLSNLYQIYAESFR, SGTVDSLDEIPPQNNNVPPR, LSHVSMFRSGFSNSSVSIIR, and NFSGTAGVIIDRFEFIPVTATLEAE, with peptide masses of 1039.8 Da, 1145.3 Da, 1238.5 Da, 1670.1 Da, 1902.3 Da, 2150.8 Da, 2213.5 Da, and 2699.4 Da, respectively.

Biological activity of Cry1Ab/Ac protein was determined in a growth inhibition assay of *Chilo suppressalis*. The Cry1Ab/Ac protein without His tag showed the lowest LD_{50} value (2.04 μg/g) followed by the purified Cry1Ab/Ac fusion protein (4.05 μg/g), and the inclusion bodies protein had much higher LD_{50} value (9.16 μg/g) than the two former. The LD_{50} values measured in the current experiment were similar to those reported by several authors. The lowest LD_{50} value of the Cry1Ab/Ac protein without His tag, in all probability, is caused by complete solubilization and accurate protein conformation. The significance of the intragroup and intergroup value differences was analyzed with SPSS v12.0 (SPSS, Chicago, IL,

USA), and for the different groups 0.55 μg/g was not significant, but results for 1.1, 2.2, 4.4, 8.8, and 17.6 μg/g were significant ($p < 0.05$).

This research successfully engineered the *Cry1Ab/Ac* gene, constructed the Cry1Ab/Ac protein expression system in *E. coli*, and established a new method to purify the protein from inclusion bodies. The protein was purified by a His TrapTM FF crude column and renatured to gain a soluble protein that when assayed by Western blot, MALDI-TOF MS, and biological activity was identified as *Chilo suppressalis*. These results open up the possibility to prepare a large amount of Cry1Ab/Ac recombinant protein by large-scale fermentation and use it as a biological insecticide in free-GM crops. The purified Cry1Ab/Ac protein could be used to evaluate the food safety of transgenic plants containing the *Cry1ab/ac* gene. In addition, the Cry1Ab/Ac protein might be used to produce the antibodies for immune-based methods in detecting the GMOs containing the *Cry1ab* or *Cry1ac* gene.

4.3.2 The Substantial Equivalence of Protein Expression

Many studies for safety assessment require grams of the transgenic proteins with high purity (>90%). It is often difficult to acquire enough purified transgenic protein from GM crops for this low amount. It is relatively easy to produce sufficient protein in fermentable microbes. A typical study for evaluation of acute oral toxicity requires a minimum of five male and five female mice each to be given a single dose of at least 5000 mg protein/kg body weight. Depending on the weight of the animals, 2 g protein is needed for this safety evaluation. Sometimes, a repeated-dose oral toxicity study is used (e.g., based on OECD guideline 407). It may be that 30 g protein is needed for this study.

When using protein produced by a microorganism, it is necessary to test its suitability as surrogate for the plant-produced transgenic proteins. Substantial equivalence does not imply that the plant and microbial proteins are identical, but that the microbial protein be sufficiently similar biochemically and functionally to the plant protein. Although there are clear advantages of microbial proteins, it is necessary to test the characters of the substances. The protein purity must be measured so that the correct amount of test substance is used to give the accurate amount of protein in a toxicity study. Information about solubility is crucial as many studies require a high concentration of protein.

4.3.2.1 Solubility

Most analytical tests require the protein in a solubilized form. The solubility determination of a protein test substance is therefore commonly the first experiment conducted during all the characterization. The solubility of protein provides important information on the delivery vehicles in animal toxicity studies and nontarget organism effects studies. For toxicology studies, water is the desirable solute for its simple structure. Unfortunately, many protein test substances have limited solubility in water, meaning that buffers such as TRIS, CAPS, or CMC are often required to dissolve test substances. For some tests, such as acute oral toxicity in mice, a homogeneous suspension of the test substance may be an acceptable alternative to a solution. When buffers are used in a toxicity study, it may be necessary to determine whether they may have adverse effects on the animals.

4.3.2.2 Purity

The purity of protein used in a toxicity study usually as high as 90%. The purity of protein is usually determined in two stages. First, the proportion of total protein is determined using standard laboratory methods such as the BAC detection kit.

4.3.2.3 Intactness and Immunoreactivity

The molecular weight is commonly called a measure of intactness. The protein of the microbial and plant proteins provides information on whether they have been truncated or degraded in a sample. Molecular weight determination can also detect whether the protein be modified, such as glycosylation, and insertions or deletions of amino acids. Immunoreactivity can detected if these two kinds of proteins have the same ability to bind to one specific antibody. Loss of immunoreactivity may indicate biochemical or functional properties changes in the protein. Western blot analysis is a convenient method for comparing the intactness and immunoreactivity of protein samples. Side-by-side comparison of the apparent molecular weight of a protein from differences sources may find a difference in mobility and modification.

4.3.2.4 Intact Mass

The molecular weight of a protein cannot be correctly determined by SDS-PAGE or Western blot. Exact protein masses can be obtained by mass spectrometry (MS). For MS analysis, quadrupole-time-of-flight (Q-TOF) analysis is better than matrix-assisted laser desorption ionization (MALDI-TOF) analysis. Q-TOF machines are able to distinguish proteins with single amino acid substitutions, or other low molecular weight modifications, such as methionine oxidation. This difference would not be found by MALDI-TOF MS.

4.3.2.5 Protein Sequence

The amino acid sequence of a protein provides useful information on its structure and function. The functional equivalence of microbial and plant proteins can be detected by amino acid sequence comparisons. Two methods are used in this functional equivalence: N-terminal sequencing and peptide mass mapping.

4.3.2.6 Glycosylation

About half the protein in plants is estimated to be glycosylated. Glycosylation is the reaction in which a carbohydrate, such as a glycosyl donor, is attached to a hydroxyl or other functional group of another molecule. Glycosylation can alter the physio-chemical characters of protein, such as its tolerance of heat, functional activity, protein folding, half-life, and transport. Glycosylation has been considered to be related to allergenicity. Increasing knowledge of plant glycosylation has recently led to the conclusion that carbohydrate moieties are probably insignificant as clinically important allergen determinants. Proteins form GMOs are not intended to be glycosylated, and recombinant proteins expressed in *E. coli* are not glycosylated. The analysis of glycosylation in protein equivalence studies is routinely accomplished using Western blot assay.

4.3.2.7 Activity

Many transgenic proteins are enzymes. It is important to hypothesize that the protein from microbial and transgenic crops have substance equivalence in activity, such as toxicity to mammalian and other nontarget organisms.

4.4 Acute Oral Toxicity of Foreign Protein

Single-dose toxicity testing, also known as acute toxicity testing, is normally used to confirm any potential acute toxicity, and this testing usually conducted in rats or mice. Acute toxicity describes the adverse effects of a substance that result either from a single exposure or from multiple exposures in a short period of time (usually less than 24 h). It is distinguished from sub-chronic or chronic toxicity, which are repeated exposures at lower levels but for longer time period to one substance. The adverse effects should occur within 14 days following exposure of animals to single doses of chemicals by ingestion, inhalation, or through the skin. Generally, acute toxicity tests in animals are not a real condition because human exposure tends to be for much longer periods with lower dosage, but this test is required in the risk assessment of many individual chemicals found in foods because it can provide the LD_{50} (the dose that causes the death of half the test animals) or the maximum

Table 4.1 In vitro genotoxicity tests as described by OECD guidelines

OECD No.	Title
471	Reverse mutation assay (*Salmonella typhimurium*)
472	Reverse mutation assay (*Escherichia coli*)
473	In vitro mammalian cytogenetic (chromosome aberration) test
476	In vitro mammalian cell gene mutation test
479	In vitro sister chromatid exchange (SCE) assay in mammalian cells
480	*Saccharomyces cerevisiae*, gene mutation assay
481	*Saccharomyces cerevisiae*, mitotic recombination assay
482	DNA damage and repair, unscheduled DNA synthesis in mammalian cells in vitro

tolerated dose for one substance. The acute toxicity tests also can determine a clearly toxic but sublethal dose, to identify major target organs for toxicity and to provide a rough guide for dose selection in subsequent repeated-dose toxicity tests or other chronic problems in food areas (Barlow et al. 2002). Acute toxicity testing has some value for proteins. In the safety evaluation of transgenic crops acute, animal toxicity testing with the purified protein is routinely conducted. The products expressed in GMOs are usually a protein. Acutely toxic proteins typically elicit their adverse effects almost immediately upon consumption of relatively high dose levels of the protein (Sjoblad et al. 1992; Genotoxicity testing of foreign protein).

To assess the potential carcinogenicity of GM foods, in vitro mutagenicity testing is an option analogous to the use of mutagenicity tests as a tool for pre-screening single substances for their carcinogenic potential. Eight testings of the OECD guideline and ten testings of the GB guideline can be followed in the genotoxicity testing (see Table 4.1). In China, according to the Food Safety Toxicology Evaluation Program, three tests can be used to evaluate the genotoxicity of one GM protein.

However, the testing of whole foods in vitro poses specific problems: whole foods cannot be satisfied at a high concentration as opposed to a single chemical. Whole foods are generally not soluble and have to be applied in a matrix that is appropriate for in vitro test systems. Freeze-drying or homogenization is often used to dispose the whole foods. Certain constituents may interfere with the test system used (e.g., histidine in the Ames test). Thus, much work is still needed to determine the value of mutagenicity tests in predictive toxicology testing of complex mixtures such as whole foods. To solve some of these problems and to concentrate any mutagenic components present, extraction procedures are needed. Thus, in normal research, the protein is expressed by bacteria or extracted from GM crops.

Animal-based in vivo assays are not usually conducted for the initial detection of a genotoxic hazard; in most conditions, the in vitro tests can be used to adequately detect the genotoxic hazard (Eisenbrand et al. 2002). Because many chemicals still keep their activity in vitro, if the activity is limited in vitro, it is usually because of limitations in the exogenous metabolism system. If a positive result is acquired in an vitro mutagenicity test, it is required to conduct further testing in vivo. In cases where it can be adequately demonstrated that the positive in vitro findings are not relevant for the in vivo situation, in vivo testing may not be necessary. In vivo testing could also be avoided where the mutagenic profile in vitro and structure–activity considerations are sufficient in themselves to argue for in vivo activity. Before undertaking any in vivo testing, a review of the in vitro test results and all available information on the toxicokinetic and toxicodynamic profile of the test substance should be carried out. A particular in vivo test should be conducted only when it can be reasonably expected from all the properties of the test substance and the proposed test protocol that the specific target tissue will be adequately exposed to the test substance or its metabolites. For a test substance with adequate systemic bioavailability (e.g., evidence is available to indicate that the substance or its metabolites reach the internal organs via the bloodstream), the in vivo tests usually undertaken are the rodent cytogenetic test and the rat liver unscheduled DNA synthesis (UDS) test, depending on the type of assay that was positive in vitro. Alternatively, for substances that show clear indications of little or no systemic bioavailability, especially for directly acting electrophilic molecules, studies at the main site of contact, such as gut tissue, may be more appropriate. The use of in vivo/in vitro studies to evaluate effects on DNA in target tissues may prove valuable in this context. Other newer test procedures, such as the 'Comet assay' for the detection of DNA damage in single cells of any tissue (McGregor and Anderson 1999), DNA covalent binding, the detection of DNA adducts by very sensitive methods such as ^{32}P-postlabeling or mass spectrometry, or tissue-specific mutations in transgenic animal models, may also be relevant for identifying a possible hazard in vivo. An improved method for detecting point mutations in ubiquitous (restriction enzyme) sites has been described and can be readily applied to any sequenced gene from various tissues (Jenkins et al. 1999). For test methods, it is generally recommended that OECD protocols for in vivo assays be used, which include tests for chromosomal aberrations, micronuclei in rodent bone marrow, rat liver UDS, and chromosomal aberrations in spermatogonia. In general, a negative outcome in bone marrow or liver assays indicates that the potential of the chemical to reach and mutate germ cells will also be low or absent, because the germ cells are less readily accessible to xenobiotics than bone marrow or liver. Conversely, in the case of positive results in somatic cells, a mutagenic hazard to germ cells cannot be excluded. In this case, the need for further studies in germ cells requires careful consideration. In most cases, in vivo somatic cell mutagens will be assumed to be both potential genotoxic carcinogens and potential germ cell mutagens, without further genotoxicity testing. In cases in which toxicokinetic and toxicodynamic data indicate germ cell exposure, in vivo testing may be deemed necessary for definitive hazard identification and eventually quantitative risk evaluation. Cytogenetic effects (metaphase analysis in spermatogonia or micronuclei in spermatocytes), dominant lethal effects, and UDS can be investigated using OECD harmonized testing protocols. None of these assays provides conclusive evidence as to whether genetic effects will be transmitted to future generations, and currently the only methods available for risk estimation are the specific locus test and the heritable translocation test, both of which require a large number of animals and are not very practical options.

We have done some studies about genotoxicity in GMO, as follows.

Research 1 The genotoxicity study of one GM lactase: this research was published in the journal *PLOS One* in 2014.

1. Introduction

Lactose, also known as lactase-phlorizin hydrolase or LPH, is one of the main sugars present in milk and other dairy products. It is helpful in improving the absorption rate of some essential trace elements such as calcium, phosphorus, and magnesium (Travis et al. 2013; Wilt et al. 2010). It is a glycoside hydrolase that hydrolyzes the disaccharide lactose to produce galactose and glucose (Poulter et al. 2003; Swallow 2003). It has been reported that lactose can be degraded by lactase and that its digestion and absorption depend on lactase activity (Travis et al. 2013). Lactase can be produced in the digestive systems of infants and of many but not all adults (Yuan 2010). The lactase gene is expressed exclusively in the colon during fetal development and declines dramatically after weaning in most of the world's populations (Kruse et al. 1988), leading to lactose intolerance in adults (Wang et al. 1998). More than 70% of people have lactose intolerance and cannot easily digest fresh milk and dairy products (Heyman 2006).

The production of lactase by microbial strains using traditional mutation breeding cannot satisfy the fast-growing market demands. The development of genetic engineering technology makes the large-scale industrial production of lactase possible through the introduction or modification of genes to promote characteristics such as higher enzyme activity or higher production. This lactase derived from recombinant microorganism needs to be evaluated in accordance with the Guideline for the Conduct of Food Safety Assessment of Foods Produced Using Recombinant-DNA microorganisms (CAC/GL 46-2003). The lactase product was subjected to a series of toxicological tests to determine its safety for use as an enzyme preparation in the dairy industry. This study was conducted to assess the toxicological properties of lactase produced by genetically modified *Pichea pastoris* for use as a food additive. The toxicity of recombinant lactase was assessed by a mutagenicity study in bacteria (*Salmonella* Typhimurium/mammalian microsomal enzyme test, Ames test; bone marrow cell micronucleus test, mouse sperm abnormality test).

2. Ames test

2.1. Materials

Lactase powder with a purity of 83.33% was used in this study. To investigate the mutagenic potential of the lactase, four histidine-requiring *Salmonella* Typhimurium mutant strains were tested: TA97, TA98, TA100, and TA102. The rat liver-derived metabolic activation system (S9) and cofactor-II were purchased from the Chinese Center for Disease Control and Prevention (CDC). The S9 mix was prepared just before use. Without S9 addition, Nexon is the positive control for TA97, TA98, and TA102; NaN3 is the positive control for TA100. With S9 addition, 2-AF is the positive control for TA97, TA98, TA100, and TA102.

2.2. Experimental design

The dosages of lactase were 5000, 2500, 1250, 625, and 312.5 μg/plate. Lactase was dissolved in distilled water at concentrations of 50, 25, 12.5, 5, and 2.5 mg/ml. The bacteria were exposed to the lactase treatments in phosphate-buffered nutrient broth with or without S9 mix. The concentrations of the positive controls were set in accordance with Chinese Standard GB15193.4. Distilled water was used as the negative control. The test was considered positive for mutagenicity if the number of reverse mutation colonies was at least twofold greater for the lactase treatments than for the negative control and if the number of colonies was affected in a dose-related fashion. Otherwise, lactase was determined to be negative for genotoxicity (Yoshikawa et al. 2013).

3. Bone marrow cell micronucleus test
3.1. Materials and animals

Lactase powder with a purity of 83.33% was used in this test. Sixty 10-week-old KM mice (30 male and 30 female) were used. Cyclophosphamide was used as the positive control.

3.2. Experimental design

The mice were divided into five groups, with six animals of each sex per group. Three groups were chosen as experimental groups and were given lactase by oral gavage in dosages of 1250, 625, and 312.5 mg/kg·BW. The negative control group was treated with physiological saline, and positive controls were treated with cyclophosphamide (40 mg/kg·BW). All mice were treated twice, with 24 h between treatments. Six hours after the second gavage, the mice were killed by cervical dislocation, and bone marrow cells were immediately collected from the femurs. The bone marrow was flushed from the femurs with fetal bovine serum, and the marrow was harvested to make smears. The smears were air-dried, fixed, and stained with 4% Giemsa. Giemsa solution can stain immature erythrocytes and erythrocytes with a micronucleus so that they turn blue, whereas mature erythrocytes without micronuclei are stained pinkish orange (Kim et al. 2010). One thousand polychromatic erythrocytes were recorded per mouse by observation under a Leica DM2500 microscope. Cells were considered to be micronucleated if they contained defined chromatin corpuscles less than one-third the diameter of a normal nucleus (Hayes et al. 2009). The frequency of micronucleated cells was calculated by counting them and dividing by the total number of polychromatic erythrocytes.

4. Mouse sperm abnormality test
4.1. Materials and animals

Lactase powder with purity of 83.33% was used in this test. Thirty 8-week-old male KM (Kung Min) mice were used in this test. Cyclophosphamide was used as the positive control.

4.2. Experimental design

Grouping and dosing were similar to the bone marrow cell micronucleus test. Lactase was given by oral gavage for 5 days, and the mice were fed with rodent basic feed for 30 days. On day 35, the mice were killed by cervical dislocation, and smears of sperm from the epididymis were fixed with methanol and stained with Eosin Y (Styrna 1991). One thousand sperm cells were counted, and the percentage of abnormal sperm was recorded for each mouse. The percentage of abnormal sperm was calculated as the number of abnormal cells number divided by total sperm cell count. The sperm were analyzed by microscopy (DM2500; Leica, USA), and the classification proposed by Krzanowska was used (Krzanowska 1976).

5. Results

5.1. Ames test

The numbers of reverse mutation colonies for the positive control substances such as sodium azide (NaN_3), 2-acetamidofluorene (2-AF), 1,8-dihydroxyanthraquinone (1,8-HAQ), and Nexon were at least twofold greater than the negative control. Recombinant lactase did not increase the number of revertants in the four *Salmonella* strains compared with their negative controls, regardless of the presence or absence of S9 mix. Additionally, no dose-dependent mutagenic effects were caused by the lactase. Lactase did not show any mutagenic activity under the experimental conditions.

5.2. Bone marrow cell micronucleus test

The micronucleus incidence rate of the positive control group treated with 40 mg/kg·BW cyclophosphamide was significantly higher ($p < 0.01$) than that of the negative control group. The micronucleus incidence rate of the experimental groups treated with 312.5, 625, or 1250 mg/kg·BW lactase showed no statistically significant differences compared to the negative control group ($p > 0.05$). Additionally, no dose-dependent genetic toxicity was caused by the lactase. No genetic toxicity was observed for lactase under the experimental conditions.

5.3. Mouse sperm abnormality test

The rate of sperm abnormalities in the positive control group was statistically higher than that of the negative group ($p < 0.01$). The experimental groups treated with different doses of lactase were all within the normal range, and there were no statistically significant differences in the lactase-treated groups compared to the negative control group ($p > 0.05$). No mutagenic activity was observed under the experimental conditions.

This recombinant lactase had the highest activity of all recombinant strains reported thus far. Genotoxic toxicity tests performed in mice showed no genotoxicity in any groups. The lactase showed no mutagenic activity in the Ames test or a mouse sperm abnormality test at levels up to 5 mg/plate and 1250 mg/kg body weight, respectively. It also showed no genetic toxicology in a bone marrow cell micronucleus test at levels up to 1250 mg/kg body weight.

4.5 Sub-chronic Toxicology Assessment of Whole GM Food or Feed in Animals In Vivo

Genetic modification technology provides a method to introduce exogenous DNA into plants, endowing the target plants with a specific characteristic. Most GM crops are being developed worldwide for tolerance of herbicides or insects, and other crops contain new genes that have improved nutritional characteristics for feed or food use. These traits make GM crops prosper all over the world. Currently, many commercial GM crops including corn, soybean, canola, cotton, and papaya are growing on a worldwide scale in nearly 30 countries. Since the first biotech crops were planted in 1996, GM crops with traits such as pest tolerance and herbicide tolerance have been grown in more than 1.8 million hectares, which is a remarkable 100-fold increase since the start of commercialization.

In recent years, remarkable concern about the safety of GM crops has reached the public view and media attention, mainly focused on the risk to human health and environment. The GM crops that are on the market are declared by WHO to have all passed through the safety assessments carried out by various national authorities. However, as indicated for the scientific field, risk assessments of the GM crops have not been well developed or systematically performed. More seriously, the information that the press and media publish is often thought to be unreliable against the available and authentic scientific evidence by professional study and research, and this has hampered the extension of GM crops worldwide.

Risk assessments of GM crops differ among countries and authorities and reflect the complexity of the subject, including feeding strategies and periods, animal models (fish, chicken, mice, rat, and pig included), and other parameters. Thus, it is no wonder that diversity among the methods and techniques used in safety assessments leads to diversity in the results. Various groups have suggested that before approval for market circulation and human consumption, long-term animal feeding studies should be carried out for all GM foods (from animals or plants). A 90-day sub-chronic toxicology study is scientific and authoritative to verify if any one substance has the potential to threaten animal health.

The European Food Safety Authority (EFSA) recommended that "the safety assessment of GM plants and derived food and feed follows a comparative approach, i.e., the food and feed are compared with their non-GM counterparts in order to identify intended and unintended (unexpected) differences which subsequently are assessed with respect to their potential impact on the environment, safety for humans and animals, and nutritional quality" (EFSA 2008). This "substantial equivalence" concept is based on this principle: "If a new food is found to be substantially equivalent in composition and nutritional characteristics to an existing food, it can be regarded as being as safe as the conventional food.". This concept can also extend to the safety assessment of GM foods and GM crops. "With different methods, key elements of the assessment procedure such as molecular, compositional, phenotypic, and agronomic traits are analyzed in both the GM line and its near isogenic counterpart" (EFSA 2008). When "molecular, compositional, phenotypic, agronomic and other analyses have demonstrated equivalence of the GM food/feed, animal feeding trials do not add to the safety assessment" (EFSA 2009; updated in EFSA 2011). Based on the concept of "substantial equivalence," safety assessments of GM crops are carried out using molecular biology analysis and related animal feeding tests.

It is acknowledged that more useful information can be provided by animal feeding studies, including the nutritional value and complement safety assessments of GM foods, especially when unintended effects are suspected and noted by the public or investigators. The EFSA experts panel recommend that "the use of 90-days studies in rodents should be considered for the detection of possible unintended effects in food and feed derived from GM plants which have been more extensively modified in order to cope with environmental stress conditions like drought or high salt conditions, or GM plants with quality or output traits with the purpose to improve human or animal nutrition and/or health" (EFSA 2008). Based on this reason, we considered that unless there is still reasonable and actual doubt about the procedure or results of a 90-day feeding test, such long-term studies clearly do not need to be carried out.

OECD Test Guideline No. 408 makes definitions of the experimental material of the protocols for in vivo toxicological studies and the 90-day rodent study and the practical conditions used for testing (including the species of target animal, number and gender of animals used, conditions of animal housing, number of doses administered, etc.). In a 90-day rodent study, body weight data of weeks, food consumption per animal, clinical biochemistry, and other phenotypic responses are also recorded and analyzed. During the past few decades, by refining an improved and reliable toxicological and safety assessment of drugs, pesticides, additives, and other low molecular weight xenobiotics, these parameters have served as fundamental data for the safety assessment of GM-based foods or feed.

Feeds are identical in the usual 90-day rodent study between the treated or control animal groups; usually, animals are provided the GM feeds and non-GM feeds. However, there are some specific questions and doubts on the strengths and weaknesses of such tests in food safety studies. For example, 30% of GM added to the animal feed is usually incorporated when one wants to assess whether if there would be potential health effects of the GM-based food or feed. There is another advantage of feeding experiments when scientists use rodent models in a 90-day feeding study because it allows the researchers to test the whole GM material, usually by the improvement of feed processing. Such safety assessments can be motivated by either a known modification in the composition of GM plant to be tested, or by indications of potential unintended effects that could also be involved (EFSA 2006, 2008).

Additionally, a 90-day animal feeding study is not carried out for one particular effect of a known and given molecule but is supposedly designed to find changes that may occur in the GM foods or feeds. These changes may include those potentially generated variations caused by some genetic modifications as well as those resulting from a compositional change that is thought to be directly or not linked to the transgene inserted into the GM crops. Based on these reasons, some 90-day feeding studies seem to be too broad and inefficient to detect some small weak effects, as EFSA (2008) has already stated: "It is unlikely that substances present in small amounts and/or with a low toxic potential will result in any observable unintended effects." However, what should be highlighted is that in the case of a chemically defined molecule for which the contact with humans might be is very low, the dosage of the classical 90-day feeding studies can be increased in further studies. The key point is that such a high dosage would not appear as a quantitatively important constituent in the diet of one person, certainly not causing potential food accidents.

Moreover, EFSA (2008) states that "the subchronic, 90-day rodent feeding study is not designed to detect effects on reproduction or development, other than effects on adult reproductive organ weights and histopathology. Thus, in some cases, testing of the whole food and feed beyond a 90-day rodent feeding study may be needed. In cases where structural alerts, indications from the subchronic study or other information on the whole GM plant derived food and feed are available that suggest the potential for reproductive, developmental or chronic toxicity, the performance of such testing should be considered."

4.5.1 Materials and Procedure

In this part, we describe the main procedure and materials in a 90-day sub-chronic study. This protocol could provide a direction and process to follow when one wanted to design and carry out a 90-day study.

4.5.2 Plant Materials

In a 90-day study, we use the GM plant and non-GM plants to make the fodder. An analysis of main nutrient contents, fatty acids, vitamins, amino acids, mineral element, and anti-nutrients in the fodder were usually carried out after the final fodder was made according the recommended methods of the State Standard of the People's Republic of China.

4.5.3 Diet Formulation and Experimental Design

Each rat diet contained a plant from GM or non-GM plants at the inclusion rate of 7.5%, 15%, or 30% (w/w). The 30% inclusion rate is the highest that would still allow proper nutritional balance. The diets were also been fortified with other ingredients such as bean pulp, wheat flour, fishmeal, and yeast powder to ensure a balanced diet for the rats. The nutritional content of these diets conformed to Chinese Standard GB14924.3-2010. All diets used in this study were vacuum-packed and irradiated with ^{60}Co by KeAoXie Li Feed Co. Ltd.; diets were then kept at 4–8 °C before feeding the rats. Finally, the nutrition of diets was analyzed according to Chinese Standard GB/T 5009-2010.

4.5.4 Animals and Management

The rats were fed with test diets for 90 days in the specific pathogen-free (SPF) animal laboratory of the Supervision and Testing Center for Genetically Modified Organisms (GMOs) Safety, Ministry of Agriculture (SYXK, Beijing, China). The temperature and humidity of the animal room ranged from 20 to 24 °C and 40% to 70%, respectively. A 12-h light:dark cycle and air exchanges 15 times/h were used to keep the environment inhabitable for the rats.

We obtained 70 male and 70 female SPF Sprague–Dawley rats from Vital River Laboratories (SCXK, Beijing, China). All rats were 4 weeks old, weighing 80–100 g. After 5 days of acclimation, rats were randomly divided into seven groups based on body weight with 10 male and 10 female rats in each group. Six treatment groups were fed with diets containing 7.5%, 15%, or 30% (w/w) GM soybean (T1, T2, and T3 groups) or non-GM soybean (N1, N2, and N3 groups). The seventh group was fed a basic rodent diet (CK group). Water and diet were supplied ad libitum. The animal study and housing procedures were carried out in compliance with the OECD Principles of Good Laboratory Practice. The animal study was approved by the Animal Experimental Welfare and Ethical Inspection Committee (No. 140032-033) in the Supervision and Testing Center for GMO Food Safety at the Ministry of Agriculture (Beijing, China).

4.5.5 Clinical Observation, Body Weight Gain, and Feed Utilization

The rats were observed daily for their behavior, hair color, and other toxicity symptoms. Feed consumption and body weight were recorded weekly. Feed utilization was determined using the following calculation:

$$\text{Feed utilization}(\%) = (\text{body weight gain} / \text{feed consumption}) \times 100\%$$

4.5.6 Hematology

On day 90, the rats fasted overnight and blood samples were collected from the orbital sinus under anesthesia using EDTA•K_2 as an anticoagulant. White blood cell count (WBC), red blood cell count (RBC), hemoglobin (HGB), hematocrit (HCT), mean corpuscular volume (MCV), mean corpuscular hemoglobin (MCH), mean corpuscular hemoglobin concentration (MCHC), red cell volume distribution (RDW), blood platelet count (PLT), and mean platelet volume (MPV) were measured using a HEMAVET 950FS (Drew Scientific, Dallas, TX, USA) automatic animal blood cell counter.

4.5.7 Serum Chemistry

On day 90, the rats fasted overnight and blood samples were collected from the orbital sinus under anesthesia. The samples were centrifuged at 4000 *g* for 15 min to separate the serum. Alanine aminotransferase (ALT), aspartate aminotransferase (AST), total protein (TP), albumin (ALB), alkaline phosphatase (ALP), glucose (GLU), blood urea nitrogen (BUN), creatinine (CREA), calcium (Ca), potassium (P), cholesterol (CHO), triglyceride (TG), lactate dehydrogenase (LDH), chlorine (Cl), and magnesium (Mg) in the serum were measured with an automatic Biochemical Analyzer 7020 (Hitachi, Tokyo, Japan).

4.5.8 Necropsy and Histopathology

On day 90, the rats were euthanized by decapitation after anesthesia, and then a gross necropsy was performed on all animals to check if there were any macroscopic pathology changes to tissues or organs. The heart, spleen, lungs, kidneys, adrenal glands, brain, liver, thymus, and testes or ovaries were separated and weighed, and the relative organ weight (organ weight/body weight in percent) was calculated. The organ tissues (including the CK group, the non-GM T3 group, and the GM T3 group) were immersed in a 10% neutral buffered formalin solution and embedded with paraffin, and then tissues were stained with hematoxylin and eosin (H&E). Pathologists of the China Agricultural University College of Veterinary Medicine conducted histopathological examinations on heart, spleen, lungs, kidneys, adrenal glands, brain, liver, thymus, and testes or ovaries.

4.5.9 Statistical Analysis

For the final statistical analysis of this study, each GM incorporation level group was compared to its non-GM counterpart, and each soybean-containing group was compared to the control diet group. The data for body weight gain, feed consumption, food utilization, organ weight, hematology, and serum chemistry between GM soybean FG72 groups and non-GM soybean Jack groups were analyzed with a *t* test using the statistical software Statistical Product and Service Solutions (SPSS) v12.0 (SPSS, Chicago, IL, USA). All the data mentioned were presented as a mean value ± standard deviation (mean ± SD), and the statistical level of significance was set at $p < 0.05$.

4.5.10 Examples

4.5.10.1 Soybean

Soybeans are an important crop throughout the whole world, being used for both human food and animal feed, and it is also one of the most commonly modified crops generated by genetic engineering. According to one report, the main soybean-producing countries are these four: the United States (US), Argentina, Brazil, and China. Also, the production of soybeans in the US reaches the highest level among the GM crops. In Asia, especially in China, soy foods are an important traditional source for primary protein and have been consumed and utilized for thousands of years. In the year 2013, the output of soybean in China was 12.0 million tons. Some studies have declared that a high soybean intake for human beings is modestly associated with a reduced risk of breast cancer (Kim et al. 2008).

4.5.10.2 GM Soybean CV127

Genetically modified (GM) herbicide-tolerant crops including soybean, rice, and maize have become an important and widespread tool in crop and food production practices. In 2011, more than 80% of the total hectares of GM crops planted all over the world had special herbicide resistance traits (Green 2012). Imidazolinone herbicides can kill the weeds by inhibiting the enzyme acetohydroxyacid synthase (AHAS), which is a important enzyme in plants associated with the biosynthesis of the branched-chain amino acids such as isoleucine, leucine, and valine.

The soybean plant (*Glycine max* L.) under investigation is a variety derived from a single transformation event designated as BPS-CV127-9. As this transformed soybean variety was introduced with an imidazolinone-tolerant AHAS large-subunit gene (*csr1-2*) from *Arabidopsis thaliana*, it is able to tolerate imidazolinone herbicides by the expression of this gene inserted into the genome of the soybean plant. The *Arabidopsis* AHAS protein belongs to a family of the class of AHAS (large subunit) proteins found ubiquitously in plants. The mutation in the *Arabidopsis ahas* gene responsible for imidazolinone herbicide tolerance is a codon change from AGT to AAT, ultimately leading to a single amino acid substitution of serine to asparagine at position 653 of the AHAS protein. This single nucleotide change of guanine to adenine makes this kind of protein as a result. A second mutation, R272K, was later identified when this gene was inserted into soybean, retaining its normal biosynthetic function in the transgenic plant and not impacting the enzymatic function of the AHAS enzyme in soybean.

Although compositional studies have shown the equivalence of CV127 to commercial soybean varieties, some animal feeding studies were carried out to find further support and evidence for commercial acceptance of this new GM soybean variety. A sub-chronic safety study in 2012 with CV127 soybean meal showed that there were no adverse effects when used to feed Wistar rats at dietary incorporation levels of 11% or 33% (Chukwudebe et al. 2012). In this following 90-day animal study, we report additionally on the safety and nutritional wholesomeness of CV127 soybeans in Sprague–Dawley rats at three dietary incorporation levels (7.5%, 15%, or 30%, respectively). Because soybeans are also known as a major protein source in the daily diets of most animals, the nutritional wholesomeness of CV127 was also investigated by feeding the processed meal to broiler chickens over a 42-day period.

The results showed that the meal was quantitatively similar in ingredient composition between CV127 and Conquista diets at each soybean meal concentration (7.5%, 15%, or 30%) and thus similar as in percentages of proximate analytes confirming the nutritional balance of the animal diets.

During the 90-day study, no rat deaths or overt signs of toxicity issues were observed. All animals appeared healthy in this study, and no adverse clinical and typical observations were found. When comparing CV127 versus its corresponding Conquista group, we found no significant differences in body weight change for males or females. The feed utilization rate of male rats fed 15% GM soybean was significantly lower compared to 15% Conquista and the standard diet, but this difference was considered to be not treatment related because such issue was found at the highest dose (30%) group. The feed intake of female rats fed 30% CV127 was found to be significantly higher when compared it to 30% Conquista, but it showed no statistically difference from the standard diet. These differences were considered not to be dose- or age related and not biologically meaningful.

Some values of hematology and serum chemistry were shown to be significantly different between CV127 and its corresponding Conquista group. However, we considered these differences were not treatment related or adverse, because such differences were not found when compared to a standard diet group. Additionally, no correlative changes in histopathology were found, and no change was indicated in the metabolism of these animals during the 90-day period.

As for the statistical differences found for relative organ weights between CV127 and Conquista groups, we considered it to be incidental and not treatment related, based on the absence of any relevant histopathology findings as just described. The weight of brain and testes in male rats fed 15% CV127 was significantly higher than for the corresponding Conquista and the standard diet. However, these changes were not considered treatment related based on the reason that no such significant differences was observed between CV127 and Conquista at the highest dose group. For female rats, the weight of the thymus glands at 7.5% and kidneys at 30% was significantly different from its corresponding Conquista, but not statistically different from the standard diet. The value of liver weight in female rats fed 30% CV127 was significantly higher than that in corresponding Conquista and standard diet groups, but it was thought to be not dose related because such changes were not found in male rats in any dose group. Furthermore, no correlative changes in liver histopathology were observed, suggesting that there was no indicated adverse effect in these animals.

Histopathology was tested and recorded on 12 animals (6 males and 6 females) in the standard diet and 30% Conquista groups and 11 animals (6 males and 5 females) in the 30% CV127 group. Data included focal lymphocytic infiltration in the liver of 10–12 rats per group and a single incidence of uterine infection in the 30% CV127 group. These findings are commonly observed in other 90-day rat feeding studies (Qi et al. 2012) and are consistent with normal background records that occur spontaneously in this rat strain. Thus, these lesions were considered to be reasonable and not considered adverse or treatment related.

In this 90-day rat feeding animal study, CV127 soybean meal was fed to Sprague–Dawley rats at dietary incorporation levels ranging from 7.5 to 30%, which exemplify two important differences (i.e., a different rat strain and a different level of dietary incorporation) in study design as compared to Chukwudebe et al. (2012). Some occasional differences were once reported for the endpoints examined, which included body weight gain, feed consumption, hematology values, serum chemistry values, and relative organ weights, between rats fed CV127 and its corresponding Conquista and the standard diet. However, none of these occasional differences or changes was thought to be treatment related or adverse, and all these differences or changes were regarded as biologically irrelevant. Moreover, the data analysis provided in this study is consistent with the interpretation presented by Chukwudebe et al. (2012). Overall, this weight-of-evidence approach made a strong conclusion that unintended adverse changes have not occurred as a result of CV127 modification.

GM soybean 3Ø5423 × 40-3-2

Oil obtained from conventional soybeans contains high concentrations of polyunsaturated fatty acids [PUFAs; linoleic acid (18:2) and linolenic acid (18:3)] that have known health benefits (Udell et al. 2005). Unfortunately, the presence of these particular fatty acids makes conventional soybean oil susceptible to oxidation and subsequent rancidity, thereby limiting the utility for food applications (Liu et al. 2007). Accordingly, soybean oils with a high proportion of monounsaturated fatty acids were developed in the interest of greater stability. However, the process of hydrogenation results in the formation of trans fatty acids as a by-product (Lichtenstein et al. 1993). Partially hydrogenated soybean oil is effective for technical applications in foods including increased stability and shelf life. However, laboratory and clinical research has established a causative relationship between dietary consumption of trans fatty acids present in partially hydrogenated oil and cardiovascular diseases, cancer, and other types of diseases (Dhaka et al. 2011; Tan 2011). In some cases, food manufacturers have minimized or eliminated the use of partially hydrogenated oils in their products (Remig et al. 2010).

DP-3Ø5423 soybeans (TREUS) are a genetically modified soybean specifically engineered to possess higher concentrations of monounsaturated oleic acid than present in conventional soybean oil and contains no trans fatty acids. A fragment of the soybean microsomal omega-6 desaturase gene (FAD2-1) was inserted into soybeans. Transcription of the gene fragment under the control of a seed-preferred KTi3 promoter acts to silence the expression of the endogenous soybean omega-6 desaturase, preventing the addition of subsequent double bonds on oleic acid. The soybean has an oil profile with a significantly higher proportion of oleic acid (70.6% of total fatty acids) and lower proportion of linoleic acid (5.50% of total fatty acids) relative to traditional soybean oil (approximately 20% oleic acid and 55% linoleic acid, respectively). Oleic acid is considered one of the healthier sources of fat in the diet. Linoleic acid is a carboxylic acid with an 18-carbon chain and two cis double bonds. Accordingly, the oil obtained from DP-3Ø5423 soybeans is more oxidatively stable than standard soybean oil without the requirement for hydrogenation. Also, a gene (gm-hra) encoding a modified version of the soybean acetolactate synthase (als gene) that confers tolerance to a sulfonylurea herbicide was used as a selectable marker for the transformation.

In this case, the high oleic acid soybean DP-3Ø5423 was crossbred with Roundup Ready (RR) soybean 40-3-2 encoding a glyphosate-resistant form of the plant enzyme 5-enolpyruvylshikimate-3-phosphate synthase (EPSPS) (CERA 2009). The stacked GM soybean DP-3Ø5423 × 40-3-2 therefore contains a high content of oleic acid and is tolerant to the field application of herbicides containing glyphosate. For the purposes of this chapter, these will be referred to as HOA-HT (high content of oleic acid–herbicide tolerant) soybeans. As a new food source, a sub-chronic feeding study in rats as part of the comprehensive safety assessment for foods was conducted.

Nontransgenic soybean (JACK) and 3Ø5423 × 40-3-2 (HOA-HT) have the same genetic background with the exception of the *gm-fad2-1*, *gm-hra*, and *cp4 epsps* genes, which are present only in the HOA-HT soybeans. Analysis of nutritional proximates (moisture, ash, fat, crude fiber, crude protein, fatty acid, minerals) and the concentrations of individual fatty acids were determined. Both JACK and HOA-HT soybeans were ground into flour that contained the oil and formulated into feed at 7.5%, 15%, or 30% (wt/wt), respectively.

Four-week-old Sprague–Dawley (SD) rats were housed five per cage and provided water and feed ad libitum. Rats of each gender were randomly divided into seven experimental groups (10/sex/group) by body weight. The seven groups were fed with balanced diets with 7.5%, 15%, or 30% (wt/wt) transgenic soybean (T1,

Table 4.2 In vitro genotoxicity te sts as described by GB guidelines (GB15193.1-2014)

GB No.	Title
GB15193.4	Reverse mutation assay (*Salmonella typhimurium*)
GB15193.5	Mammalian erythrocytes micronucleus test
GB15193.6	Mammalian bone marrow cell chromosome aberration test
GB15193.8	Chromosome aberration test of spermatogonium or spermatocyte in mice
GB15193.9	Rodent dominant lethal test
GB15193.23	Vitro mammalian cells chromosome aberration test
GB15193.20	In vitro mammalian cell gene mutation test
GB15193.11	Sex-linked recessive lethal test
GB15193.10	Test method of DNA damage and repair/unscheduled DNA synthesis with mammalian cells in vitro
GB15193.12	V79/HGPRT gene mutation test

T2, T3) or traditional soybean (N1, N2, N3) and the control diet, respectively (Table 4.2). During the exposure period, body weight and food consumption were determined weekly. Food efficiency was calculated from the food consumption and body weight. Behavior and appearance of all rats were observed at least every 2 days throughout the study. Hematology and serum chemistry were conducted on the 45th day (mid-stage trial) and 90th day (the end of the trial). Heart, liver, spleen, lungs, kidneys, adrenal glands, thymus, brain, ovaries and uterus (female), or testes and epididymides (male) were weighed. These organs and stomach and intestine (duodenum, jejunum, ileum) were fixed in 10% neutral buffered formalin and examined microscopically.

All rats survived for the duration of the 90-day study and there was no sign of adverse clinical effects observed in any of the rats. No differences in body weight, weight gain, food consumption, or food efficiency were observed. Some statistically significant differences in mean relative organ weight were observed in rats fed the 3Ø5423 × 40-3-2 diet. However, the differences were not considered treatment related and commonly fell within the normal ranges of the control group consuming the commercial diet. Hematology values were not different between GM soybean groups and non-GM soybean groups. Incidental tumid uterus and focal lymphocytic foci in the liver and lung inflammatory cell infiltrates were observed in a number of rats, but these are common findings among experimental rats of this strain and age (Appenzeller et al. 2008; Delaney et al. 2008b; McNaughton et al. 2008).

New stacked GM crops have not revealed safety issues (Pilacinski et al. 2011). In this case, both individual events have been assessed for food safety individually in sub-chronic rodent feeding studies with no evidence of adverse effects (Delaney et al. 2008b; Hammond et al. 1996). The results indicated that GM HOA-HT soybean 3Ø5423 × 40-3-2 is as safe as the non-GM conventional isogenic line.

GM soybean MON87708

At present, there are 23 single or stacked herbicide-tolerant soybeans that account for approximately 80% of the genetically modified (GM) soybean (ISAAA 2014a).

However, populations of several troublesome weeds may gradually increase and acquire tolerance or resistance to herbicides currently used (Behrens et al. 2007). Dicamba-tolerant crops developed because few dicamba-resistant weeds have appeared after more than 50 years of use. The advent of dicamba-tolerant crops provides a broader choice for agricultural production and weed management. Dicamba (2,6-dichloro-2-methoxybenzoic acid) is an herbicide that mimics a plant growth hormone that is effective at killing broadleaf plants. The purpose of the present study was to evaluate the potential toxicity of dicamba-tolerant MON87708 compared to that of the near-isogenic non-GM soybean line A3525.

The GM soybean grain MON87708 and its near-isogenic non-GM conventional soybean grain A3525 (lacking the dmo gene) were provided by Monsanto. A standard commercial rat diet, namely, SPF rodent maintenance feed, was also included as a blank control. Soybean MON87708 can biosynthesize dicamba mono-oxygenase (DMO); DMO rapidly catalyzes the oxidative demethylation of dicamba to 3,6-dichlorosalicylic acid (DCSA) and formaldehyde (Dumitru et al. 2009). The amino acid sequence of the DMO enzyme does not share significant amino acid identity with any known protein toxins or allergens (FSANZ 2012). The soybean MON87708 has already obtained regulatory approval in the US in 2011, in Canada and New Zealand in 2012, in the European Union in 2015, and so on, and has even been approved for direct use and processing in food (ISAAA 2014b)

The full-fat, unhulled milling soybean powder from the GM and the non-GM soybeans was heated 40 min at 70 °C to inactivate enzyme inhibitors. The soybean powder (full-fat, unhulled) used in our study was also different from many other processing methods of soybean (Appenzeller et al. 2008; Delaney et al. 2008a), and is more consistent with the Chinese habit of consumption (crude soybean without oil-extraction, hulling, and toasting). To make the respective experimental rodent diets, the soybean powder was incorporated at proportions of 7.5%, 15%, or 30% and formulated with the other diet ingredients by Ke Ao Xie Li Feed Co. (Beijing, China). The feed of the commercial diet rat group was from Ke Ao Xie Li Feed also, namely, SPF rodent maintenance feed. The 30% dietary incorporation of the GM soybean in our study would be generally dozens of times higher than the recommended intake for humans. Three dietary doses (7.5%, 15%, 30%) groups of the GM and non-GM soybean are a greater variety of doses than those tested in previous studies (Brake and Evenson 2004; Chukwudebe et al. 2012; Daleprane et al. 2010).

The health status, morbidity, mortality, and cage-side clinical signs were observed and monitored at least once per day throughout the whole dietary exposure. Blood samples were collected after overnight fast (16 h) in the morning at the mid-stage and end of the study for hematology and serum chemistry studies. Some organs, including the brain, thyroid, heart, lungs, thymus, liver, spleen, kidneys, adrenals, stomach, small intestine (duodenum, jejunum, and ileum) and paired testes or ovaries, were fixed with 10% neutral buffered formalin and weighed, except stomach, small intestine, and thyroid. Then, a histopathological examination was conducted.

Some significant differences were observed in food consumption and utilization, hematology, and serum biochemistry by analysis of variance (ANOVA). It is particularly necessary to point out that the statistical analysis of the food consumption data has limited power because there are only two values (cages)/sex/group, and the following differences should be especially mentioned. The level of HCT was observed to be slightly higher in the 15% MON87708 male group than that in 15% A3525 and commercial diet groups. Considering the differences of relative parameters RBC and HGB, it was attributed to slight hemoconcentration, and the differences in the parameters were in the range of historical studies (Qi et al. 2012). Slightly elevated TP and ALB were also observed in the male soybean (GM and non-GM) groups when compared with the commercial diet group, but these changes were also in the range of historical reference values. A slightly lower value of GLU was observed in some GM and non-GM soybean treatment groups, but the differences were not dose related. A relatively higher value of LDH was observed in the female soybean treatment groups, which was attributed to biological variability rather than a result of the consumption of transgenic soybean based on the analysis of the dose-related effects. No differences in relative organ weights in either male or female rats were attributed to consumption of MON87708 soybean. All organ weight differences were observed only between groups fed soybean diets (GM or non-GM) and the commercial diet group. In general, very few differences were observed, and these differences were always incidental, because they were minor (within historical ranges), not dose related, and in isolation.

The results of this study demonstrated no indication of adverse effects associated with the consumption of dicamba-tolerant soybean MON87708 in Sprague–Dawley rats during the entire exposure period. The test substance, MON87708, is as safe and wholesome as conventional non-GM isogenic soybean and presents the same nutritive value to animals at dietary inclusions up to 30%.

4.5.10.3 Maize

Maize, rich in carbohydrates, is one of the most widely grown cereal crops, accounting for more than one-half of global cereal production. World maize production for 2007–2008 was 791.6 million tons (Barros et al. 2010). Currently, about one fourth of the global maize yield is produced using biotechnology(Coll et al. 2011). However, weeds and insects can lead to a decrease in its quality and yield. Since the production of the first genetically modified (GM) plant, which occurred in 1983, applying genetic engineering techniques to improving the production of crops has been adopted rapidly (Song et al. 2014). GM maize is the second most dominant GM crop, covering 30% of the global GM planting area and 24% of the total maize production in 2008 (Qaim 2009). The presence of different added traits in GM maize, including insect resistance (Li et al. 2013), disease resistance (McComas et al. 2014), and herbicide resistance (Kawahigashi et al. 2007), has effectively improved the yield.

GM maize GH5112E-117C

The *Cry1Ah* gene is a novel insecticidal gene that is cloned from *Bacillus thuringiensis* BT8 (Xue et al. 2008). The active Cry1Ah protein with a molecular weight of 65 kDa (Li et al. 2013) is highly toxic to *Lepidopteran* insects, and is more toxic to *Helicoverpa armigera*, *Ostrinia furnacalis*, *Chilo suppressalis* (when compared with the Cry1Ac protein), and *O. furnacalis* (when compared with the Cry1Ab protein) (Li et al. 2013). The Cry1Ah toxin is highly specific to its target insect, and is innocuous to humans, vertebrates, and plants (Bravo et al. 2007). Glyphosate (N-phosphonomethylglycine), as a potent herbicide, is widely used to control weeds in agricultural fields by inhibiting the enzyme 5-enolpyruvylshikimate-3-phosphate synthase (EPSPS; EC 2.5.1.19) (Cao et al. 2012a). Having a low activity or no activity from the target enzyme, EPSPS is used as a basic strategy for developing the glyphosate-resistant maize. At present, multiple *zm-2mepsps*, two *cp4-epsps* are widely used to develop glyphosate-resistant maize (Green 2012). Inserting the *aroA* gene (located in *Salmonella typhimurium*) could inhibit the biosynthesis of aromatic amino acids by suppressing the activity of the EPSPS enzyme (Zhu et al. 2013), which shows an enhanced glyphosate-resistance in transgenic crops (Dun et al. 2014).

Compared with the mono-trait (insect-resistant or glyphosate-resistant) GM maize, the stacked insect-resistant and glyphosate-resistant traits can satisfy various needs of consumers and producers. However, in comparison with the mono-trait GM maize and the conventional non-GM maize, the insertion of the double genes into the genome of maize might lead to unintended impacts on nutrition and toxicology.

Seventy male and 70 female SPF Sprague–Dawley rats (5 weeks old, weighing 80–100 g) were fed with the GM maize GH5112E-117C or the non-GM maize Hi-II. The GM maize GH5112E-117C, in which the expression levels of the proteins Cry1Ah and G2-aroA are 1.47 ± 0.17 (μg/g) and 0.34 ± 0.06 (μg/g), respectively, and the near-isogenic non-GM maize Hi-II were grown in the same isolated field plots in China. During the test, the rats were observed and recorded daily for their behavior, clinical signs including skin, eyes, excretion, behavior, mortality, and other toxicological clinical signs. Additionally, body weight and feed intake were measured and recorded weekly. Blood samples from fasting (overnight) rats were collected from the orbital sinus under anesthesia on the 45th and 90th days. White blood cell count (WBC), red blood cell count (RBC), hemoglobin (HGB), mean corpuscular hemoglobin (MCH), mean corpuscular hemoglobin concentration (MCHC), red cell volume distribution (RDW), blood platelet count (PLT), and mean platelet volume (MPV) were analyzed using an automatic animal blood cell counter, HEMAVET 950FS (Drew Scientific, Dallas, TX, USA). The sera were obtained, and serum chemistry including alanine aminotransferase (ALT), aspartate aminotransferase (AST), total protein (TP), albumin (ALB), alkaline phosphatase (ALP), glucose (GLU), blood urea nitrogen (BUN), creatinine (CREA), calcium (Ca), potassium (P), cholesterol (CHO), triglyceride (TG), lactate dehydrogenase (LDH), chlorine (Cl), and magnesium (Mg) was measured with an automatic Biochemical Analyzer 7020 (HITACHI, Tokyo, Japan). At the end of the 90-day feeding study, rats were sacrificed by decapitation under anesthesia and a complete gross necropsy was performed on all animals by visual inspections. Brain, liver, heart, spleen, lung, kidney, adrenal gland, thymus, and testis (or ovary) were isolated and weighed. The relative organ weight (the percentage of the organ weight/ body weight) was calculated. The tissues of the organs from the negative control group and the high dose (50% maize in the diet) group were immersed in 10% neutral buffered formalin. After paraffin-embedded tissue sections, all tissues were stained with hematoxylin and eosin (H&E). Histopathological examinations were conducted on cerebrum, cerebellum, heart, spleen, liver, lung, kidney, stomach, duodenum, jejunum, ileum, cecum, colon, rectum, thymus, thyroid, adrenal gland, testis, epididymis, ovary, and uterus.

In this research, a 90-day sub-chronic feeding study of the stacked trait GM maize GH5112E-117C (obtained from Beijing Origin Seed Technology, China) with *Cry1Ah* and *G2-aroA* was carried out, according to the Organization for Economic Cooperation and Development (OECD) guidelines no. 408, as well as the Chinese standard safety assessment of genetically modified plant and derived products 90-day feeding test in rats (NY/T 1102-2006). The obtained results of the study will provide a valuable reference for the food safety evaluation of other stacked trait GM crops and can greatly promote the adoption of stacked trait GM maize GH5112E-117C. At the same time, the obtained results were similar to that of the GM maize with the *G2-aroA* gene (Zhu et al. 2013), indicating that the simultaneous insertion of *Cry1Ah* and *G2-aroA* did not result in unintended effects on nutrition and toxicology. Based on these facts, a 90-day sub-chronic toxicology feeding study in rodents was one of the effective methods to assess the food safety of stacked trait GM crops. Additionally, the results of this study demonstrated that the stacked trait GM maize GH5112E-117C can be further developed so that it can be widely adopted.

The risk assessment of GM stacked events is on the agenda of the development of agricultural biotechnology. The science-based food safety evaluation of stacked traits GM crops will ensure the sustained development and timely introduction of new GM varieties, which are needed to meet future agricultural demands. Based on the experiences obtained from the safety evaluation of mono-trait GM events, stacked trait crops are considered as safe as traditional crops on condition that there are no significant differences in nutritional and agronomic characters.

4.5.10.4 Rice

Rice is one of the most important food resources for humans and feed resources for animals in the world, especially popular as planted in Asia. Rice is also a good bioreactor to produce pharmaceutical proteins and other nutritional elements. Rice is now consumed by more than 3.5 billion people, which means that more than half the world's population depends on rice,. Some studies have declared that the GM rice can promote human health and improve nutritional levels; most importantly, GM rice is environmentally safe by a series of safety assessments (He et al. 2008). Scientists have successfully expressed lactoferrin and human lysozyme in rice seeds

(Nandi et al. 2005; Yang et al. 2003). The former can be a nutritional supplements for human beings and the latter has a function of protecting humans against some microbial attacks.

GM rice 4-114-7

Human serum albumin (HSA) is known as a famous protein because it can be widely used to treat patients beset with hemorrhagic shock, serious burn injuries, hypoproteinemia, fetal erythroblastosis, and ascites (Belgacem et al. 2007). Annual use of HSA worldwide reaches 500 metric tons at a high cost of more than $1.5 billion (Daniell 2003). Nowadays, the main production of HSA for commercial purposes is extracted from human plasma, which is highly costly, and may have the potential risk of transmission of blood-derived infectious pathogens such as human immunodeficiency virus (HIV) (Chamberland et al. 2001; Erstad 1996). However, such problems may be solved as HSA can be expressed in rice, in which the accurate expression level can be a productivity rate of 2.75 g/kg in one kind of brown rice, achieved by Wuhan University, China (He et al. 2011). This production rate can certainly meet the huge needs of large-scale commercial production once it were spread among the world. He et al. carried out many experiments related with "substantial equivalence" between rice-extracted HSA (rHSA) and plasma-derived HSA (pHSA) in 2011. The amino acid sequence and crystal structure of rHSA were exactly the same as those of pHSA. Both these proteins have equivalent biological functions such as efficiency of promoting cell growth and immunogenicity in vitro and vivo (He et al. 2011). Although the GM rice will not be sold in the food market, safety assessments of the GM rice must be carried out before large-scale applications because of some concerns and doubts about the food safety of GM rice: this is of particular concern for the potential and future release of some GM crops such as rice, soybean, maize, and potatoes into conventional plant populations of the markets (Halford and Shewry 2000).

The GM rice 4-114-7 and its nontransgenic isogenic control "Taipei 309" were both planted in Taoxian City, Hubei Province, China under an identical and same climate and environment condition in the proper growing season of 2010. A specific polymerase chain reaction (PCR) assay was used to identify the GM and non-GM rice based on the presence of the HSA gene. The results demonstrated that the gene was found in the 4-114-7 rice although it was not detected in the non-GM isogenic control rice. rHSA accumulates a final expression at up to 4.93 g/kg in the GM brown rice (He et al. 2011).

By utilizing a stratified randomization procedure, 5-week-old Sprague-Dawley rats (70 male and 70 female rats) free from specific pathogens were randomly distributed into seven experimental groups by body weight (10 rats/sex/group). The animals were fed with balanced rodent diets with commercialized rodent diet (CK) or a content of 12.5%, 25%, or 50% (wt/wt), GM rice (T1, T2, T3, respectively), or non-GM rice (N1, N2, N3, respectively). The rats used in this study were housed (5 rats/cage) in a specific pathogen-free (SPF) animal laboratory meeting the r standard laboratory conditions. The housing condition were controlled with adequate fresh air exchange all day, temperature mainly 20–25 °C and relative humidity

40–70%. A 12-h light:dark automatic cycle of artificial illumination was constructed to meet the rats' living habits. All animals were provided with sterile drinking water. The results of heat stability and in vitro digestion of rHSA protein were shown in accordance with a published report (Cao et al. 2010). Clinical observation, feed efficiency, hematology, serum chemistry, organ weights, histopathology, and other ranges of indexes were examined and recorded as regulated.

The levels of amino acids, reflecting protein content, were also analyzed. No statistically significant differences were observed in the levels of moisture, ash, crude fiber, fat, phytic acid, vitamin B_1, and most of the minerals between GM rice and non-GM rice. However, some components of the GM rice exhibited statistically significant differences ($p < 0.05$) compared with the non-GM rice. Most of the differences remained within the ranges reported in the literature (USDA 2004).

In the males, the mean body weight of the T1 group was lower ($p < 0.05$) than that of the CK group. Some individual rats in the T1 group were observed wasting feed, which is a habit for some rodents, which may account for the higher feed use and lower body weight for this group. Furthermore, this significant difference was not considered adverse or related to the addition of GM rice in the rodent diets because the mean body weight in the T2 and T3 groups was not different from the CK group. The mean body weight of the female rats in the GM group was comparable with both the control group and the non-GM group. No significant differences in total food consumption or food efficiency were observed.

A few statistically significant differences between the control group and other treatment groups were found in some hematology parameters on the 45th and 90th day. Most of these variations are within historical control data ranges. Although some statistically significant differences were observed in a few serum chemistry parameters on the 45th and 90th days, there were no treatment-related biologically adverse effects. All the variations are within historical control data ranges and considered as incidental and toxicologically insignificant as there was no dose-dependency and corresponding changes in histopathology.

The mean relative weight of the brains of the male rats that consumed 25% GM rice was slightly lower ($p < 0.05$) than that of the non-GM group and the CK group. However, the change was within historical control data range and considered incidental because no significant differences were observed in the female rats, and the microscopic histology was normal. Any changes noted in organ weights were incidental, with a lack of corresponding changes in histopathology. At necropsy, there were no gross or microscopic anatomic pathological changes that were considered to be test article related.

Although the two rice varieties were cultivated under identical environmental and climatological conditions, statistical analyses indicated that several parameters of the GM rice were significantly different from the isogenic non-GM variety. In this case, additional field trials are required to confirm whether the differences measured are caused by the genetic modification process or biological variability in the field. Most of the compositional parameters of the GM rice were within the ranges reported in the literature (USDA 2004), except for calcium, lysine, tyrosine, and vitamins B_1 and B_2. However, the existing food composition database does not necessarily reflect the full range of natural variation (Burlingame 2004). Amino acids, especially alanine, arginine, proline, and serine, in the GM rice were all higher compared with the isogenic non-GM line, which probably reflects the increased level of protein caused by the high expression of HSA in the GM rice. The expression of the HSA protein did have some influences on the composition of the rice. However, relevant unintended side effects could not be determined from the analyses of specific compounds or intermediates in important nutritional and anti-nutritional pathways (Kok and Kuiper 2003).

Some parameters of hematology, serum chemistry, and organ weights in the GM groups demonstrated statistically significant differences compared with the control group and/or the non-GM groups. In the male rats, a lower level of WBC, an indication of hypoimmunity, was detected only in the 25% GM group. This change was amplified by individual rats in this group, but no histopathological symptoms in the liver or thymus were observed. The absolute and relative decreases of ALT and AST were observed only in the male rats. This change was not considered to be clinically significant to the health of the animals (Anderson et al. 2000); it may actually reflect beneficial effects such as membrane stabilization in hepatocytes (Lee et al. 2005). The HGB and HCT values decreased in the 12.5% GM male group, symptoms often related to anemia. Some individual rats of the group were observed wasting feed, causing higher feed use and lower body weight. Anemia often follows dietary restriction (Chow and Lee 1964), so low mean HGB and HCT values of some individuals may be associated with anemia. Hence, these changes were considered biologically insignificant, incidental, and not associated with the test article administration. Some statistically significant differences were also observed in blood biochemistry, hematology, immunological parameters, and histopathology between the GM groups and non-GM groups, but none of these differences were considered to indicate adverse effects on animal health.

Plasma-derived HSA has been widely used clinically for a long time and there is a significant history of safe human exposure (Anderson and Anderson 2002). "Substantial equivalence" between rHSA and pHSA was certified through a series of experiments. Physical and biochemical characterization of rHSA revealed it was equivalent to pHSA (He et al. 2011). Recombinant protein production in plants to provide botanical therapeutics is responsible for launching a new generation of improvements in human health in the twenty-first century (Raskin et al. 2002). Research on pharmaceutical GM crops is an exciting field of genetic engineering with great advantages, such as low cost, high safety, and facilitated mass production. However, food safety assessment for pharmaceutical GM corps is necessary before the development of their commercialization. The present study indicates no evidence of potential allergenicity of the protein, and the GM rice 4-114-7 had no adverse or toxic effects on the test animals. The GM rice appears to be safe as a dietary ingredient when it is used at up to 50% in the diet on a sub-chronic basis.

GM rice T2A-1

Bacillus thuringiensis (Bt) is a naturally occurring soil bacterium that produces proteins active against certain insects. Bt crops provide highly effective control of major insect pests including the European corn borer, southwestern corn borer, tobacco budworm, cotton bollworm, and pinkworm, reducing reliance on conventional chemical pesticides (Betz et al. 2000). Many transgenic Bt rice plants have been successfully produced since the mid-1990s (Tyagi and Mohanty 2000). A number of laboratories developed different local varieties with Bt genes including *Cry1Ab*, *Cry1Ac*, *Cry1Aa*, *Cry2A*, and *Cry1B*, or a combination of these genes for resistance against lepidopteran pests (Manimaran et al. 2011). The first field testing of Bt rice was conducted in China in 1998. However, no Bt rice has been commercialized legally yet (Manimaran et al. 2011).

T2A-1 rice is a kind of GM rice that was produced by insertion of one novel synthetic *Cry2A** gene from *Bacillus thuringiensis* into the elite indica rice restorer line Minghui 63 by *Agrobacterium*-medicated transformation. The Bt gene Cry2A has low homology with Cry1A (<45%). It produces a δ-endotoxin crystal protein that is active against both lepidopteran and dipteran insects (Chen et al. 2005). Researchers have confirmed that Cry2A is toxic to several of the main lepidopteran pests such as yellow stem borer, striped stem borer, and two species of rice leaf folder (*Marasmiapatnalis* and *Cnaphalocrocis medinalis*) when the insects are fed the toxin in an artificial diet (Chen et al. 2005; Karim and Dean 2000). A modified novel Cry2A* gene designed on the basis of rice preference codons was introduced into the elite indica rice restorer line Minghui 63 (Chen et al. 2005). This transgenic line is significantly resistant to lepidopteran rice pests, can be used to produce insect-resistant hybrids, and serves as a resistant source for the development of two-toxin Bt rice (Cry1Ac plus Cry2A, or Cry1C plus Cry2A) (Chen et al. 2005). The average protein concentration of this GM rice reached approximately 10 μg/g leaf fresh weight, which is adequate to meet the requirement of the "high-dose/refuge" strategy (Cohen et al. 2000).

Numerous genetically modified (GM) crops expressing proteins for insect resistance have been commercialized following extensive testing demonstrating that the foods obtained from them are as safe as those obtained from their corresponding non-GM varieties. In this chapter, we report the outcome of safety studies conducted on a newly developed insect-resistant GM rice expressing the *cry2A** gene by a subchronic oral toxicity study on rats. Based on the compositional analysis, the transgenic T2A-1 rice and its parental control line Minghui 63 rice were formulated into a basic rodent diet at inclusion levels of 30%, 50%, and 70%, respectively. The high concentrations of rice used in toxicology study are usually set as 60–80% (Cao et al. 2012c; Wang et al. 2014a; Zhang et al. 2014b). In this study the highest supplementation of rice was set to be 70%. The 70% dietary consumption was equivalent to an intake of 58 g/kg BW/day in rats. The concentration of Cry2A* in T2A-1 rice is 10 μg/g, which means that the maximum daily exposure of rats to Bt toxin was 583 ug/kg·BW, which is far lower than the "no-effect level" (8400 mg/kg·BW) of pure Bt toxin in chronic experiments with rats (Betz et al. 2000). A commercial rodent diet with comparable nutritive values incorporated was applied as a negative control diet. One hundred and forty male and female Sprague–Dawley rats, 4 weeks after birth with an average body weight of 80–100 g, were killed for this study. All animals were housed individually with ad libitum access to water and diets. The rats were acclimatized for 5 days, then divided into seven groups randomly with 10 rats/sex/group. The test groups were fed with diets formulated with GM T2A-1 rice, compared with the control groups fed with diets containing Minghui 63 rice. The group fed with the AIN93G diet were the negative control group.

Cage-side observation was conducted twice daily for all rats for mortality, morbidity, and any noteworthy signs of toxicity during the 90-day duration. Body weight and feed intake were measured at weekly intervals throughout the experiment. Body weight gain and feed utilization were calculated weekly, Hematology and biochemistry evaluation all were recorded at the 45th and 90th days. Pathology examination included brain, heart, kidney, liver, lung, spleen, thymus, and testes or ovaries. Tissue sections from brain, heart, liver, lung, spleen, kidney, thymus, stomach, intestine, and testes or ovaries were fixed with 10% buffered formalin.

No treatment-related adverse or toxic effects were observed based on an examination of the daily clinical signs, body weight, food consumption, hematology, serum biochemistry, and organ weight or based on gross and histopathological examination. These results demonstrate that the GM rice with the *cry2A** gene is as safe for food as conventional non-GM rice. Several statistical differences were observed among rats consuming the GM diets and non-GM rice diets. The statistical differences were randomly observed among the different groups, were not dose related, and were not considered to be biologically significant. All these differences were also observed in other 90-day feeding studies (Qi et al. 2012; Yoshikawa et al. 2013; Zhu et al. 2013; Zou et al. 2014) and were considered to be normal for rats of this strain and age. The studies described here indicate that the T2A-1 rice is not likely to causes toxicity to rats. Collectively, the comprehensive results from these studies demonstrate that rice from lepidopteran insect-protected GM rice T2A-1 is as safe and wholesome as rice from non-GM strains.

4.6 Chronic Toxicology Study of Whole GM Foods or Feed in Animals In Vivo

Chronic toxicity is the development of adverse effects as a result of long-term exposure to one substance. Adverse effects associated with chronic toxicity can be directly lethal but are more commonly sublethal, including changes in growth, reproduction, or behavior. Chronic toxicity is in contrast to acute toxicity, which occurs over a shorter period of time with exposure to higher concentrations. Various toxicity tests can be performed to assess the chronic toxicity of different contaminants that usually last at least 10% of an organism's lifespan.

The principal objective of chronic toxicity is to identify substances that may cause an increase in cancer at any site by any mechanism, by observing test animals for the incidence of neoplastic lesions as a consequence of exposure to one substance, during the growth period of their lifespan, to various doses of the test substance by an appropriate route of administration. For food chemicals and food components, this implies the oral route. Initially, because of lack of knowledge of specific mechanisms, the testing of individual chemical substances for chronic toxicity virtually ignored mechanistic considerations, and any statistical increase in tumor incidence or numbers and sizes of tumors was considered to be a potential risk factor to humans. Before chronic toxicity testing progressed, a number of tests may be conducted in evaluation of the allergic or acute toxicity and carcinogenicity. It is worth noting that if a new food or food additive were evaluated, and any adverse effects including genotoxicity or other toxicities were detected, these foods or food additives would not normally be introduced into human food and thus would not be a candidate for chronic toxicity testing. The use of MTD (methadone) in carcinogenicity testing and the potential problems with this approach are discussed elsewhere (Dybing et al. 2002).

For carcinogenicity testing, most regulatory systems continue to accept the ‘two species’ (rat and mouse) approach. This approach has been used extensively over the years by the U.S. National Toxicology Program. These two species are commonly used in the evaluation of any adverse substance, and the information or data about their responses in carcinogenicity bioassays are accepted by researchers, including the knowledge of specific strains and their background incidences of tumors. A guideline for such studies has been provided by the OECD, intended for all chemicals, including food-related substances. In essence, protocols for chronic toxicity studies have not undergone any significant changes since this time. However, the past two decades have seen an increase in our knowledge of the mechanisms of carcinogenesis, of species specificity, and in the development of other testing regimens, which has enabled a number of scientific bodies to propose alternative or modified approaches for the testing of substances for potential carcinogenicity to humans, which are next. The OECD guideline discusses the sufficient numbers of 50 female and 50 male animals to be killed in one testing, and sets the duration of testing as 18 months for mice and 24 months for rats. Fundamentally, the protocols are designed not simply to maximize the detection of a positive result but also to give a sense of reassurance and confidence in a negative finding. It is important to note that particular attention and checking should be given to animal growth, necropsy, histopathology, and statistical evaluation.

The International Agency for Research in Cancer (IARC) provided detailed guidance on these aspects. Of particular importance in carcinogenicity testing of food substances and dietary components is consideration of the possible peculiarities of animal models, especially when the test material is a nutrient, such as a commercially treated protein or starch, where the product may be incorporated into the diet at levels as high as 20–60% at the expense of the corresponding nutrient. Usually, for non-nutrient materials that are being tested via the oral route, the highest concentration should not exceed 5%. An important factor in rodent dietary studies is the finding that a decrease in calorific intake per se, without a deficiency in essential nutrients, extends the maximum lifespan and retards the onset of both spontaneous and induced tumours and that this effect is broadly consistent across gender, strain, and species (Hart and Turturro 1997; Hart et al. 1995). Overall, investigations into the effects of dietary restriction in rodents have concluded that the total amount of energy intake per animal has the prime effect on longevity and that differing macronutrient compositions of the diet are of lesser importance. In contrast, most studies of energy-providing components of the diet in relation to human carcinogenesis have focused on dietary composition rather than on the total amount eaten.

In 2012, Seralini published a study on one Roundup-tolerant genetically modified maize (NK603). The health effects of NK603 (from 11% in the diet), cultivated with or without Roundup, and Roundup alone (from 0.1 ppb in water), were studied for 2 years in rats. In females, all treated groups died two to three times more than controls and more rapidly. The results were shown to be hormone- and sex dependent, and the pathological profiles were comparable. Females developed large mammary tumors almost always more often than and before controls, and the pituitary was the second most disabled organ: the sex hormonal balance was modified by GMO and Roundup treatments. At about the same time my research team conducted the same study about one glyphosate-tolerant maize with the G2-aroA gene. In our study, "A 90-day feeding study of glyphosate-tolerant maize with the G2-aroA gene in Sprague–Dawley rats" (Zhu et al. 2013), the material was not treated with herbicide. The paper on the 90-day rodent feeding study with the whole genetically modified (GM) food aims to assess potential unintended effects of toxicological and/or nutritional relevance and to establish whether the GM food is as safe and nutritious as its traditional comparator (Delaney 2007); therefore, this study can lead to the conclusion that "the herbicide-tolerant maize is as safe and wholesome as its isogenic non-GM control." It is true that the herbicide will be involved in our food chain when we consume the herbicide-tolerant crops. The toxicity of herbicides has been evaluated by expert scientific bodies and regulatory agencies, such as the World Health Organization (WHO) and the U.S. Environmental Protection Agency (EPA), before their commercialization but not after using them on GM crops. In fact, many traditional non-GM crops are also using various kinds of herbicides, obviously, which should not be the reason to doubt the safety of GM crops. For example, glyphosate [*N*-(phosphonomethyl) glycine] is used widely in agricultural, forestry, and residential markets. It was concluded that the glyphosate had low oral acute mammalian toxicity with the lethal dose for 50% of test animals (LD_{50}) being greater than 5000 mg/kg, and it was not considered to be a mutagen, carcinogen, teratogen, or reproductive or developmental toxicant. The maximum residue limit (MRL) for glyphosate in dry soybean was recommended as 20 mg/kg by the Codex Alimentarius Commission (CAC) (Codex pesticides residues in food online database, http://www.codexalimentarius.net/pestres/data/index.html). Additionally, Williams et al. (2000) reviewed the studies about the safety of Roundup herbicide, including its active compound glyphosate, the major breakdown product [aminomethylphosphonic acid (AMPA)] of glyphosate, and the predominant surfactant [polyethoxylated tallow amine (POEA)] used in Roundup formulations worldwide. According to the studies performed for regulatory purposes as well as published research reports, it was concluded that the use of Roundup herbicide did not result in adverse effects on development, reproduction, or endocrine systems in humans and other mammals. A similar conclusion was drawn again 12 years later (Williams et al. 2012). We also noticed that some studies indicated adverse effect of the glyphosate or the surfactant of Roundup herbicide on mammals or cells in recent years. These data might be very helpful for the WHO or regulatory agencies of different countries to reconsider the safety and MRL of Roundup. It should be kept in mind throughout the discussion that the herbicide was widely used in traditional agricultural and forest products, not only in the GM products. The other question is whether the 3-month (or 90-day, 13-week) feeding study was longer enough for the detection of potential adverse effect of GM crops, which was one of the main issues between the study of Zhu et al. (2013) and Seralini et al. (2012). According to literature data from studies investigating whether toxicological effects are adequately identified in 3-month sub-chronic rodent studies by comparing findings at 3 and 24 months for a range of different chemicals, it is concluded that rodent animal feeding studies of 90-day duration appear to be sufficient to pick up adverse effects of diverse compounds that would also give adverse effects after chronic exposure. As recommended by the FAO/WHO Expert Consultation (FAO/WHO 2000), the minimum requirement to demonstrate the safety of long-term consumption of a food is a sub-chronic 90-day study. Longer-term studies may be necessary if the results of a 90-day study indicate adverse effects, such as proliferative changes in tissues. Specific research protocols need to be established for further studies (Kuiper et al. 2001), which is the greatest regret of the study of Seralini et al. (2012): they designed the 24-month chronic feeding study according to the protocols of a 90-day sub-chronic feeding study, which did not meant that more time was more scientific. According to the observation in Zhu et al. (2013), it was not necessary to conduct chronic studies of the GM maize because no adverse effect was found in the sub-chronic study.

Thank you for your attention to the safety assessment of GM food. The debate on the safety of GM foods has continued along with their development. The major responsibility of toxicologists was to detect and predict the potential adverse effect of novel foods to protect the health of consumers with science-based methods. We call for scientists around the world to work together to explore the safety of GM foods, and we welcome anyone to propose more suggestions to improve existing studies.

4.7 Prospect: Safety Assessment of Toxicology in the Future

4.7.1 The Protocols for the Safety Assessment of GMOs Come from the Protocols for Chemicals: These Studies Have Limitations for the Testing of Whole Foods and Feed

No dietary proteins are known to date to be directly associated with teratogenic, mutagenic, or carcinogenic effects in animal models (Pariza and Foster 1983; Pariza and Johnson 2001). The current performance of the safety assessment of whole foods is mainly based on the protocols for low molecular weight chemicals such as pharmaceuticals, industrial chemicals, pesticides, food additives, and contaminants. However, without adaptation, these protocols have limitations for the testing of whole foods and feed, primarily because defined single substances can be dosed to laboratory animals at very large multiples of the expected human exposure, thus giving a large margin of safety. In contrast, foodstuffs are bulky, lead to satiation, and can only be included in the diet at much lower multiples of expected human intakes. When testing whole foods, the possible highest concentration of the GM food and feed in the laboratory animal diet may be limited because of nutritional imbalance of the diet, or by the presence of compounds with a known toxicological profile.

4.7.2 The Development of Alternative Testing to Animal Experiments

An underexposed ethical issue raised by tissue engineering is the use of laboratory animals in tissue engineering research. Even though this research results in suffering and loss of life for the animals, tissue engineering also has great potential for the development of alternatives to animal experiments. With the objective of promoting a joint effort of tissue engineers and alternative experts to fully realize this potential, this study provides the first comprehensive overview of the possibilities of using tissue-engineered constructs as a replacement for laboratory animals. Through searches in two large biomedical databases (PubMed, Embase) and several specialized 3R databases, 244 relevant primary scientific articles, published between 1991 and 2011, were identified. By far most articles reviewed were related to the use of tissue-engineered skin/epidermis for toxicological applications such as testing for skin irritation. This review article demonstrates, however, that the potential for the development of alternatives also extends to other tissues such as other epithelia and the liver, as well as to other fields of application such as drug screening and basic physiology. This review discusses which impediments need to be overcome to maximize the contributions that the field of tissue engineering can make, through the development of alternative methods, to reduction of the use and suffering of laboratory animals.

There are various areas in the development of alternative testing approaches in environmental hazard and risk assessment that require future action and research. As outlined here, they appear to apply to many endpoints, regardless of whether testing or nontesting approaches are considered. These goals include the need for a detailed mechanistic understanding, such as for the exploitation of the AOP concept, as well as knowledge on bioavailability and ADME of chemicals in vivo and in vitro. However, equally important for the development of alternatives is the availability of high quality in vivo reference data. For the final acceptance of a scientifically valid approach, successful validation and availability of internationally accepted test guidelines is a prerequisite. It is now generally accepted that it is difficult to replace an animal test by a single experimental or nontesting approach. Therefore, ITS can provide the operational means to combine the different promising alternative methods in a powerful and predictive approach that allows significant reduction of animal testing. To achieve this goal it will be necessary to develop formalized ITSs that allow industry and regulators to provide potential hazard information with high reliability with at least the confidence of a vertebrate test or endpoint. Finally, a serious problem in the implementation of alternatives approaches in environmental risk assessments arises from the fact that discussions about alternatives are frequently based upon 'perceptions' and not exclusively based on scientific arguments. The problem is that such perceptions may interfere with an objective discussion of the pros and cons of various alternative approaches. The best (and only) option to overcome emotions and conservatism are, of course, an even more convincing outcome from validation projects and more striking scientific arguments. Moreover, continuous information and education of a broad public including academics, industry, and regulators is required to create a better understanding of the opportunities and potential applicability of new alternative methods. If a broad consensus on these issues cannot be achieved, the future for widespread adoption of alternative approaches in environmental risk assessment will remain an admirable, but distant, goal.

Chapter 5
Human Trials

5.1 The Debate on the Necessary of Human Trials on GM Foods

5.1.1 Can Animal Tests and In Vitro Cell Tests Reflect the Safety of GM Foods?

Experimental animals are useful alternatives for humans in the safety assessment of foods, chemicals, and drugs, due to their clear genetic backgrounds, good controls, and reduced expense, and for ethical reasons. And animal tests were widely accepted around the world for one century. However, the current approach of directly adopting the results of animal tests as the endpoint of risk evaluation has been recognized as not being appropriate due to the following limitations: uncertainty in the correlation between animal models and human responses, the safety margin, and the choice of multiple endpoints (Hartung 2009). Therefore, the results of animal tests cannot fully satisfy the consumers' continually increased expectations about the safety of GM foods.

Nowadays, the use of human cell tests in vitro substitutes the animal tests in vivo that are emerging. Cell tests are considered easy to be cultured, treated, and detected and to represent the response of the human body and reduce animal usage and charge. There have also been in vitro experiments on the Bt protein in human cells, with the results indicating that the Bt protein had some impact on the metabolism of cells (Mesnage et al. 2012). Considering that in vitro cultured cells were too sensitive to any slight changes in the culture conditions and cannot reflect the real situation of the complex organism, some people still insist that a human diet exposure trial is necessary to further examine the safety of GM foods.

K. Huang, *Safety Assessment of Genetically Modified Foods*,
DOI 10.1007/978-981-10-3488-6_5

5.1.2 The Debate of Human Trials for GM Foods

Human clinical trials are widely used in the development of drugs to make sure of effectiveness and to detect the side effect. Human diet intervention trials or diet exposure trials are also very commonly used in the disease therapies, investigation of the nutritious or functional effect of diet or food ingredient, and sensory study of new types of foods. While the regulatory requirements of different international organizations, such as FAO, WHO, CAC, OECD, and governments of various countries, none of them ask for the safety or nutritional assessment of GM foods on humans. There are some reasons: first of all is the ethic reason for human trials. According to the principle of safety assessment of GM foods, if the GM organism is "as safe as" traditional food, it should not have undergone more safety assessment, and even some scientists considered that the 90-day feeding study on rats is unnecessary. Some people doubt the representation of animal tests on safety assessment of GM food and ask for long-term test on human, even for several generations. While as we know, it is difficult to control a human's diet for a long time, especially mainly consuming one kind of foods or foods derived from one kind of crops. And the subjects are easily influenced by the environment, emotion, disease, and so on. It is impossible to do a long-term human diet intervention trial on one kind of GM food.

So how about doing a short-term safety assessment trial on human for each kind of GM food? The answer is still *not*, because hundreds of GM organisms are commercialized and thousands of GM organisms are developed and the genetically engineering techniques are thought not more risky than traditional breeding technology and hybridization techniques. The GM organisms have been ensured for food safety by nutritious, toxicological, and allergenic strategies, which is not necessary to spend more money and time on human tests.

Furthermore, the limitation of human trials itself restricts the application on safety assessment of GM foods. Rather than the clinical trials for drugs, which choose special subjects with the designated disease, the general definition of safety assessment test on human should choose different representatives for human, considering sexes, age groups, race, economic position, healthy situations, and special groups, such as pregnant or lactating women. It is hard to define how many types of groups and how much numbers of subjects are enough for the safety assessment tests.

For these reasons above, it is not required to do human trials on the safety of GM foods. While, does it mean that no human trials are necessary for any kinds of GM foods? In fact, some special GM organisms do need supported evidences of human trials, such as nutritionally improved GM food and pharmaceutically purposed GM organisms. For these special-purpose GM organisms, the metabolic situation and nutritious effect, curative effect, or side effect on humans should be verified on human subjects after valid safety assessment on animals or in vitro tests. While for the safety assessment, only few special human tests have been conducted such as allergenic test on human subjects.

5.2 The Principles and Requirements of Human Trials

5.2.1 What Is a Science-Based Human Trial?

In China, there are more and more concerns about the safety of GM foods. The public are expected to see more valid evidences on the safety of GM foods, especially some results from human trials. On the other hand, the nutritious experiments on children with "golden rice" (Tang et al. 2012) made some people felt scared, except for the reasons of little knowledge on golden rice and GM foods, lack of evidences of safety assessment on animals, and the illegal process. Meanwhile, some scientists working on the GM breeding work want to show the safety of their products, so they organized some foretaste activities for their products, such as Bt maize and Bt rice. It is also not science-based human trials. So what is a science-based human trial?

The definition of human clinical trials is as follows: experiments done in clinical research on human participants are designed to answer specific questions about biomedical or behavioral interventions, including new treatments (such as novel vaccines, drugs, dietary choices, dietary supplements, and medical devices) and known interventions that warrant further study and comparison. Clinical trials generate data on safety and efficacy (from Wikipedia, https://en.wikipedia.org/wiki/Clinical_trial).

There are several international consensus rules and requirements on human trials, including the *Nuremberg Code*, the *Declaration of Helsinki* (DoH) by the World Medical Association (WMA), *International Ethical Guidelines for Biomedical Research Involving Human Subjects* (also referred to as *CIOMS Guidelines*), the WHO and ICH Guidelines for *Good Clinical Practice* (GCP), *Operational Guidelines for Ethics Committees That Review Biomedical Research* by the World Health Organization (WHO), etc. Compliance with these guidelines helps to ensure that the dignity, rights, safety, and well-being of research participants are promoted and that the results of the investigations are credible (WHO 2000). Basic principles of human clinical trials require informed consent which should be reviewed by ethics committees for ethical and scientific concerns. And the appropriate protection was required to protect participants who are unable to consent (WHO 2000).

According to the *Declaration of Helsinki* (DoH), the fundamental principle of human trials is respect for the individuals. They have the right to determine and decide whether to participate in the research, both at the beginning and during the research. The welfare of subjects should always be beyond the scientific and social interests, and the laws and regulations should always be inferior to the ethics.

Except for the ethical review and inspection by the committee, the operational principles of human trials also include the scientific knowledge, evaluation of risks and benefits, the reason to the subjects studied, and proper protocols performed by trained investigators.

5.2.2 The Protocol Consideration of a Human Trial

For a science-based human trial, there should be a detailed planned protocol, which describes the scientific background, requirements for participants, plan and methods, statistical methods, and the trial's organization. The details should be provided in the protocol of the experiment and distributed to the subjects (https://en.wikipedia.org/wiki/Clinical_trial). The most important thing is that the study protocol should be approved by the ethics committee, and all of the subjects or their guardians should read, understand, and sign the informed consent form for the study.

When a human trial was conducted on GM foods, there are some special considerations such as the following:

1. The source of samples should be clear and legal. The genetic background of the GM organism should be verified. And a suitable non-GM isogenic control should be considered. Both of the GM and non-GM organisms should be cultivated in the similar conditions and be undergone a compositional analysis to make sure the comparable nutritious efficacy except for the target nutrients.
2. The choice of subjects should satisfy the objective of the study. In general, for nutritional or metabolic analysis, the health subjects should be involved. If detecting the effectiveness of the nutrient to improve the nutritional situation of humans, some subjects who lack this kind of nutrient or with malnutrition will be considered.

For example, the recruited standard in the study on golden rice in a health volunteer was as follows:

> The study protocol was approved by the Tufts Medical Center Institutional Review Board (*EC review—by author of this book*). Persons who had not taken vitamin A or b-carotene supplements within the past month (*because this experiment was associated with vitamin A—by author of this book*) and who were not disqualified based on several exclusion criteria were eligible to become study volunteers. Potential subjects were accepted into the study if they had none of the following conditions (*the detailed requirements on the health situation—by author of this book*): severe or symptomatic cardiac disease or hypertension; history of bleeding disorders; chronic history of gastric, intestinal, liver, pancreatic, or renal disease; any portion of the stomach or the intestine removed (other than an appendectomy); history of intestinal obstruction, malabsorption, or use of antacid drugs; cancer (active or use of medications for a history of cancer treatment within the past 5 year); history of chronic alcoholism; a convulsive disorder; or abnormal results in screening blood or urine samples. (Tang et al. 2009)

The above requirements are good examples for investigators who want to define the health situations of subjects.

3. The schedule of the test should be well planned, including how many meals to be provided, just once or several weeks, and when to collect the biological samples, such as blood, feces, urine, and other health parameters, according to the metabolic situation of the nutrient or food.

4. Later observation should also be considered. After the test period, if it is necessary, the follow-up observations should be conducted to make sure of the persistent effect on the health of subjects.
5. Before the study, the protocol should be reviewed and approved by registered ethics review committee, and all of the subjects or their guardians should read, understand, and sign the informed consent form of the trial.

For example, the paper of the study of "golden rice" on Chinese primary school students was retracted for ethical reasons such as the following: the trial was not reviewed by a local ethics committee in China and the children and their guardians involved in the trial were not supplied with the completed consent form (Am J Clin Nutr. 2015 Sep.;102(3):715).

5.3 The Application of Human Trials on GM Foods

Compared with widely employed human trials in disease research and drug assessment, there have been few dietary exposures or clinical experiments involving GM food in humans due to ethical issues. There are no standard methods for demonstrating the food safety and nutritious effects of a novel food by a human exposure trial. To date, several human dietary intervention tests on GM food have been performed, mainly on nutritionally improved food. The investigations on human subjects to detect the safety or efficacy of GM foods include food allergen or immunology detection, metabolic or nutritious effectiveness, and epidemiological investigation.

5.3.1 Food Allergenic Assessment

In the guideline for evaluating the food safety of GM crops, the FAO/WHO report did not involve any experiments on humans, such as skin prick testing and double-blind, placebo-controlled food challenges, which are "gold standard" in allergenic clinical diagnoses and were recommended by the original ILSI strategy (FAO 2001; Metcalfe et al. 1996). While in the FAO allergen trees, these two clinical tests were replaced by target IgE screening test and special IgE screening test for ethical considerations. Even though, if special food allergen risk was found in the GM crops, the clinical tests should be considered. The most persuasive evidence is the case of GM soybean introduced by 2S albumin from Brazil nuts (Nordlee et al. 1996) (details in Sect. 5.3.1.1). There is also another case to detect the immunogenicity of proteins introduced into GM food by human subjects, ice structuring protein (ISP) type III. Healthy volunteers consumed ice structuring protein (ISP) for 8 weeks (5 days a week). The general situation of health was recorded during the test, and the levels of specific IgG and IgE antibodies were detected to monitor the immunogenicity of the protein. During the 8-week study and the 4-week follow-up period, the health situation of the subjects was good. No ISP-specific IgG or IgE antibodies were detected in the blood of the subjects indicating little allergenic risk of ISP (Crevel et al. 2007). Such human clinical trials were not commonly used in the detection of allergenicity of GM food which also doesn't have a standard protocol. The most common method used to confirm the cross active of a new protein with known food allergens is sera screening method or IgE-binding test, which was widely used in the detection or prospection of newly introduced proteins; among them, the most famous case is Cry9C protein (Raybourne et al. 2003).

5.3.1.1 Transgenic Soybeans with 2S Albumin from Brazil Nut

The only documented case where a human allergen was introduced into a food component by genetic engineering occurred when attempts were made to improve the nutritional quality of soybeans using a Brazil nut protein, the methionine-rich 2S albumin from Brazil nut.

The nutritional quality of soybean seeds (*Glycine max*) is deficient in methionine in the protein fraction (Utsumi S 1992). For this reason, animals' diets based on soybean meal must be fortified with methionine or protein sources of methionine. To improve the nutritional quality of soybean, a transgenic soybean was developed by introduced 2S albumin from the Brazil nut (*Bertholletia excelsa*) which was rich in methionine. Allergies to nuts are one of the most common food allergies (Taylor 1992), and Brazil nut has been reported to arouse food allergy (Arshad et al. 1991). Immunoblotting and autoradiography were conducted with serum from subjects allergic to Brazil nuts to detect the IgE-binding ability of proteins in GM and non-GM soybeans, Brazil nuts, and purified 2S albumin. Skin prick test with extracts of GM and non-GM soybeans and Brazil nut was also conducted on three subjects. The results showed that proteins extracted from the GM soybean inhibited binding of IgE to Brazil nut proteins in the radioallergosorbent test with pooled serum from four subjects allergic to Brazil nuts. On immunoblotting, purified 2S albumin and proteins with similar molecular weight in the Brazil nut and the GM soybean bound to serum IgE from eight of nine subjects. Moreover, all of the three subjects had positive reactions to proteins of Brazil nut and GM soybean and negative reactions to the non-MG soybean proteins in the skin prick test (Nordlee et al. 1996). This study indicated that the 2S albumin is probably a major Brazil nut allergen. It also shows that an allergen from a food known to be allergenic can be transferred into another food by genetic engineering. Once this was discovered, further development of the new soybean variety was halted, and it was never marketed.

5.3.1.2 StarLink Corn Containing a Potential Allergenic Protein Cry9C

The US EPA has approved the StarLink corn, which was developed by Aventis Corporation, as an animal feed source. The approval of StarLink corn only for animal feed was because it contained a *Bacillus thuringiensis* (Bt) protein, Cry9c.

When compared with other Bt proteins, Cry9c protein appeared to be heat stable and maybe an allergen to human according to an earlier version of the decision tree (Bucchini and Goldman 2002). Since StarLink was planted on over 300,000 acres and it is hard to segregate corn grain after harvest, some of it ended up processed for food in corn chips and taco shells. After media reports about the StarLink corn, the US FDA received reports from 18 states and territories about health effects potentially related to consuming a corn product. Fifty-one individuals reported signs and symptoms that include gastrointestinal illness and anaphylactic shock (Bernstein et al. 2003). The Centers for Disease Control and FDA investigated and tested claims of food allergy. An ELISA test was developed to investigate the possibility of allergic reactions due to Cry9C by detecting IgE antibodies to Cry9C in blood samples from 18 people who self-reported allergic reactions (Raybourne et al. 2003). Tests were negative, and the investigators conclude that there was no scientific basis to assume that Cry9C caused allergic reactions. The StarLink corn episode was a good example for researchers to investigate the utility of the allergic decision tree as well as the ability to conduct post-market surveillance for allergic reactions of GM foods (Bernstein et al. 2003).

5.3.2 Nutritional Study on Nutritionally Improved GM Foods

According to the data of WHO, ~250 million preschool children are deficient in vitamin A. About 250,000–500,000 children become blind every year because of vitamin A deficiency, and about 50% die within 1 year after losing their sight (Micronutrient deficiencies; available at http://www.who.int/nutrition/topics/vad/en/; accessed October 30, 2015). Those children have restricted diets with deficient sources of nutrients, such as rice (National Academies of Sciences, Engineering, and Medicine 2016). Many efforts have been tried over the past 10 years to develop nutritionally improved food crops by biotechnology. To benefit the world's poor population, many kinds of biofortified staple crops such as rice and corn are developed to improve essential micronutrients (Hefferon 2015). The nutritionally improved crops including ascorbate enriched in corn and potato; β-carotene enriched in rice, potato, corn, wheat, cassava, and sorghum; folate enriched in rice and corn; γ-tocopherol enriched in corn; α-tocotrienol in rice, corn, and barley; iron enriched in rice, corn, and barley; and zinc enriched in rice and barley (De Steur et al. 2015). Among them, "golden rice" with high β-carotene content (Ye et al. 2000; Paine et al. 2005) is the most famous one. These improved micronutrients not only should be generated in foods at high levels; furthermore, they should be bioavailable to improve human's micronutrient situation (Hefferon 2015).

For this reason, several human clinical trials or diet intervention trials have been conducted to detect the bioavailability of the micronutrients in nutritionally improved GM foods. Studies on GM low phytic acid maize and the absorption of iron in healthy subjects showed that the consumption of such GM maize might improve iron absorption in humans (Mendoza et al. 1998; Mendoza et al. 2001). A study on GM soybeans with high oleic acid oil (18:1) indicated that, similar to other types of blended oil, the GM soybean oil resulted in more favorable lipoprotein profiles than partially hydrogenated soybean oils in moderately hyperlipidemic subjects (Lichtenstein et al. 2006). Moreover, a series of human dietary intervention studies on biofortified foods, such as golden rice and orange-fleshed sweet potato with β-carotene, indicated that β-carotene derived from the GM-biofortified crops helped improve the vitamin A status in healthy adults and primary school children (Tang et al. 2009, 2012; Van Jaarsveld et al. 2005). These human trials demonstrated that the nutritional bioavailability or effectiveness of nutrients improved by biotechnology in GM foods.

5.3.2.1 Bioavailability of β-Carotene from Golden Rice

In more than 50% of the world's countries, Vitamin A deficiency causes significant public health problems. It is estimated that vitamin A deficiency may bother approximately 1/3 of preschool-age children and 15% of pregnant women (Li et al. 2010). Metabolic engineering has been used to increase vitamin content in staple crops. In them, "golden rice" is most famous, which was introduced with genes from daffodil and the bacterium *Pantoea*. And then, other β-carotene-enriched staple crops followed the golden rice steps, such as corn, cassava, potato, and wheat (De Steur et al. 2015).

It is important to determine the vitamin A conversion of β-carotene from golden rice because the conversion of provitamin A carotenoids in vegetables to vitamin A is in the range of 10–27 to 1. Three women and two men were given 130–200 g golden rice expressing 0.99–1.53 mg β-carotene. The results showed the conversion factor of β-carotene from golden rice to retinol is 3.8 ± 1.7 to 1, indicating that 100 g uncooked golden rice may provide 500–800 μg vitamin A. As recommended by the US National Academy of Science, this amount of vitamin A represents 80–100% of the estimated average requirement (EAR) and 55–70% of the recommended dietary allowance (RDA) for men and women (FNB 2001). This study not only demonstrated that β-carotene derived from golden rice is effectively converted to vitamin A in humans but also indicated the probable safety of golden rice by monitoring subjects for any possible adverse effects after the consumption of golden rice, including allergic reactions or gastrointestinal disturbance. A much longer exposure study may be more helpful to make definitive assertions regarding the inherent safety of golden rice for human use (Tang et al. 2009).

The author also considered that additional study of golden rice is essential on children. In 2012, 68 Chinese children were involved in the experiments comparing the vitamin A conversion of β-carotene in golden rice and spinach with pure β-carotene. The children (aged 6–8 years) were randomly assigned to three groups to consume golden rice or spinach or pure β-carotene for 21 days. Serum samples collected from subjects on dose-day 1, 3, 7, 14, and 21 were analyzed. The results showed that children can be provided vitamin A nutrition from these three provitamin A sources. Among them, golden rice was as effective as the pure β-carotene, and both were much more effective than spinach. According to the Chinese recommended nutrient intake, 50 g uncooked golden rice can provide ~60% of vitamin A for 6–8-year-old children (Tang et al. 2012). (*This paper was retracted on 29 July 2015 for ethical reasons.)

5.3.2.2 The Human Trials on the Low-Phytic Acid Maize

In many cereals, phytic acid is played as an inhibitor of mineral absorption by binding minerals such as zinc and iron to form an insoluble complex that prevents mineral absorption. For this reason, iron absorption in the human body was very low for many plant foods have high phytate contents (Mendoza et al. 2001).

As early as 1994, a *lpa-1-1* mutant maize with low-phytic acid was developed (V Raboy, K Young, P Gerbasi, unpublished observations 1994), and much of the phytic acid in maize can be removed without affecting the total phosphorus content of the grain or the health of the plant. Fourteen non-anemic men ingested tortillas made by low-phytic acid GM maize (the *lpa-1-1 mutant*) and wild-type strain non-GM maize (nonmutant "flint" maize) for 2 weeks to detect the iron absorption. The results of the red blood cells showed that iron absorption was 49% greater from low-phytic acid GM maize (8.2% of intake) than from wild-type strain (5.5% of intake). It indicated that low-phytic acid maize may be helpful on iron absorption in populations with maize-based diets ((Mendoza et al. 1998).

Then 14 non-anemic women were detected with iron absorption by consuming porridges prepared from the GM (*lpa-1-1 mutant*) and non-GM maize for 2 days. The results showed that when porridge was fortified with either sodium iron EDTA or ferrous sulfate, the phytate contents had no significant effect on iron absorption. Thus, when maize porridge was fortified with additional iron, the previously identified benefit of low-phytic acid GM maize was no longer detectable (Mendoza et al. 2001).

5.3.3 Horizontal Gene Transfer Study on GM Food

Horizontal gene transfer is defined as "the stable transfer of genetic material from one organism to another without reproduction or human intervention" (Keese 2008). Some people were concerned that GM plants with foreign DNA might transfer the foreign genes to the human intestinal microflora and human somatic cells, especially those antibiotic-resistant marker genes (National Academies of Sciences, Engineering, and Medicine 2016).

Gene transfer from GM plants to the intestinal microflora requires survival of the plant DNA within the gastrointestinal tract. There were several studies on the degradation of plant DNA in rumen, ovine, chickens, mice, etc., but few studies in the human gastrointestinal tract, except for a study on GM soybeans with the *epsps* gene that assessed the survival of the transgene in the human gut tract (Netherwood 2004).

5.3.3.1 The Digestion of GM Soya DNA in the Human Gastrointestinal Tract

If a gene could transfer from GM plants to resident intestinal microbiota or host cells, the DNA would need to survive through the digestive system of the gastrointestinal tract. In a previous study, the degradation of foreign genes of GM soybean and maize was evaluated in simulated gastric and intestinal system. The results showed that 80% of the transgene in DNA extracted from GM soybean was degraded in the simulated gastric fluid, while the transgenes contained in GM soybean and maize didn't degrade (Martín-Orúe 2002). In the simulated intestinal fluid, the foreign genes in the DNA extracted from soybean were degraded at a similar rate to those in the soya protein. In contrast, 85% of the maize nucleic acid was rapidly degraded in the simulated intestinal fluid, while the rest of the DNA was degraded at a similar rate to that in the soya material. These results suggested that some transgenes may go through the small intestine in the material of GM foods. This experiment was a simulated situation, which could not reflect the real situation of transgene in diet in the human digestive system.

To detect the persistence of DNA from dietary GM plants in the human digestive system, the survival of the transgene *epsps* from GM soya in the small intestine of human ileostomists, where the terminal ileum was resected and digesta were diverted from the body via a stoma to a colostomy bag, was compared with healthy volunteers. All of the subjects fasted overnight, and DNA was extracted from their stoma effluent before the start of the trial. The test meal consisted of GM soya burgers (containing 150 g texturized soya protein) and a GM soya milk shake (containing 100 g soya protein). Each subject was fed 190 g of the GM soya burger and 264 g of the GM soya milk shake. The contents were collected every 30 min for 6 h from the ileostomists' stoma bags, and the healthy volunteers' stools were collected for over 72 h.

The results showed that the amount of transgene that survived passage through the small bowel varied among individuals, with a maximum of 3.7% recovered at the stoma of one ileostomy, while the transgene could not survive from the intact gastrointestinal tract of human subjects consuming GM soya, which was completely degraded in the large intestine. Even the author found that before involving in this experiment, low-frequency gene transferred from GM soya to the intestinal microflora in three subjects, while no complete foreign gene had been transferred to the microflora. So, the gene transfer from GM foods seen in this study is unlikely to alter gastrointestinal function or cause a risk to human health (Netherwood 2004).

5.3.4 *Epidemiological Investigation and Post-marketing Surveillance of GM Foods*

Considering the food allergen risk and other food safety issues associated with GM foods, there were some suggestions on the post-market surveillance of GM foods, such as a previous report *Safety of Genetically Engineered Foods* pointing out that "there were important limitations in allergenicity predictions that could be done before commercialization" (NRC 2004), which also call for a two-step process of pre-commercialization testing and post-marketing surveillance. In the case of StarLink corn, existing post-market surveillance resources were used by the US FDA and the CDC to follow up people who self-reported as experiencing allergic reactions (CDC 2001). Recently, in EU, Commission Implementing Regulation (EU) No 503/2013 on applications for authorization of genetically modified (GM) food and feed legalized the requirement for a post-market monitoring (PMM) plan if the outcome of the risk assessment has identified a need to do so. Three specific examples are given in this regulation: (1) when the GM food or feed has altered nutritional composition, (2) when the nutritional value of the GM food or feed differs from the conventional ones, and (3) when there is a likelihood of increased allergenicity due to the genetic modification. Such testing recognized would be logistically challenging, as described in a scientific report to EFSA for the following reasons: under the legislative requirements, PMM should confirm that specific recommendations of use are followed by the consumer/animal owner, qualify the predicted consumption of GM food and feed, and further characterize any intended or unintended effects highlighted during the pre-market risk assessment. The "unrecognizable" property of GM traits in processing and the absence of proper consumption data were recognized as the two largest barriers to conducting PMM of GM food and feed in the EU (ADAS 2015).

For these reasons, it is difficult to conduct a post-marketing surveillance for the safety of GM food. There was a controversial sample of investigation on epidemiology on the persistence of Bt protein and herbicides in nonpregnant women, pregnant women, and their fetuses, which can give some inspiration for the post-marketing surveillance of GM foods (Aris and Leblanc 2011). After this study, the researchers in the USA and Germany have done similar investigations for the persistence of glyphosate in human milk and/or urine (McGuire et al. 2016; Steinborn et al. 2016).

5.3.4.1 Maternal and Fetal Exposure to Pesticides Associated to Genetically Modified Foods (Aris and Leblanc 2011)

Pesticide-associated GM foods allow the protection of desirable crops by introducing herbicide-resistant genes such as *CP4 epsps* or *bar/pat* to tolerate herbicide glyphosate and glufosinate or by providing insect-resistant gene such as the *Bacillus thuringiensis* (Bt) genes. Few studies have examined the kinetics of absorption, distribution, metabolism, and elimination (ADME) of glyphosate in humans.

Glyphosate [N-(phosphonomethyl)glycine] is a nonselective herbicide used for the control of a wide range of weeds and was widely used in the herbicide-tolerant GM crops (Hori et al. 2003). Glyphosate concentrations in urine among people living in Iowa of the USA have been detected (Curwin et al. 2007). The results showed that, between farm and nonfarm mothers, there were no significant difference and no positive association between the mothers' urinary glyphosate levels and glyphosate dust concentrations. It was suggested that other sources of exposure such as diet may be involved.

The levels of glyphosate and its metabolite aminomethylphosphonic acid (AMPA), glufosinate and its metabolite 3-methylphosphinicopropionic acid (3-MPPA), and Cry1Ab protein were detected in blood samples of 30 pregnant women and 39 nonpregnant women from Eastern Townships of Quebec, Canada. The results showed that glyphosate was not detected in the maternal and fetal blood but present in the blood of some nonpregnant women (5%), whereas its metabolite AMPA was not detected in all analyzed samples. Glufosinate was detected in 18% of nonpregnant women's blood and not detected in the maternal and fetal blood. On the other hand, 3-MPPA (the metabolite of glufosinate) was detected in 100% of maternal and umbilical cord blood samples and in 67% of the nonpregnant women's blood samples. This highlights that this metabolite is more detectable than its precursor and seems to easily cross the placenta to reach the fetus. Cry1Ab toxin was detected in 93 and 80% of maternal and fetal blood samples, respectively, and in 69% of tested blood samples from nonpregnant women.

Even with the controversy on the sensitivity and the selectivity of detection methods used in this study and the unknown diet exposure to GM foods of the subjects, this is the first study to investigate the presence of pesticide-associated genetically modified foods in maternal, fetal, and nonpregnant women's blood. It called for more attention on the post-marketing surveillance of GM foods.

5.3.4.2 Detection of Glyphosate and Aminomethylphosphonic Acid (AMPA) in Human Milk and Urine

Glyphosate (N-(phosphonomethyl)glycine) is among the most frequently used active ingredients of plant protection products worldwide. The rapidly increasing amount of glyphosate is considered resulting from the widely acceptance of herbicide-resistant GM crops, such as soybean, corn, and cotton. Published data indicated that glyphosate is quite common to be detected in human urine because of different exposures or intake pathways (Niemann et al. 2015; Krüger et al. 2014).

In America, it was indicated that glyphosate in three of ten breast milk samples was detected ranging from 76 to 166 ng/mL in 2014 (Honeycutt et al. 2014). In 2015, it was reported that 16 breast milk samples from Germany were analyzed with glyphosate levels between 0.2 and 0.4 ng/mL (Steinborn et al. 2016). Because of the high concern of the potential health risks for breastfed infants, the researchers in the USA and Germany performed detection of persistence of glyphosate in breast milk in greater scope, respectively.

In America, glyphosate and AMPA were detected using highly sensitive liquid chromatography-tandem mass spectrometry methods on breast milk and urine samples from healthy lactating women living in two states. It was not surprising that the urine samples were found with glyphosate and AMPA according to previous studies. Subjects consuming organic foods were also found with urine glyphosate and AMPA, with concentrations similar to those consuming conventionally grown foods. Furthermore, there were no differences between women living on or near a farm and those living in an urban or nonfarm area. No detectable glyphosate and AMPA were found in any breast milk samples for both analytes (McGuire et al. 2016).

In Germany, two independent analytical methods for the determination of glyphosate in breast milk were newly developed and are able to quantify glyphosate residues at concentrations as low as 1 ng/mL. The methods were applied to quantify glyphosate levels in 114 breast milk samples collected from Lower Saxony and Bavaria of Germany. None of the investigated samples were detected with glyphosate residues above the limit of detection (Steinborn et al. 2016). "The result shows how important professionally conducted scientific studies are to ensure that consumers are not unnecessarily confused in the emotional debate on pesticide residues", says the Federal Institute for Risk Assessment (BfR) President, Professor Dr. Andreas Hensel.

Both of the two independent investigations demonstrated that dietary glyphosate exposure is not a health concern for breastfed infants. It was recommended that mothers should not be confused by unfounded claims and should continue to breastfeed as before.

5.4 Conclusion

Even though there is still a great controversy on human clinical trials or diet intervention trials on the safety assessment of GM foods, the existing trials on food allergenic assessment studies, nutritional effectiveness studies, and horizontal gene transfer studies show that the human trials can provide powerful evidences on the safety or nutritional effect of GM foods, which can further improve the confidence of customers on GM foods. Moreover, more and more emerging kinds of GM products are developed, such as pharmaceutical GM plants, which need more safety and/or efficacy detection on human subjects. Meanwhile, along with the development of modern biotechnology, the genomics, proteomics, metabonomics, and intestinal microbiota methods are widely used in safety and nutritional studies on diets, drugs, etc. It is time to consider to employ these new technologies to detect the safety and/or nutritional effect of emerging GM foods.

For the post-marketing surveillance of GM foods, it is recognized that it's difficult to implement such a progress on a special type of GM food for the complicated food matrix of a society or country. The investigations on persistence of glyphosate in the human blood, milk, or urine are more likely to be a pesticide post-marketing surveillance rather than studies on GM foods, since the glyphosate is also widely used in non-GM crops, grass, and forestry. There is still a need for more careful consideration on the possible way of post-marketing surveillance of special GM foods, such as nutritionally improved or pharmaceutical GM foods.

Part II

Chapter 6
New Technology Used in GMO Safety Assessment

6.1 The Definition of Unintended Effects

Genetically modified (GM) food is designed with a specific demand in our life. The normal plants or animals are being transferred with foreign DNA and get the specific changes within it. For instance, after the specific gene transfer, the plants probably become herbicide resistant and pest resistant, contain more amount of certain substance, or express some new substances. These are the intended (target) effects. FAO/WHO defined the unintended effect: "In achieving the objective of conferring a specific targeted trait (intended effect) to the host organism by the insertion of defined deoxyribonucleic acid (DNA) sequences, additional traits could, theoretically, be acquired or existing traits lost" (Mally et al. 2004). European Food Safety Authority (EFSA) defined the unintended effect: "Unintended effects are considered to be consistent differences between the GM plant and its appropriate control lines, which go beyond the primary expected effect(s) of introducing the target gene(s)" (Gross-Steinmeyer et al. 2002).When the foreign genes are integrated into genome of receptor plants, unintended effects caused by random insertion might disrupt endogenous gene expression and alter levels of micronutrients and anti-nutritional factors or even produce new proteins which we do not expect. These changes are called unintended effects. Unintended effects are those experimental results that appear in addition to or, in some cases, instead of the intended results. Sometimes the inserted gene does express what we want; however, it affects other characteristics as well. It is suggested that the interaction of "foreign" DNA sequences with the existing genes of the plants may disrupt, modify, or silence active genes or activate silent genes leading to disruption of metabolic and biochemical pathways in unpredictable ways, produce new toxic compounds, or improve the content of the already existing ones (Janczyk et al. 2007). Unintended effects are especially vital to potential and long food safety issues in the process of safety assessment of GM crops and their products. Unintended effects of GM food are more unpredictable than toxic, allergenic, and anti-nutritional effects, leading to

K. Huang, *Safety Assessment of Genetically Modified Foods*,
DOI 10.1007/978-981-10-3488-6_6

the hard control status. One of the main reasons that genetically modified food cannot be accepted by consumers is that unintended effects probably lead to some adverse influences to health. The Food and Agriculture Organization (FAO) of the United Nations, the World Health Organization (WHO) of the United Nations, the Codex Alimentarius Commission (CAC), the Organization for Economic Co-operation and Development (OECD), the European Community (EC), and several other international agencies emphasized the importance of unintended effects in GM food safety assessment, because it is possible to bring food safety issues in the long term. In many countries, safety assessment of unintended effects is an essential part under genetically modified organism (GMO) administration regulations, safety assessment principles, and technological schemes. The Ministry of Agriculture and Ministry of Health Regulations in China consider that unintended effect in safety assessment of GMOs is of great value (Deng et al. 2008).

Unintended effects include both predictable and unpredictable effects. Predictable unintended effects indicate those effects which are not target but can be predicted in advance. For example, we can predict some effects through the knowledge of gene interaction and metabolism pathways. The metabolites that probably are influenced and the allergenicity of the proteins can be predicted as well. Metabolomics is often used for analyzing some of these effects. Unpredictable unintended effects are difficult to predict based on current knowledge. In transgenic systems, well-known unintended effects are mainly two types, position effects and pleiotropic effects. Position effects are caused by insertion of a transgene at a particular locus in the genome, and some interference occurs. Due to the different site of integration, these effects are various, therefore acquiring unique plant line. Position effects can be easily eliminated by screening for plant lines that have no or little position effects. In contrast, pleiotropic effects are independent of the site of transgene insertion and are the sum of all the phenotypic effects caused by expression of the transgene. Some of traits are possibly the intended ones, while others may occur through certain interactions of the gene with plant processes and constitute the unintended pleiotropic effects. These effects are more difficult to eliminate, so they can more easily create safety issues.

We know the chemical composition, nutrients, anti-nutritional factors, toxins, and allergens about traditional food, so we can manage our intake to control the limit of those adverse effects on our body. We can balance the nutritional part and the harmful part according to long experience so that we would not get hurt from these foods. The unintended effects of genetically modified organisms are random events, so it is difficult for us to prevent them. It will be harmful to our health in case of new toxins or allergenic proteins produced in genetically modified organisms (Cellini et al. 2004).

6.2 Why Unintended Effects Happen?

In spite of the fast development of the transgenic technology, genetic engineering has some problems. Exogenous genes were randomly inserted to the carrier; therefore, it prefers the location that genes are enriched. Unintended effects may result from disruption of a functional gene at the point of insertion, rearrangements of the gene cassette, or expression change of neighboring genes (Cheng et al. 2008). Overexpression of transcription factors, alterations of biosynthetic pathways, and expression of transgenes to increase tolerance to biotic or abiotic challenges all carry the potential to give rise to unexpected interactions, including increasing or decreasing the availability and activity of other components. Some unintended changes might also be caused by pleiotropy or result from the alterations of intended traits, leading to the composition, metabolism, or physiology changes of the plant. In the process of recombination event, all can happen including gene fracture, change of gene sequence, unexpected new protein being produced, and other situations. Does unintended effect happen to only genetically modified organisms? No. Hybridization, a traditional way of breeding, can also bring the unintended effect. It takes a lot of time to eliminate the adverse unexpected effect in the process of conventional breeding. So the unexpected effects of genetically modified organisms are not a new event. We have faced to unintended effects for a period of time.

6.3 Characters of Unintended Effects of Genetically Modified Organisms

Up to now, we do not have enough knowledge about the appearance of the unintended effects of GMOs. How to assess the unintended effect of GMOs? We discuss from the following aspects.

6.3.1 Change of Targeted Phenotypic Characters

We expect that GMOs acquire some features such as tolerance to herbicides, heavy metal, insects, dry land, and other factors. However, sometimes they do not show the features what we want. Sometimes, the features will gradually disappear. GMOs which have certain tolerance can more easily grow in adverse environment, but are probable to accumulate more pollutant compared with traditional food. Some GMOs are designed to reduce harmful components or increase the quality, quantity, and function of nutritional ingredients, while it does not reach the amount we designed or does not reduce harmful components or increase nutritional components anymore as passages increase.

6.3.2 Change of Targeted Hereditary Characters

The expression amount of target genes, the amino acid sequence of new proteins, and the biological activity of the new product sometimes are deviated with the designed target. The expression activity of the targeted gene possibly decreases; the amount of targeted gene expression possibly decreases after a period of time; or the function or regulatory pathway of the targeted product probably has some difference. Also, it is possible that insertion, deletion, or replacement of amino acids in targeted expression products can happen. Apart from change of nucleotide sequence, posttranscriptional modification of the targeted gene possibly has an impact on its protein structure. And the insertion of exogenous gene could influence the expression of the endogenous genes as well due to the change of structure and function, further changing the endogenous expression product or metabolism.

6.3.3 Variations of Nontargeted Characters

When the exogenous gene was inserted to the recipient genome, it is possible to change the transcription, translation, and functions of endogenous genes in the recipient. These effects possibly result in the change of the hereditary and phenotypic characters which are not targeted. Due to the large amount of genes and the unknown integration site of exogenous genes in the recipient plant genome, it is difficult to predict the nontargeted hereditary and phenotypic characters. Because the insertion site cannot be controlled, the exogenous gene can interrupt the expression of the endogenous gene. That means the endogenous genes cannot express in a normal way. If nontargeted phenotypic and hereditary characters have some variation, for example, some novel allergens and biotoxins, pesticides, heavy metals, or anti-nutritional factors are produced, and some nutrients are lost, these factors will damage animal's and human's health. In some other circumstances, some silent genes will be activated because of the insertion of the exogenous gene. When the expression of the endogenous genes changes, their regulated pathways could also be changed. Their metabolism level and the metabolites would change as well. Most times we can only know the change until it happens.

6.3.4 Gene Horizontal Transfer

DNA horizontal transfer can happen among different species and among different cells in the same species. Exogenous DNA is able to embed into the recipient genome by DNA recombination, including homologous recombination, transposition, site-specific recombination, and DNA repair. What people worry about gene horizontal transfer is mainly that after the GMOs are released to the environment, the exogenous genes of genetically modified organisms that express the products related with antibiotics may transfer to the environment or to the animals' gastrointestinal tract, so that the recipient will have antibiotic resistance (Deng et al. 2008).

Antibiotic resistance genes do not necessarily cause bacterial drug resistance. Novel drug resistance is usually dependent on the effective expression and integrity of function of transferred genes. Therefore, in the safety assessment of exogenous gene horizontal transfer, expression activity of transferred genes in bacteria should be finally taken into account.

6.4 Assessment Criteria of Unintended Effects of Genetically Modified Organisms

The aim of the safety assessment of genetically modified food is to demonstrate that the novel food is as safe as its traditional counterpart. And it does not bring any new risks to consumers' health. The occurrence of unintended effects is a dynamic way and might be recognized or obscure. Even in the same GMO, different unintended effects could occur in different periods during their growth. Also, different generations are possible to change along with growth conditions. Sometimes, different unintended effects can be detected even at the same period. Therefore, all unintended effects are hardly identified in GMOs by an experiment. It is urgent to build a system to assess unintended effects of GMOs. Based on their extent of risk to microorganisms, plants, animals, humans, and the environmental ecology, GMO unintended effects are grouped into four levels in China:

Level one: It has been proved that neither phenotypic nor hereditary feature variation occurs and the horizontal transfer of antibiotic resistance gene does not occur.

Level two: It has been proved that either phenotypic or hereditary characteristic variations occur, but the variations have no adverse effects on edible safety, and the horizontal transfer of antibiotic resistance gene does not occur.

Level three: It has been proved that both phenotypic and hereditary characteristic variations occur, but the variations have no adverse effects on edible safety, and the horizontal transfer of antibiotic resistance gene does not occur.

Level four: It has been proved that one of the following two statuses occurs—one is that some phenotypic and/or hereditary characteristic variations occur and the variations have a harmful impact on edible safety and the other is that the horizontal transfer of antibiotic resistance gene occurs (Deng et al. 2008).

We need to focus on which level the unintended effects arrive at when food safety of genetically modified organisms is assessed.

6.5 Detection Approaches for Unintended Effects of GMOs

It has some challenges to assess unintended effects due to two parts. On one hand, it is difficult to determine unintended effects on technical level. On the other hand, it is hard to determine the possibility of the unintended effects on consumers' health. It is a forefront topic on the research of unintended effects for GM food safety in the world. The related techniques include modern instruments analysis techniques and modern molecular biology techniques. Traditional methods for detecting the unintended effects of GM plants include environmental adaptability evaluation, analysis of the chemical composition, and agronomic character comparisons between the transgenic line and its non-transgenic control. These methods have been used to investigate unintended effects in a large number of transgenic plant systems and evaluate their potential safety issues. This is also called targeted detection. In recent years, omics methods are applied to detect unintended effects, which are called nontargeted detection.

6.5.1 Targeted Methods

At present, targeted approaches are mainly used to detect nutrient composition, toxins, and anti-nutritional factors. As a part of "substantial equivalence," it has been accepted by international communities and applied to safety assessment of GMOs. GMOs are usually compared with their parental lines to see whether the composition of the nutrients is significantly different. If the chemical composition of new crops is substantially similar to that of traditional crops, it is not considered to pose a health risk to humans and animals. The key of targeted detection is to choose the compounds that you want. They include proximate analysis (e.g., protein, fat, moisture, ash) as well as the analysis of key compounds, such as naturally occurring toxicants and essential nutrients (e.g., minerals, vitamins) and anti-nutrients (e.g., phytic acid, trypsin inhibitor), which may have a substantial impact on the overall diet. When some nutrient compositions are predicted to be changed, they can be detected as well. For example, GMOs with herbicide resistance usually change the level of alanine, tyrosine, and tryptophan. Tang et al. compared the nutritional composition between the transgenic lysine-rich maize lines and conventional maize. They found an increase of total protein and lysine and decrease in phytate phosphorus, resulting in an increase of nutritional value (Tang et al. 2013). This research was published in the *Journal of the Science of Food and Agriculture* in 2013: *Nutritional assessment of transgenic lysine-rich maize compared with conventional quality protein maize*.

Maize is the main food for humans and livestock in China. However, lysine and tryptophan are not enough, and human beings and monogastric animals cannot synthesize these two amino acids. There have been many attempts to improve lysine content of maize seeds using plant biotechnology in these years (DiBaise et al. 2008; Hammer et al. 2001; Waters and Bassler 2005; Zhang et al. 2014a). When assessing the safety of genetically modified (GM) crops, the concept of substantial equivalence is the basic role (FAO/WHO 1996; OECD 1993b). But for the nutritionally improved genetically modified food (GMF), the nutritional profiles could be affected. In this study, the transgenic lysine-rich maize lines (an inbred line 642 and a hybrid line Y642) were compared with a conventional quality protein maize Nongda 108, which has been used safely for more than 20 years in China. Then the composition data from GM maize were compared with the data set from ILSI crops composition database (Kuiper-Goodman and Scott 1989) to provide a comprehensive overview of the results.

The maize used in this study was acquired from the experimental field of Beijing farm. The parents of hybrid Y642, line 642, and conventional maize Nongda 108 were planted adjacent to the transgenic field under the same climate conditions. The following composition was analyzed, including fat, protein, ash, fiber, moisture, vitamins (vitamin B1, vitamin B2, and vitamin E), fatty acids, total carotene, antinutrient (phytate phosphorus), and selected minerals (potassium, sodium, magnesium, phosphorus, calcium, copper, zinc, manganese, and iron) (AOAC 2000). Different solubility fractions of maize protein (globulin, albumin, zein, G1-glutelin, G2-glutelin, and G3-glutelin) of 642, Y642, and Nongda 108 were extracted (Alexander et al. 2006).

The proximates of transgenic inbred maize (642) and hybrid line (Y642) and the conventional QPM (Nongda 108) indicated that the moisture and fiber levels of the two transgenic varieties were not significantly different from Nongda 108. The protein, lipid, ash, and carbohydrate contents of 642 were significantly different from those of Nongda 108, but for the hybrid transgenic line Y642, only protein content was significantly higher than the QPM Nongda 108. This result, in agreement with prior research ((JECFA) 1991), indicated that the protein content of maize with introduction of sb401 gene was higher than conventional QPM. Meanwhile, the statistic changes of transgenic maize didn't beyond the range of ILSI composition database apart from the carbohydrate content of the three varieties all lower than the minimum data.

The amino acid analysis indicated that, compared to Nongda 108, the 17 tested amino acids of 642 and Y642 all showed higher levels, resulting in higher content of total essential and nonessential amino acids, which suggested a more nutritive value of the transgenic maize. There were nine kinds of amino acids of 642 maize and eight kinds of Y642 maize significantly different from those of Nongda 108. Most of the contents of amino acid of transgenic maize lines fell within the range of the database; nevertheless, three of them (threonine, glycine, and phenylalanine) were higher than the reported range. According to the proposed amino acid scoring pattern of WHO technical report 935 ((JECFA) 1995), the first three limiting amino acids of all of the maize varieties are lysine, leucine, and valine. The amino acid score of Nongda 108 was 0.61. The amino acid scores of high-lysine transgenic maize 642 and Y642 were improved to 0.76 and 0.70, respectively, which were notably improved compared with Nongda 108.

The results of minerals showed that compared to Nongda 108, the content of iron, zinc, and manganese of inbred line 642 and zinc of hybrid line Y642 was significantly higher. All of the data of transgenic maize were covered by the range of the ILSI database, except for the lower content of phytate phosphorus which may result from different measurement methods.

The vitamin B1, vitamin B2, vitamin E, and total carotene content of 642 were all higher than Nongda 108; especially, the content of vitamin B2 and total carotene of 642 was significantly increased. The vitamin B2 of Y642 was higher than the conventional QPM, but other nutritional compositions were comparable to Nongda 108.

Maize is one of the protein sources for human beings and livestock. Grain protein can be separated into six fractions according to their solubility (Alexander et al. 2006), namely, albumin, globulin, zein, G1-glutelin (or zein-like protein), G2-glutelin (glutelin-like protein), and G3-glutelin (true glutelin). In comparison with the conventional QPM, the water-soluble protein of transgenic maize lines was significantly increased by 60%. But as one of the easily digested protein fractions, the salt-soluble protein of transgenic maize was just about 20% of the content of conventional QPM. As a result, the sum content of the albumin and globulin of transgenic maize was comparable to that of the conventional one. The content of the other kind of nutritional available protein G2-glutelin (glutelin-like protein) of Y642 was about 12-fold of that of Nongda 108, which was statistically significantly higher. The G3-glutelin (true glutelin) of the transgenic lines was slightly higher than that of the conventional line. The sum of the nutritional available glutelin of transgenic maize was 20.8%, which was higher than that of the conventional one. The alcohol-soluble proteins including zein and G1-glutelin of the two cultivars were comparable. The higher lysine content appeared to be a result of a generally higher content of total protein or free amino acids.

The transgenic lysine-rich maize showed a good balance of amino acids with improved total content of essential amino acids and will have a protein-sparing effect (Benford et al. 2001). And due to the higher protein level, every kind of amino acid of the transgenic maize was higher than that of conventional QPM Nongda 108. It was demonstrated that the insertion of lysine-rich gene did not affect the amino acid compositional balance and the transgenic lysine-rich maize lines were proved to be superior with higher content of lysine and amino acid to the conventional QPM widely cultivated in China. The analysis of mineral elements demonstrated that transgenic lysine-rich maize was higher in phosphorus than Nongda 108 but lower in phytate phosphorus, which might indicate more usable phosphorus in transgenic lysine-rich maize. This could promote the utilization of phosphorus, iron, and zinc (Bendele et al. 1985; Pfohl-Leszkowicz and Manderville 2007). These results are meaningful for the nutritional improvement of maize.

Wang et al. compared the nutritional composition of GM rice and non-GM rice from two places and found nutritional quality of transgenic rice seed was substantially equivalent to that of non-transgenic counterparts and that no detrimental changes related to potential unintended effects of the transgene were observed (Wang et al. 2012). This research was published in the *Journal of Cereal Science* in 2012: *Comparative analysis of the proteomic and nutritional composition of transgenic rice seeds with Cry1ab/ac genes and their non-transgenic counterparts.*

Rice (*Oryza sativa* L.) is the main crop and staple food that serves nearly half of the world's population (Xin et al. 2008). Genetically modified crops are becoming an increasingly important feature of the agricultural landscape. *Bacillus thuringiensis* (*Bt*) is an already useful alternative or supplement to synthetic chemical pesticide applications in commercial agriculture, forest management, and mosquito control (Kuiper et al. 2001). The Cry protein encoded by the *Cry1ab/ac* gene, a fusion of *cry1ab* (GenBank accession no. X54939) and *cry1ac* (GenBank accession no. Y09787) into a single gene, is toxic to three important lepidopteran pests of rice, *Chilo suppressalis*, *Scirpophaga incertulas*, and *Cnaphalocrocis medinalis* (Xu et al. 2009). In order to solve the questions about whether exogenous DNA sequences could induce intended or unintended effects, it is necessary to assess the safety of genetically modified crops to address both effects (Huff 1991; Maaroufi et al. 1995; Zaied et al. 2011). According to the concept of substantial equivalence for the risk assessment of transgenic foods, nutrients, including proximates, amino acids, fatty acids, minerals, and vitamins, are measured. The anti-nutritive components, such as phytic acid, lectin, and trypsin inhibitors, are also examined. The protein, fiber, ash, carbohydrate, and moisture contents were similar among these four samples. Although the lipid content in rice harvested at Xuancheng (WT01 and T01) was slightly lower than that at Jingzhou (WT02 and T02), there were no significant differences between the transgenic and non-transgenic counterparts. It showed no statistically significant differences in the levels of amino acid, fatty acid, micronutrients, and anti-nutrients, such as phytic acid, lectin, and trypsin inhibitors. The oleic acid (C18: 1) and gamma linoleic (C18: 3) contents of the WT01 and T01 samples were slightly lower than in WT02 and T02.

The nutrients and anti-nutritive components examined in this study demonstrated that the nutritional quality of transgenic rice was substantially equivalent to that of the non-transgenic counterparts. The levels of nutrients and anti-nutritive components were comparable to the non-transgenic counterparts. Since all significant difference mean values for transgenic rice were within rice reference ranges (OECD), these differences were not considered biologically relevant.

Qi et al. compared the stacked GM soybean with its parental line and found that the concentrations of proximates and minerals were approximately the same, except for higher oleic acid and lower linoleic acid in GM soybean (Qi et al. 2012). This research was published in the journal *Food and Chemical Toxicology* in 2012: *Subchronic feeding study of stacked trait genetically-modified soybean (3Ø5423 × 40–3-2) in Sprague-Dawley rats.*

Oil obtained from conventional soybeans contains high concentrations of polyunsaturated fatty acids (PUFAs; linoleic acid [18:2] and linolenic acid [18:3]) that have known health benefits (Udell et al. 2005). Unfortunately, the presence of these particular fatty acids makes conventional soybean oil susceptible to oxidation and subsequent rancidity thereby limiting the utility for food applications (Liu et al. 2007). Accordingly, soybean oil with a high proportion of monounsaturated fatty acids was developed because of more stability. Dietary consumption of trans-fatty acids has some relationship with cardiovascular diseases, cancer, and other types of diseases (Dhaka et al. 2011; Tan 2011).

DP-3Ø5423 soybeans (TREUS™) are a genetically modified soybean specifically engineered to possess higher concentrations of monounsaturated oleic acid than present in conventional soybean oil and contain no trans-fatty acids. The soybean has an oil profile with a significantly higher proportion of oleic acid (70.6% of total fatty acids) and lower proportion of linoleic acid (5.50% of total fatty acids) relative to traditional soybean oil (approximately 20% oleic acid and 55% linoleic acid, respectively). Oleic acid is considered one of the healthier sources of fat in the diet. Linoleic acid is a carboxylic acid with an 18-carbon chain and two cis double bonds. Accordingly, the oil obtained from DP-3Ø5423 soybeans has more stable oxidation than standard soybean oil without the requirement for hydrogenation. Besides, a gene (gm-hra) encoding a modified version of the soybean acetolactate synthase (als gene) that confers tolerance to a sulfonylurea herbicide was used as a selectable marker for the transformation.

In this case, the high oleic acid soybean DP-3Ø5423 was crossbred with Roundup Ready® (RR) soybean 40-3-2 encoding a glyphosate-resistant form of the plant enzyme 5-enolpyruvylshikimate-3-phosphate synthase (EPSPS) (CERA 2009). The stacked GM soybean DP-3Ø5423 × 40-3-2 therefore contains a high content of oleic acid and is tolerant to the field application of herbicides containing glyphosate. For the purposes of this paper, these will be referred to as HOA-HT (high content of oleic acid-herbicide tolerance) soybeans. As a new food source, a subchronic feeding study in rats as part of the comprehensive safety assessment for foods was conducted.

Non-transgenic soybean (JACK) and 3Ø5423 × 40-3-2 (HOA-HT) have the same genetic background. Analysis of nutritional proximates (moisture, ash, fat, crude fiber, crude protein, fatty acid, minerals) and the concentrations of individual fatty acids were determined.

The concentrations of proximates and minerals (Ca and P) of the HOA-HT and JACK soybeans were approximately the same, indicating that they are substantial equivalent. As expected, the fatty acid composition of HOA-HT soybeans contained a higher proportion of oleic acid (18:1, 76.56% of total fatty acids) and lower proportion of linoleic acid (18:2, 2.84% of total fatty acids) compared with the JACK soybeans (18:1, 21.65%; 18:2, 52.75%).

Apart from composition, unintended effect on agronomic trait is another part to detect. For instance, wheat with aluminum resistance overexpresses phosphatidylserine synthase which can lead to the development of necrotic lesions in wheat (Delhaize et al. 1999).

Targeted detection focuses on those known compounds. It plays an important role in nutrient analysis for genetically modified organisms although it is one-sided.

6.5.2 Nontargeted Approaches

Compared with targeted approaches, untargeted approaches are usually used to analyze a large amount of unknown compounds. Current new methodologies "omics" were developed to detect unintended effects, such as genomics, transcriptomics, proteomics, and metabolomics.

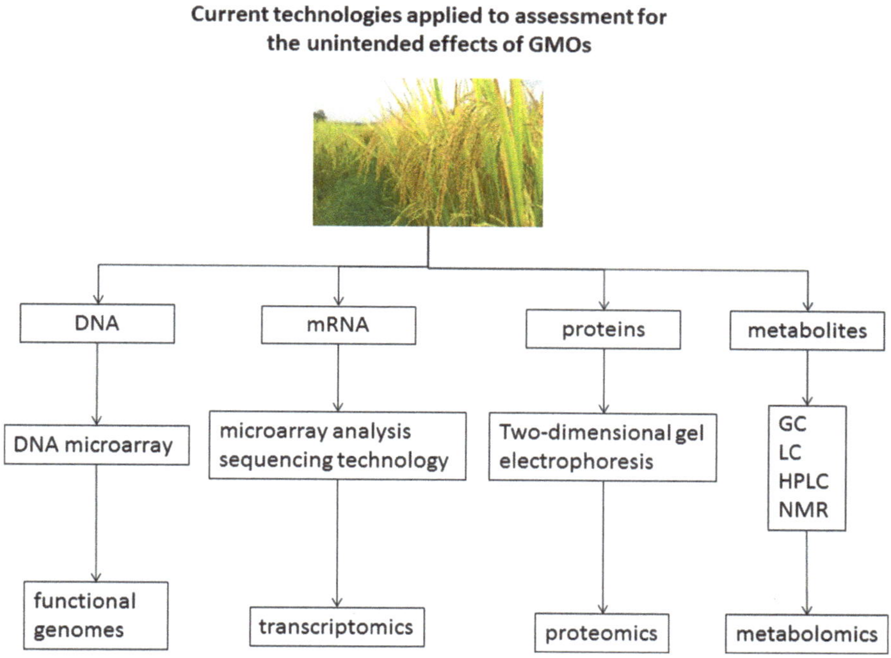

These results are really exciting because they also give profound insights into the extent of variation within conventional cultivars. They have been proved to be promising methods to investigate unintended effects of GM plants and reduced the uncertainties in the safety assessment. The use of nontargeted global profiling technologies, such as microarray analysis, to identify unintended effects in plant systems has been proved an effective means of determining the "substantial equivalence" of a transgenic plant to its traditional counterpart. Such approaches have been used to investigate unintended effects in a lot of transgenic plant systems (Baker et al. 2006; Baudo et al. 2006; Catchpole et al. 2005; Zhang et al. 2004b). As transgenic crops with more complex traits will soon be entering the market, it is important to extend these analyses to investigate their potential risk in such systems. Omics approaches can screen the potential changes at cellular, tissue levels and metabolism level in a nonselective and unbiased manner. It can be a supplement to targeted approaches in safety assessment of GMOs. This would definitely result in a more complete compositional profile of the crop and increase the understanding of complex metabolic pathways. At present, few such data are available. A lack of comparative compositional profiles may lead to misunderstanding of data and lack of concern. We can build a database for the GM crops using omics methods. Omics approaches make us to understand the whole story that how one part change leads to the change of another part, resulting in the entire change of the organism. We will discuss several nontargeted approaches in the following part.

6.5.3 Functional Genome Detection

Genomes research focuses on the function of transcribed genes and regulatory element. Actually, functional genomes research the direct expression product of genes. DNA microarray technology is used in this research. A DNA microarray is a collection of microscopic DNA spots attached to a solid surface. DNA microarrays can be used to measure the expression levels of large numbers of genes simultaneously. Comparing the gene expression map of GM crops with the parental lines, if some significant differences were detected, unintended effects may happen. Functional genomes platform is able to screen the gene expression differences from the different tissues. It also can be used to analyze the effect of environmental stress on organisms, providing us a method to deeply know pathways.

6.5.4 Transcriptome Detection

In omics studies, microarray analysis and sequencing technology are important methods to detect differentially expressed genes. They have been used to investigate unintended effects in several transgenic plants (Cheng et al. 2008). For instance, microarray analysis was conducted on drought-tolerant transgenic *Arabidopsis* overexpressing the transcription factor ABF3 (Abdeen et al. 2010) and glufosinate-tolerant transgenic *Arabidopsis* expressing the bialaphos resistance (*bar*) gene (Abdeen and Miki 2009), respectively, which found few differences in gene expression profiles, suggesting no unintended effects were discovered. The Affymetrix GeneChip Rice Genome Array was used to search for transcriptome changes in resistant genetically modified and conventional rice; the results indicated that only half of the transcriptomic differences are related to the transgene, and these changes are not enough for causing any obviously unintended effects (Montero et al. 2011).

6.5.5 Proteomics Detection

Proteomics is becoming an important tool in biological sciences and will contribute greatly to the understanding of gene function, particularly as it is known that mRNA level does not correlate well with protein level (Gygi et al. 1999). Proteomics researches all the proteins that genomes express. It is applied to three main fields: the confirmation of proteins, protein precursor, and posttranslational modification; the show of the protein difference; and protein interaction. Proteomics plays a vital role in evaluating unintended effects of GM crops. Wang et al. compared the proteomics between Cry1Ab/Ac rice and traditional rice and found that the growth environment made more influence than the insertion gene (Wang et al. 2012). This research was published in the *Journal of Cereal Science* in 2012: *Comparative analysis of the proteomic and nutritional composition of transgenic rice seeds with Cry1ab/ac genes and their non-transgenic counterparts.*

The Cry protein encoded by the *Cry1ab/ac* gene, a fusion of *cry1ab* (GenBank accession no. X54939) and *cry1ac* (GenBank accession no. Y09787) into a single gene, is highly toxic to *Chilo suppressalis*, *Scirpophaga incertulas*, and *Cnaphalocrocis medinalis*, three important lepidopteran pests of rice (Xu et al. 2009). A major concern of these genetic modifications is the possible occurrence of unintended effects. The evaluation of crop composition is sometimes referred to as "targeted analysis." However, concerns are seldom reported regarding the analytical procedures used to detect the unintended effects of transgene. In principle, proteomic technologies can allow the visualization of all of the changes that take place when the genetics, nutrition, or environment of an organism is altered. To date, proteomic technologies have not been formally applied to the food safety assessment of transgenic crops. We analyzed the proteomic profiles of a transgenic rice variety (*cry1ab/ac*) and its isogenic controls (WT01 and WT02) in two different growth environments (T01 and T02).

By comparing proteomic profiles, we found that 21 proteins were up- or down-regulated as a consequence of environmental influence (WT01 vs. WT02). Similarly, 20 to 22 protein levels were differentially modulated in transgenic rice seeds in comparison to their non-transgenic counterparts (T01 vs. WT01; T02 vs. WT02). These latter changes may be due to the influence of growth environment and the insertion of a single gene into the rice genome. We observed many vertical and horizontal position shifts when comparing the proteomic profiles. A number of seed storage proteins, such as globulins and putative globulin proteins, exhibited truncated forms with molecular weights significantly lower than the native ones. A possible explanation for this observation is that frequent mutations lead to changes in protein molecular weight (Gekle and Silbernagl 1994). However, the above results would not be observed in a traditional substantial equivalence assessment. The storage proteins are abundant in seeds and include albumins, prolamins, glutelins, and globulins, all of which play important roles in determining the nutrition value of rice seeds (Gekle et al. 2005). The truncated forms of storage proteins might affect the nutrition value or digestion and absorption in the intestinal tract of humans and animals in an unknown way. In agreement with our data, Zolla et al. (Gekle and Silbernagl 1994) found frequent molecular weight changes in the major storage proteins of transgenic maize lines. In this study, a significant number of proteins were shown to be differentially regulated by both environment and genetic modification. The levels of approximately 21 proteins were differentially modulated as a consequence of environmental influence (WT01 vs. WT02), whereas 20 and 22 differentially expressed proteins were found in transgenic seeds in comparison to their non-transgenic lines (T01 vs. WT01; T02 vs. WT02). These differentially expressed protein spots were related to starch synthesis, glycolysis, gluconeogenesis, and the glyoxylate cycle. However, no significant differences were observed between the different genomic and environmental groups. Therefore, they consider that the differences in protein spots that exist between transgenic and non-transgenic lines fall within the range of natural variation. Interestingly, MS/MS identification of modulated spots revealed that the proteins belong to several different metabolic pathways. No new or toxic proteins unique to individual transgenic lines were observed. The expression of the target *Bt* gene is below the limit of detection (LOD) of the method employed here. This is expected because proteomics is a useful method for comprehensive analyses but not if the level of a target protein is extremely low. There is little difference in the proteomic profiles of the transgenic rice seeds and their non-transgenic counterparts. This also includes the different environment group comparisons, which indicate that both single gene insertion and slightly different growth environments have only a moderate effect on the rice proteome. To be more precise, the effects of environmental growth conditions on the rice proteome are not significantly different than effects caused by a single gene insertion. Single gene insertions may affect only a limited number of proteins; in contrast, a greater number of potential protein changes may be expected as a result of significant environmental changes.

Based on the proteomic profile analyses, the transgenic rice seeds containing Cry1Ab/Ac protein were comparable to the non-transgenic counterparts. Moreover, protein spots with more than twofold differences in abundances were monitored; however, the differentially expressed proteins identified were storage proteins or proteins involved in basic metabolism, not allergen or toxic proteins. The environmental effects were also not significantly different from the transgenic effects.

Two-dimensional gel electrophoresis has high resolution and sensitivity. It is relatively easy to further analyze the isolated proteins. Around 1000 proteins can be dissolved in one gel. However, proteomic technology has some limitations for detecting unintended effects. Each step, including protein abstracting, sample treatment, electrophoresis, dyeing, can have an impact on the results. Up to now, the number of detected proteins is still much less than the actual number. In conclusion, proteomic technology is vital for our understanding about structure and metabolism.

6.5.6 Metabolome Detection

Metabolome refers to the total metabolites in cell. Metabonomics is a science to research metabolome. This novel way is to measure "everything at once" rather than isolating a single compound of interest. The results could reflect a specific composition measured at a specific moment in time. The researches on metabolism are mainly divided to four classes. The first is the analyses of target compounds, that is, analyzing the specific compounds that we are interested in the experiment. The second is to analyze a group of compounds those connected to the known metabolic analysis. The third is to analyze large amounts of compounds extensively rather than some specific compounds. The fourth is to analyze all compounds. We try to analyze biological systems of all the metabolites as much as possible to reflect the overall metabolic product information. Metabolic networks are usually complicated, overlapping each other. It is reported that 50% of the metabolites are involved in more than two reactions, and more than 67% of all reactions involve more than one substrate and product (Nielsen and Oliver 2005). However, we cannot say that introduced genes would result in more unintended effects because of the complex metabolites.

At present, major techniques for researching metabolomics are gas chromatography (GC), liquid chromatography (LC), high-performance liquid chromatography (HPLC), and nuclear magnetic resonance (NMR). These techniques are complementary because of their intrinsic limitations. One of the key challenges of metabolite analysis is to find an optimal balance between accuracy and coverage of metabolite measurements. GC has a high resolution. It can be used with flame ion detector (GC/FID) or mass spectrometry (GC/MS), which can quantify the metabolites with unknown structure. HPLC is the most commonly method to analyze plant material target compounds. NMR can provide information of structure. In principle, ^{1}H NMR can detect any metabolites containing hydrogen.

Cao et al. analyzed rat urine fed with GM rice using NMR. They found that the urine metabolome did not have significant difference between two groups fed with GM or non-GM rice (Cao et al. 2011, 2012c). These researches were published in the journals *Molecular BioSystems* and *IUBMB Life* in 2011 and 2012: *Metabonomics study of transgenic Bacillus thuringiensis rice (T2A-1) meal in a 90-day dietary toxicity study in rats.*

In this study, the urine from the rats fed with transgenic *Bacillus thuringiensis* rice (T2A-1 rice) meal for a 90-day dietary was detected first using metabonomics. T2A-1 rice and the non-transgenic near-isoline control rice (MH63 rice) were provided by Huazhong Agricultural University (Wuhan, China). Rice flours from T2A-1 and MH63 rice were used at 70% w/w to produce diets similar to the nutrient requirements of AIN93G diets. Twenty-four Sprague-Dawley male and female rats were divided into groups randomly with each group consisting of six male and six female rats. Rats in the experimental group were fed with diet formulated with 70% (w/w) T2A-1 rice flour, while those in the control group were fed with diet formulated with the same concentration of MH63 rice flour. Twenty-four hour urine samples were collected from each rat each month: 0, 1, 2, and 3 months (Appelgren and Arora 1983).

In this study, they aimed at proof of the substantial equivalence of GM rice and non-GM rice at the level of metabonomics. The results of the multivariate analysis showed that the α-ketoglutarate in rat urine has different changes in directions between the non-GM and GM groups for the three time points and the GM rats' urine had higher creatinine, acetic acid, sugar, and hippurate than the non-GM rats' urine on all of the three time points. As the multivariate analysis could not tell whether the differences were significant, ANOVA was employed. The ANOVA results showed the α-ketoglutarate in rats' urine on the first month from the GM group was significantly different from that from non-GM group, but on the second and the third month, the α-ketoglutarate in rats' urine from the GM group was comparable with that from non-GM group. It is another single factor of statistical difference that GM group had only a significant difference of the hippurate in rats' urine from the non-GM group on the first and third month. In view of no stability and no continuity, these changes cannot be regarded to be associated with the GM rice.

In view that the multivariate analysis could not tell whether the differences were significant or not, ANOVA was also employed in this study. Although the differences between the groups may be very tiny, the distinguishing metabolites can be separated to differentiate the groups by the multivariate analysis. ANOVA was then used to tell the significance of difference. Thus, correct results were obtained, and we saved our time by using both of them. The results of the metabonomics were comparable with that from the traditional method in this study.

This is the first time to use NMR to detect the rat urine from a 90-day dietary experiment in a non-invasive dynamic monitoring way. The method will be a complementary way with the traditional methods of safety assessment of GM foods. Metabonomics is regarded as a simple operation, as short-time-consuming, to be of low cost, and to have no damage in experimental animals when compared with the traditional methods of safety assessment of GM foods. Because of dynamically reflecting the changes in metabolism without damage in experimental animals, the metabonomics should be recommended to be a useful in vivo evaluation method for genetically modified foods. In this study, a system was established for other food safety assessment of GM foods (Cao et al. 2012c): *Safety assessment of transgenic Bacillus thuringiensis rice T1c-19 in Sprague-Dawley rats from metabonomics and bacterial profile perspectives*.

T1c-19 is transgenic rice that harbors genes encoding the Cry1C protein from *Bacillus thuringiensis* and the phosphinothricin acetyltransferase (PAT) protein from *Streptomyces viridochromo* genes. Expression of the Cry1C and PAT proteins in rice confers resistance to *stem borers* and *leaf folders* and tolerance to herbicides containing glufosinate ammonium, respectively.

Experimental rodent diets were composed of flour from T1c-19 rice at 70% wt/wt. The control diet was composed of rice flour from its transgenic parent, MH63, and at the same concentration. The urine metabonomics was analyzed at four time points in this paper. The relationships among the groups, metabolites, rat sex, and rat age were discussed.

The metabonomics data revealed that the GM rice did not induce continuous metabolites changes in rat urine. Several metabolites were correlated with the age and sex of the rat. TMAO, TMA, and DMA are derived from choline, which can be derived from the diet (Ferrufino-Guardia et al. 2000) or catabolized from phospholipids (Denli et al. 2008). TMAO is produced by the N-oxidation of trimethylamine, a degradation product of choline in the gut, in the presence of flavin-containing monooxygenase (FMO) enzymes (Elaroussi et al. 2008). Testosterone has been previously shown to inhibit the expression of FMO enzymes in the liver (Sava et al. 2007), which, along with the hormonal variation caused by the estrus cycle, may account for the higher concentrations of TMAO detected in female urine. The creatine–phosphocreatine system is crucial for cellular energy transportation. Most of creatine is phosphorylated by the enzyme creatine kinase, creating phosphocreatine in the cell. This reaction is reversible and can act as an energy buffer to maintain ATP homeostasis. Additionally, approximately 2% of all the creatine and phosphocreatine are nonenzymatically converted to creatinine (Zhang et al. 2009). As the rats mature and age, energy demand decreases. Thus, creatine is likely diverted to creatinine rather than phosphocreatine production. Hormones may regulate the increase in urinary taurine as the rats grow. A sex hormone influence on taurine excretion has been suggested by Chatagner and Bergeret (46), where the decarboxylation of cysteine sulfinic acid was two to three times faster in liver homogenates from male than from female rats. Sloane-Stanley (Al-Anati and Petzinger 2006) observed that the decarboxylation of cysteic acid was twice as fast in male rats as in females. These decarboxylations are steps in the conversion of sulfur amino acids to taurine; thus, taurine formation may occur similarly. Consistent with previous results, this study demonstrated higher levels of taurine in male rat urine than in female rat urine. All in all, metabonomics of GM rice-fed rats were comparable to the non-GM rice-fed rats. They conclude that the GM rice is as safe as non-GM rice. This method offers a novel approach to extensively study the impacts of GM foods. Urine can be acquired without damaging the animal. Metabonomics is regarded as nondestructive, dynamic, simple, quick, and inexpensive. These methods will protect the animals' welfare and allow for their reuse.

Qi et al. studied on the rat urine metabonomics fed with genetically modified rice expressing human serum albumin (Qi et al. 2014). They found that the urine metabolites of GM rice-fed rats were more similar to control diet group. This research was published in the journal *Food and Chemical Toxicology* in 2014: *Safety assessment of genetically modified rice expressing human serum albumin from urine metabonomics and fecal bacterial profile*.

In recent years, a third generation of genetically modified (GM) crops has risen up, called "pharma plants" (Alderborn et al. 2010). "Pharma plants" can produce drugs, such as antibodies and vaccines, to prevent and cure diseases (Pandey et al. 2010). They are inexpensive and highly scalable and do not support human pathogens (Paul and Ma 2011).

Human serum albumin (HSA) has been widely used clinically to treat serious burn injuries, hemorrhagic shock, hypoproteinemia, fetal erythroblastosis, and ascites caused by cirrhosis of the liver (Hastings and Wolf 1992). In addition, HSA is used as an excipient for vaccines or therapeutic protein drugs as well as a cell culture medium supplement (Marth and Kleinhappl 2001). Approximately more than 500 tons of HSA is produced each year worldwide. Currently, globally commercial HSA production is primarily based on collected human plasma, which is limited in supply, leading to an increase in price. Furthermore, there is a potential risk of contact with blood-derived infectious pathogens during the process of utilizing plasma-derived HSA (Chamberland et al. 2001). Due to the above limitations, He et al. developed GM rice that can stably express recombinant HSA (rHSA) (He et al. 2011b); rHSA can be produced on large scale, and we can avoid the potential above-mentioned problems. Although the GM rice is for nonfood purposes, its food safety assessment should be conducted because several episodes of accidental mixture of conventional food plants with GM food plants have been reported (Lu 2008; Li 2008; (Holst-Jensen 2008). Therefore, food safety assessment for the GM rice expressing HSA is necessary.

Analyses for the metabolites excreted in urine are supplement for traditional methods for safety assessment of GM food. Metabolites are regarded as simple to conduct, as non-time-consuming, to be of low cost, and to not cause damage to experimental animals compared with traditional methods. Urine metabolites can reflect unintended effects. Currently, no standards in regard to food safety assessment have been established for pharmaceutical GM crops.

GM 4-114-7 rice and its non-transgenic parent "Taipei 309" were provided by Wuhan University, China. The GM rice 4-114-7 and its non-transgenic parent "Taipei 309" brown rice were used at 50% wt/wt (Domon et al. 2009; Schroder et al. 2007; Zhou et al. 2011) to produce diets. Four-week-old Sprague-Dawley male and female rats were randomly divided into three groups (10 rats/sex/group). The three groups were respectively administered diets with 50% (wt/wt) GM rice, diets with 50% (wt/wt) non-GM rice, or commercialized rodent diet. Urine samples were collected from the rats in each group (ten rats/sex/group) for 24 h on days 30 and 90 (Cao et al. 2012c). A urinalysis is an array of tests performed on urine, including ketones, bilirubin, protein, glucose, urine-specific gravity, occult blood, pH, bacteria, erythrocytes, white blood cells, squamous epithelium, and clarity. Also urine samples were collected for NMR spectroscopy.

Throughout the feeding trial, all of the test animals appeared to be healthy; no adverse effects in clinical appearance were observed. All variables, except clarity of female rat urine on day 90, observed were not significantly different between the GM and the non-GM rice group on day 30 and 90. The control, non-GM, and GM groups overlapped with each other in the score plots of principal component analysis (PCA), suggesting that no obvious profile separations were found between the three groups of both male and female rats on days 30 and 90. DGGE was used to provide an overview of intestinal bacterial composition of each rat on day 30 and 90 of each group.

The metabolism of GM rice-fed rats observed through the urine metabolites was comparable with the non-GM rice-fed rats, more similar to control diet group; thus, it can be concluded that the GM rice expressing HSA is as safe as the traditional daily diet. Apart from traditional food safety assessment methods for GM foods, such as detecting some physiological indexes and pathological examination, unintended effects should be considered. We are the first to use the metabolite profile of rat urine and feces to evaluate the food safety of GM rice expressing HSA. In addition, we constructed a metabolite model for the food safety assessment of GM crops for nonfood purposes. Compared with traditional methods, they are easy to perform, do not bring damage to animals, and can realize a long-term continuous monitoring on the animals. Urine is easy to collect.

These profiling techniques offer the capacity to perform broad screening for possible changes of the food product at different integration levels in a nonselective, unbiased manner. However, there are still limitations associated with the use of these new technologies that have to be addressed. One of these limitations is the interpretation of the observed differences with respect to their biological relevance and toxicological significance. It is crucial to have information on sources of natural variation that can result from crop management practices, from interactions between genotype and environment, and from breeding systems (Davies 2010).

6.5.7 Gut Health

GM food probably has unintended effects on animals' gut health. Apart from toxicology and physiology parameters, intestinal bacteria and related gut indexes can be detected. The change of intestine microenvironment can influence body health to some extent.

Xu et al. analyzed the rat caecum fed with GM rice using RT-PCR method (Xu et al. 2011). This research was published in the *Journal of Food Science* in 2011: *Analysis of caecal microbiota in rats fed with genetically modified rice by real-time quantitative PCR*.

Rice is one of the most important crops and is supplied as main food to over half of the word population. In order to meet the increasing demands, the appearance of genetically modified rice (GMR) has shed light on this target. In this research, they focused on the effects of GMR on intestinal microbiota. Intestinal microbiota is very complex and can be influenced by the food you have taken for a long time (Sokol et al. 1988). Moreover, the changes of the intestinal ecosystem could influence human health in various aspects (Dahlmann et al. 1998; Zingerle et al. 1997).

Conventional microbiological methods have many limitations, and only 20% microflora could be isolated and identified on plates (Schwerdt et al. 1997; Welborn et al. 1998). So they applied RT-PCR which is a very sensitive and precise technique to quantify certain bacterial species and groups in intestinal content samples.

A total of 12 groups of rats, which included male and female, were fed with the basal diets containing 30%, 50%, and 70% GMR (B1, B2, B3) or 30%, 50%, and 70% non-GMR (D1, D2, D3) for 90 days. Total DNA was extracted from the cecal content. All the PCR experiments were performed using the Applied Biosystems 7000 instrument (Applied Biosystems Inc., Norwalk, CT, USA). The targeted bacteria, annealing temperatures, amplicon sizes, primer sequences, and literature references are given in Table 6.1. Standard curves were constructed using tenfold serial dilutions of DNA extracted from the six pure strains. Logarithms of the average estimate of target bacterial genomes per 1 g of feces (wet weight) were used to achieve normal distribution, and the mean ± standard deviation (SD) of each target was calculated. The percentage of bacterial group in relation to total bacteria (relative abundance) was calculated for each individual, and the median was determined for each subject. All statistical calculations on data obtained from the feeding study were carried out using SAS release 9.0 (SAS Institute Inc., Cary, NC, USA).

According to RT-PCR, there is no statistically significant difference ($P > 0.05$) observed in total bacteria of all samples. The genome copies of *Lactobacillus* group from the cecum of male rats fed with 70% non-GMR were higher than those fed with 70% GMR, and the relative abundance of *Lactobacillus* group is also higher for group D. This result was in contrast with the *E. coli* subgroup, which was more numerous in proportion of group B, except D2 and B2 for male rats. The *Clostridium perfringens* subgroup was numerically more abundant in group D than group B of the same level, also except D2 and B2 for male rats. These results suggested that GMR had a complex effect on cecal microflora that may be related to the health of the host. Our conclusion was that these results suggested that GMR had a complex effect on cecal microflora that may be related to the health of the host.

There are some notices about the results and conclusion. Only the comparison of *Lactobacillus* group in male rats is significant, and thus we used the expression of "be higher than." The description of *E. coli* and *Clostridium* is based on the numerically abnormal appearances, and thus we used the expression of "be more numerous

Table 6.1 16S rDNA RQ-PCR primers used for amplification of cecal samples

Target bacteria	Amplicon size (bp)	Annealing temperature (°C)	Sequence (5′–3′)
All bacteria	466	58	F: TCCTACGGGAGGCAGCAGT
			R:GGACTACCAGGGTATCTAATCCTGTT
Bifidobacterium genus	437	60	F: GGGTGGTAATGCCGGATG
			R: TAAGCCATGGACTTTCACACC
Bacteroides–Prevotella	418	59	F:GAAGGTCCCCCACATTG
			R:CAATCGGAGTTCTTCGTG
C. 16S rRNA cluster I	120	57	F: ATGCAAGTCGAGCGAKG
			R: TATGCGGTATTAATCTYCCTTT
Enterococcus genus	144	61	F: CCCTTATTGTTAGTTGCCATCATT
			R: ACTCGTTGTACTTCCCATTGT
E. coli subgroup	340	61	F: GTTAATACCTTTGCTCATTGA
			R: ACCAGGGTATCTAATCCTGTT
Lactobacillus group	341	58	F: AGCAGTAGGGAATCTTCCA
			R: CACCGCTACACATGGAG

in" or "be numerically more abundant in." Due to the public concern on the safety problem of GMO, we think it is our duty to exhibit any abnormal phenomenon during the researches. There are no doubt "effects" indicated by the significantly different (*Lactobacillus*) and numerically different (*E.coli* and *Clostridium*) results. However, it's a "complex effect" as we don't know these effects are beneficial or harmful, and we don't know the mechanism by which the GMO influences the gut microbiota. So "complex effect" is a general conclusion and an appropriate one in our opinion. However, it's a "complex effect" as we don't know these effects are beneficial or harmful, and we don't know the mechanism by which the GMO influences the gut microbiota. So "complex effect" is a general conclusion, and an appropriate one in our opinion.

This research was conducted before 2010, and the article was published on 2011, when the health of gut microbiota had not received enough attention as today. This article brought the pioneering research on the effects of GMO food on the intestinal bacteria. The development of next-generation sequencing (NGS) and metagenomics hasn't been mature at that time; thus, real-time quantitative PCR was used to detect the main bacterial group of the gut microbiota. Both the viewpoint and the method were advanced at that time. The conclusion of this article was the reflection of the results, although it may be limited by the methods and techniques at that time; however, it was valuable. Any truth is not accomplished at one stroke.

Cao et al. studied on the rat fecal bacterial profile and found that the change of the bacteria had nothing to do with GM rice (Cao et al. 2012c). This research was published in the journal *IUBMB Life* in 2012: *Safety assessment of transgenic Bacillus thuringiensis rice T1c-19 in Sprague-Dawley rats from metabonomics and bacterial profile perspectives*.

The composition of feces can reflect the health and metabolism of the animal. Feces can be acquired without damaging the animal. Bacterial profiles are regarded as nondestructive, dynamic, simple, quick, and inexpensive; moreover, experimental animals are not damaged, unlike when traditional methods of GM food safety assessment are used.

Experimental rodent diets were composed of flour from T1c-19 rice at 70% wt/wt. The control diet was composed of rice flour from its transgenic parent, MH63, and at the same concentration. The bacterial profiles were analyzed at four time points in this paper. The relationships among the groups, bacterial profiles, rat sex, and rat age were discussed.

Urine samples were collected from each rat from 09:00 to 09:00 the next day every month, including a pre-dose time point (Appelgren and Arora 1983). Urine samples prepared and 1H NMR spectra acquired were done as previous paper (Cao et al. 2011). Approximately 0.2–0.25 g of fecal sample was taken for DNA extraction with a stool extraction mini kit (Qiagen, Germany). A quantitative PCR core kit (Toyobo) by ABI Prism SDS 7500 instrument (Applied Biosystems) was used to analyze the shift and composition of fecal bacteria (*Clostridium*, *Lactobacilli*, *Bacteroides*, *Bifidobacterium*, *Escherichia coli*, *Enterobacter*, and total bacteria).

The V3 region of the 16S rDNA gene was amplified with primers 518R and GC-338F. DGGE was performed using the DCode™ universal mutation detection system (Bio-Rad, USA). Cluster analysis was performed using the unweighted pair group method with an arithmetic mean clustering algorithm (UPGMA) to calculate the dendrogram of each DGGE gel.

In this study, even though GM rice did not alter the bacterial profiles of rats, a relationship between the metabolites and gut bacteria was identified. Dynamics in gut microbiota, citric acid, DMA, TMA, creatinine, taurine, and acetic acid correlated with the changes in the metabolites and the age and sex of rats, indicative of interactions among the gut microbiota, metabolites, and the age and sex of the rat.

Because the bacterial profile of GM rice-fed rats was comparable to the non-GM rice-fed rats, we conclude that the GM rice is as safe as non-GM rice. The feces from animals reveal the physical condition of those animals. We used them to establish safety assessment methods, which are comparable to the traditional method. Without experimental animal bias, this method offers a novel approach to extensively study the impacts of GM foods.

Liu et al. studied on the effect of GM *Lactobacillus plantarum* 590 on the gut health of rats (Liu et al. 2012a). They concluded that no significant differences between GM Lp 590- and non-GM Lp-fed groups. This research was published in the journal *IUBMB Life* in 2012: *The effect of genetically modified Lactobacillus plantarum 590 on the gut health of Sprague-Dawley rats.*

Genetically modified microorganisms (GMMs) provide a new way to develop food production, pharmaceutical industries, and so forth (Hagenbuch and Meier 2004). As gene technology progresses from modifications aimed at relatively simple traits to addressing more complex attributes, the challenges of evaluating the unintended effects of GMMs on human health will increase. Gastrointestinal (GI) microbiota acts as an essential "organ" and comprises a complex system. Gut health is inseparable from intestinal microbiota and their metabolites. Gut permeability and mucosal immunity can reflect the function of the intestinal barrier, further present intestine and body health.

Genetically modified *Lactobacillus plantarum* 590 (GM Lp 590) derived from electroporation of pLEB590 as a food-grade vector into *L. plantarum* showed that nisI functions as an immunity protein in *Lactobacilli*, enabling nisin resistant. In this study, they observed intestinal bacteria, their main metabolites, intestinal permeability, and intestinal mucosal immunity by administration of *L. plantarum* and *L. plantarum* 590 to Sprague Dawley (SD) rats for 4 weeks.

The analysis of fecal microbiota was used by real-time quantification and denaturing gradient gel electrophoresis method. Also, fecal enzyme activity, fecal SCFAs, ZO-1, occludin, and SIgA were analyzed. The analysis of ZO-1 and occludin was used by immunohistochemistry localization and RT-PCR. And the analysis of SIgA was also used by RT-PCR.

Bifidobacterium and *C. perfringens* are the most important bacteria in the GI tract, and some researchers considered them to be beneficial and harmful bacteria, respectively (Tsuda et al. 1999). At all five time points (days 0, 7, 14, 21, and 28), no statistically significant differences in the numbers of specific bacteria were found between the GM Lp 590 group and its parental group (non-GM Lp). Compared with the PBS-fed group, significant changes of the two bacteria were observed in microorganism-fed groups, with higher levels of *Bifidobacterium* and lower levels of *C. perfringens* ($P < 0.05$) at day 7 and day 14. After 14 days, there was no notable difference in cell numbers of the two bacteria in all groups. *L. plantarum* has long been considered as a probiotic and been used for food fermentation. Probiotic increased the number of beneficial bacteria *Bifidobacterium* and reduced the number of harmful bacteria *C. perfringens* in the early phase of treatment, an indisputable fact. The results indicated that *L. plantarum* do play a role in improvement of health. The similar result can be found in many references about assessment of *Lactobacillus* strains. From the result of DGGE at the time point of day 28, the bacterial composition in Lp- and Lp 590-fed rats shows high similarities to those in PBS-fed rats. Consistency of data changes in Lp and Lp 590 groups demonstrated that both GM and non-GM *L. plantarum* affected intestinal microbiota in a similar way.

In contrast to the control group, Lp- and Lp 590-fed rats exhibited a statistically significant decrease in the activities of β-glucuronidase, β-glucosidase, and nitroreductase, which are known to be associated with the conversion of procarcinogens into proximal carcinogens (Jung et al. 2001) before 14 days. Then, the three enzyme activities showed no remarkable change in all groups at day 21 and day 28. According to previous studies, a decrease in β-glucuronidase, β-glucosidase, and nitroreductase, which had positive correlation with beneficial *Bifidobacterium* (Jung et al. 2001), can promote intestinal health (Babu et al. 2002). Our results just proved the above correlation. And lower three enzyme activities in Lp- and Lp 590-fed groups at day 7 and day 14 ($P < 0.05$) further suggested that there is no different effect of non-GM and GM *L. plantarum* on intestinal microbial metabolites.

In contrast to the control group, the level of butyric acid of Lp and Lp 590-fed groups significantly increased at day 7 and day 14 ($P < 0.05$). At other time points, for the level of butyric acid, there was no prominent change in all groups. It is known that *Bifidobacterium* produce more SCFAs especially butyric acid (Takeuchi et al. 2001), which was in agreement with our results. Compared with control group, the level of both acetic acid and propionic acid is a not significant increase in microorganism-fed groups. However, in view of no statistically difference in non-GM or GM groups, it cannot be regarded as being associated with GM Lp 590 consumption.

Relatively large changes of the fecal bacteria and their metabolites occurred in the first 2 weeks, which were both related to the fast growing of the rats during this period and the different diets. And gender cannot affect any test result. Both Lp 590- and Lp-fed rats showed a twofold lower plasma DX-4000-FITC area under the curve compared with the PBS-fed rats. Gut permeability is controlled by several specific tight junction proteins (Leier et al. 2000). Among these, ZO-1 and occludin have been proposed as key markers of tight junction integrity (Leier et al. 2000). Lp and Lp 590 treatment increased ZO-1 and occludin mRNA in the jejunum segment. ZO-1 and occludin immunohistochemical staining appeared to be increased uniformly and continuously distributed in Lp- and Lp 590-fed rats tissues. In accordance with the mRNA analysis, compared with the control group, Lp- and Lp 590-fed rats have improved the tight junctions ($P < 0.05$), because the immunohistochemical staining of both proteins localized along the apical cellular border was greater. Previous studies on the relationship between intestinal beneficial bacteria and permeability illustrated that *Bifidobacteria* can augment two proteins of ZO-1 and occludin and then promote gut permeability (Schrickx et al. 2006). The tendency also appeared in this study. To identify whether the changes in the gut microbiota and the expression of tight-junction proteins were associated with in vivo gut permeability, multiple correlation analysis between these parameters was performed. Indeed, a negative correlation between tight-junction protein mRNA level and plasma DX-4000-FITC was found in the jejunum segment.

The results of the RT-PCR analysis showed that the level of SIgA in the ileum was higher in the Lp and Lp 590 groups than in the control group ($P < 0.05$). The results of previous studies with animals and volunteers as object all suggested that live bacteria are especially efficient mucosal immunogens (Youngblood and Sweet 2004). Moreover, the synthesis and excretion of SIgA depends mainly on healthy and stable beneficial intestinal bacteria especially *Bifidobacterium* (Yokoyama et al. 2008). It was once again verified from our results that the number of *Bifidobacterium* on day 28 and SIgA mRNA level had positive correlation.

In this study, there are no significant differences between GM Lp 590- and non-GM Lp-fed groups. Evaluation of gut health including intestinal microbiota, their metabolites, gut permeability, and gut mucosal immunity will be a new approach of assessing the safety of GMM.

Qi researched on the fecal bacterial profile of the rats fed with GM rice expressing human serum albumin (Qi et al. 2014). The bacterial profile of GM group rats was comparable with the non-GM rice-fed rats. This research was published in the *journal Food and Chemical Toxicology* in 2012: *Safety assessment of genetically modified rice expressing human serum albumin from urine metabonomics and fecal bacterial profile*.

In recent years, "pharma plants," a third generation of genetically modified (GM) crops, have risen up (Alderborn et al. 2010). "Pharma plants" can produce drugs, such as antibodies and vaccines, to prevent and cure diseases (Pandey et al. 2010). The drugs produced by plants have several advantages as they are inexpensive, are highly scalable, and do not support human pathogens (Paul and Ma 2011).

Human serum albumin (HSA) has been widely used clinically to treat serious burn injuries, hemorrhagic shock, hypoproteinemia, fetal erythroblastosis, and ascites caused by cirrhosis of the liver (Hastings and Wolf 1992). In addition, HSA is used as an excipient for vaccines or therapeutic protein drugs as well as a cell culture medium supplement (Marth and Kleinhappl 2001). Approximately more than 500 tons of HSA is produced each year worldwide. Currently, globally commercial HSA production is primarily based on collected human plasma, which is limited in supply, leading to an increase in price. Furthermore, there is a potential risk of contact with blood-derived infectious pathogens during the process of utilizing plasma-derived HSA (Chamberland et al. 2001). Due to the above limitations, He et al. developed GM rice that can stably express recombinant HSA (rHSA) (He et al. 2011b). Although the GM rice is for nonfood purposes, its food safety assessment should be conducted because several episodes of accidental mixture of conventional food plants with GM food plants have been reported (Lu 2008; Li 2008; Holst-Jensen 2008).

Analyses for the metabolites excreted in urine and feces are supplements for traditional methods for safety assessment of GM food. Metabolites are regarded as simple to conduct, as non-time-consuming, to be of low cost, and to not cause damage to experimental animals compared with traditional methods. Fecal metabolites can reflect unintended effects. Currently, no standards in regard to food safety assessment have been established for pharmaceutical GM crops.

GM 4-114-7 rice and its non-transgenic parent "Taipei 309" were provided by Wuhan University, China. The GM rice 4-114-7 and its non-transgenic parent "Taipei 309" brown rice were used at 50% wt/wt (Domon et al. 2009; Schroder et al. 2007; Zhou et al. 2011) to produce diets. Four-week-old Sprague-Dawley male and female rats were randomly divided into three groups (10 rats/sex/group). The three groups were respectively administered diets with 50% (wt/wt) GM rice, diets with 50% (wt/wt) non-GM rice, or commercialized rodent diet. Feces samples are used for analyses of the bacterial profile, short-chain fatty acids (SCFAs), and enzyme activity.

Throughout the feeding trial, all of the test animals appeared to be healthy; no adverse effects in clinical appearance were observed. DGGE was used to provide an overview of intestinal bacterial composition of each rat on day 30 and 90 of each group. From the perspective of bacterial composition, no obvious profile separations were found between the three groups on day 30. Additionally, no apparent separation was observed in the female group on day 90. From the perspective of bacterial diversity, β-diversity was significantly different between the GM and non-GM group in male rats on day 30 and in the female rats on day 90 ($p < 0.05$). However, it was not significantly different in both genders at the same time and thus not time dependent. *Lachnospiraceae* bacterium was the dominant bacteria throughout the feeding trial in all three groups. For male rats, *Clostridiales* bacterium and *Ruminococcus* sp. were dominant in both the control diet and GM group on day 30. In addition, *Lactobacillus* sp. accounted for a predominant position in the rice diet groups on day 90. For female rats, *Bacteroidales* bacterium, *Clostridiales* bacterium, and *Lactobacillus* sp. were predominant in all three groups on day 30. Additionally, *Bacteroides* sp. and *Clostridiales* bacterium were predominant in the control diet and GM group on day 90. As for SCFAs, butyrate was significantly higher in the control diet group than the rice diet groups (both GM and non-GM) for only male rats. In regard to metabolic enzymes, they are not different in GM group from the control group. Serum chemistry was investigated by Sheng et al. (Sheng et al. 2014). To assess whether there is a possible association between the structural changes of the gut microbiota and blood and urine, the bacteria acquired from DGGE were employed to predict the change in blood and urine by using PLS regression models. Urine creatinine was significantly correlated to *Ruminococcus* sp. By using PLS regression models, it showed that the urine metabolites cannot be used to predict the blood changes.

The metabolism of GM rice-fed rats observed through the bacterial profile and urine metabolites was comparable with the non-GM rice-fed rats, more similar to control diet group. Thus, it can be concluded that the GM rice expressing HSA is as safe as the traditional daily diet. Apart from traditional food safety assessment methods for GM foods, such as detecting some physiological indexes and pathological examination, unintended effects should be considered. We are the first to use the metabolite profile of rat urine and feces to evaluate the food safety of GM rice expressing HSA. In addition, we constructed a metabolite model for the food safety assessment of GM crops for nonfood purposes. Compared with traditional methods, they are easy to perform, do not bring damage to animals, and can realize a long-term continuous monitoring on the animals. The samples (urine and feces) are easy to collect, the amount of the sample is more than blood, and the approach can reduce harm to the animals, so it would be better for animal welfare. PLS regression modeling performed well in predicting some blood biochemical indexes and urine metabolites with the bacteria profile, suggesting a possible close association between blood biochemical indexes, urine metabolites, and rat intestinal bacteria. Gut microbiota has a potential to be used to predict the change of blood indexes in the future as an unharmful method to rats.

6.6 Conclusion

Both genetically modified organisms and normal organisms can have unintended effects. We cannot say that unintended effects only appear in GMOs. Not all the unintended effects will bring risk to humans. When it exceeds the variation range, it will be taken into consideration in safety assessment. However, we lack a database of natural variation, so it becomes difficult for us to detect and assess the unintended effects. Omics methods are not official request to assess the unintended effects of GMOs, but it can be developed. In consideration of the safety assessment of GMOs, unintended effects of GMOs on human and animal health should be thought over. Therefore, it is essential to build a relative complete assessment system which can be applied to safety assessment in unintended effect of GMOs.

Chapter 7
Safety Assessment of Transgenic Microbiology

7.1 Definition of Transgenic Microorganism

"Transgenic microorganism" means bacteria, yeasts, or filamentous fungi in which the genetic material has been changed through in vitro nucleic acid techniques including recombinant deoxyribonucleic acid (DNA) and direct injection of nucleic acid into cells or organelles (CAC/GL 2003).

7.2 Outline of Transgenic Microorganism

7.2.1 Background of the Use of Microorganisms in Food

Food-grade microorganisms are widely used to improve and keep safety quality and nutritional value of perishable foodstuffs in food industry. The types of food microorganism differ in different foods (Cooke et al. 1987; Nout 2001). Nowadays many processes in the food industry are conducted using defined starter strains with predictable and known properties, but spontaneous fermentation or poorly characterized starter cultures with unknown composition are still used in many products, especially in artisanal production or home.

7.2.1.1 Lactic Acid Bacteria

The microorganism is capable of lactic acid fermentation. The lactic acid bacteria (LAB) include a variable group of bacteria. They share certain common characteristics (Axelsson 1998). They are Gram-positive rods or cocci. Lactic acid is the main fermentation end product with a microaerophilic metabolism. The common food-related LAB genera are *Lactococcus*, *Leuconostoc*, *Lactobacillus*, and

K. Huang, *Safety Assessment of Genetically Modified Foods*,
DOI 10.1007/978-981-10-3488-6_7

Pediococcus. Their applications in foods are greatly limited and usually restricted to certain types of products, for example, enterococci in many endogenous food products and oenococci in red wine ripening, *Leuconostoc*, and *Streptococcus thermophilus* in cheese and yogurt starters. But from the physiological perspective, *Enterococcus*, *Streptococcus*, *Vagococcus*, *Carnobacterium*, *Tetragenococcus*, *Oenococcus*, and *Weissella* are also representatives of LAB. From the ecological perspective, LAB are usually confirmed in energy-rich habitats, including milk, fruit, decaying animal and plant material, the gastrointestinal tract, and so on.

They are widely and traditionally used in the food manufacture, such as the production of pickled vegetables, ripened cheeses, fermented milks, brined olives, sausages, and sourdough sand cured meats (Nout 2001). Beside traditional food applications, LAB are being increasingly used in promoting health in functional food products (von Wright and Salminen 1999).

7.2.1.2 Other Bacteria

Bifidobacterium and anaerobic Gram-positive bacteria, which do not have traditional food uses, belong to the normal intestinal microflora. The bacteria comprise another group of probiotic bacteria. Nowadays the bacteria are increasingly being used in various functional formulations (Ballongue 1998).

Compared with LAB, other bacteria have a less common but still important role in food production processes. Propionic acid bacteria are used in Swiss-type cheeses to produce gas and aroma compounds (Shaw and Sherman 1923; Lemee et al. 1994; Stlie et al. 1995). In meat fermentations, certain *Kocuria* (formerly *Micrococcus*; Montel 2000) are also utilized, as well as other Gram-positive cocci, such as staphylococci (*Staphylococcus carnosus*, *S. xylosus*). The genus *Brevibacterium* are important for the proper ripening of surface texture and pigmentation of certain types of cheeses (Weimer 2000).

7.2.1.3 Yeasts and Filamentous Fungi

In contrast with bacteria, yeasts and filamentous fungi are eukaryotic microorganisms. They have cellular organizations that resemble that of animal or plant cells. While molds are aerobic microorganisms, yeasts are facultative anaerobic bacteria with carbon dioxide and ethanol as the main fermentation end products. As we all know, main applications of yeasts are in wine making, brewing, and baking, which are based on the fermentative capability of these microorganisms.

While wild yeasts associated with fruit are responsible for the fermentation of many traditional drinks and wines, it is the brewers' yeast, *Saccharomyces cerevisiae* with various subspecies and variants, that is widely used in production of beer, wine, fruit juice, bread, cake, and other bakery products. Other yeasts associated with food include *Kluyveromyces* (Batt 2000) and *Candida* (Hommel and Ahnert 2000). The former are able to resolve lactose, and they can be used in dairy products (e.g., kefir; O¨ zer and O¨ zer 2000). The making of sourdough and kefir (Salovaara 1998) gives good cases of the complex microbiological interactions between LAB and the yeasts, which produce the desired technological properties and flavor of the final products. Beside food fermentations, *Saccharomyces boulardii* is used as a probiotic (Elmer et al. 1996).

Because of aerobic metabolism, the food processing application of molds is based on the action of the different enzymes and their metabolic end products. In some countries, *Penicillium*-coated fermented sausages, meat and hams, and especially surface-ripened cheeses, made with the typical starter microorganisms, *P. camemberti* and *P. roqueforti*, are probably the best known examples of the use of molds in food production (Blank 2000). In some countries people are accustomed to using various types of molds (mainly of genus *Rhizopus*) to produce peanut, bean, or soybean-based products, for example, different types of soy sauce and tempeh (Gandjar 2000).

7.3 Transgenic Technology in Microorganisms Used in Food Industry

Application of microorganisms in food industry can be traced back thousands of years ago (Mckay and Baldwin 1990). Now, microorganisms play increasingly significant role in food-associated industry. Microorganisms (bacteria, filamentous fungi, and yeasts) can be utilized in food processing both in the preparation of different fermented foodstuffs and in producing food additives (organic acids, food enzymes, flavoring agents, etc.) (Adams 1999).

7.3.1 Bacterial Starter Cultures

Conventionally, bacterial starters for certain products and processes have been selected from among spontaneous mutants and natural variants. Now, molecular and genetic biological studies and the level of sophistication of the available genetic techniques change largely among different bacterial groups used in food fermentations. Many researches have been concentrated on LAB, especially on the lactococci (von Wright and Sibakov 1998). According to the series of detection, many of lactococcal metabolic plasmids code important functional properties, for example, lactose fermentation, aroma production, proteinase activity, phage resistance, restriction-modification systems, production of exopolysaccharides, and so on (McKay 1983). Since then, studies have culminated in the sequencing of the general genome of *Lactococcus lactis* IL1403 (Bolotin et al. 2001). The possibilities of real genetic modification of lactococci started at the progress of transformation techniques: first, protoplast transformation (Kondo and McKay 1982; von Wright et al.

1985) followed by electroporation (Harlander 1987; Holo and Nes 1989). Electroporation also was proved to be the method of choice to transform other species of LAB, for example, *Lactobacillus plantarum* (Scheirlinck et al. 1989; Bates et al. 1989), *Streptococcus thermophilus* (Mercenier 1990), and even the significant but poorly transformable *Lb. delbrueckii* subspecies *bulgaricus* in food industry (Serror et al. 2002).

Along with the introduction of transformation methods the progress of cloning vectors has experienced a flying progress. The small lactococcal plasmids had been cloned on antibiotic resistance markers. The first-generation vectors were based on the small lactococcal plasmids (Gasson and Anderson 1985; Kok et al. 1984). Followed, integration vectors which based on either homologous integration have been developed (Bhowmik et al. 1993) or phage attachment sites (Bronsted and Hammer 1999). Because of unacceptable antibiotic resistance markers in any last food applications, several food-grade cloning systems, which often associated with limited host-range plasmid replicons, have been designed. The selection is based on thymidylate synthase (thyA) gene (Ross et al. 1990), on bacteriocin resistance (Froseth et al. 1988; von Wright et al. 1990), on components of the proteolytic system (X-prolyl dipeptidyl aminopeptidase; Mayo et al. 1991), or on multicomponent systems (Emond et al. 2001). Just like ocher or amber suppressors against a background of nonsense mutants in purine biosynthesis genes (Dickely et al. 1995), because of being based on plasmids carrying LacF in the background of LacF-deficient host strains, complementation systems have also been introduced and LAB-signal sequences and specific promoters including bacterial secretory pathways have been definite and utilized to generate and secrete homologous and heterologous proteins in lactobacilli and lactococci. Examples involved in the product of bacillar alpha-amylase in lactococci and of clostridial endoglucanase in different lactobacilli (Bates et al. 1989). Obviously, the S-layer signal sequences could be potentially helpful for the optimum scheme of heterologous gene expression in LAB (Kahala and Palva 1999).

Compared with LAB, fewer genetic studies have been conducted on other food-related bacteria. Vectors and transformation systems have been represented for both brevibacteria (Bonnassie et al. 1990) and propionic acid bacteria (Jore et al. 2001). A few studies on the genetic modification in bifidobacteria have been reported, too (Kullen and Klaenhammer 2000).

7.3.2 *Yeasts and Filamentous Fungi*

Saccharomyces cerevisiae is perhaps genetically and biochemically the most famous eukaryote. The application of recombinant DNA technologies to this organism has long history, and its integrated genomic sequence is well known (Goffeau et al. 1996). The selection markers and vectors applied in yeast genetics differ in many aspects from those applied in bacteria (Sandhir et al. 2000). However vectors based on the episomal yeast are similar to prokaryotic plasmids; vectors which carried the yeast origin of a chromosomal centromeric region (the YCp plasmids) and DNA replication are considered by the yeast mitotic spindle, in cell division, ensuring precise partitioning. Yeast artificial chromosomes (YAC) including telomeric sequences could be linearized after the cloning event, giving a way to clone lengthy DNA inserts. Yeast selection markers representatively either complement auxotrophic or suppress mutations in the host or are based on the CUP1 gene, which conveys copper resistance (Henderson et al. 1985). From the safety point of view, these kinds of markers are less questionable than bacterial antibiotic resistance genes. While, a lot of yeast vectors are *Escherichia coli* yeast shuttle vectors, the *Escherichia coli* yeast shuttle vector commomly contains an antibiotic resistance gene and a prokaryotic origin of replication, in view of selection in the background of bacteria.

In filamentous fungi, vectors are available (Bowyer et al. 1994; Gems et al. 1994), while transformation efficiencies are usually low, and the resulting genetic constructs show integrations of the vector into the host genome. Linear, telomere containing plasmids have been suggested as more effective vectors (Barreau et al. 1998).

7.4 Potential Impact of Food-Associated GMMs on Human Body

7.4.1 Interactions with Human Microbiota

A lot of the final applications of GMMs in food would result in a condition in which viable GMMs would be consumed in great quantities by humans. This necessitates factor of the possible interactions between resident human microbiota and the GMMs. Each human being possesses a lot and multifaceted microbial community, connected with the skin and many mucous membranes. The main interactions between microorganisms connected with endogenous microbiota and food exist in the gastrointestinal tract, where microbial counts also come to the highest quantities. So as to estimate interactions, their significance to the host and the physiological role of microbiota below are first reviewed.

Human gastrointestinal microbiota are estimated to consist of 400 species, a lot of which are difficult or impossible to be cultivated commonly (Mikkelsaar et al. 1998; Willis and Gibson 2000). The species and the numbers vary largely in different locations in the gut. Because of its low pH, the stomach is regularly almost short of actual resident microorganisms. The existence of relatively fast transit times and bile acids maintains bacterial numbers low in the small intestine. In the colon, bacterial densities attain up to 1012 bacterial cells per gram luminal content; yeasts or other fungi generally show only an unimportant fraction of the general microbial count. The major bacterial genera, coming to densities of 1010 or more, are *Bifidobacterium*, *Bacteroides*, and *Eubacterium*. But numbers of groups, for example, *Enterococcus*, *Enterobacteriaceae*, and *Lactobacillus*, are some orders of magnitude lower.

Metabolic activities associated with microbiota are obviously important but highly hard to estimate or study. The saccharolytic activity of the colonic microbiota leads to the formation of short chain fatty acids (SCFAs), for example, butyrate, propionate, and acetate, making a significant contribution to the energy metabolism of the host (Willis and Gibson 2000). Intestinal bacteria can change both ingested xenobiotics and endogenous compounds. Bile acids and their derivatives could be dehydroxylated or deconjugated to the secondary bile acids (Eyssen and Caenepeel 1988). Microbial beta-glucosidases, beta-glucuronidases, azoreductases, and nitro-reductases activate xenobiotics or deconjugate to more poisonous derivatives (Mallet and Rowland 1990). Small molecular weight microbial metabolic products, many of which may have or have bad consequences, include hydrogen sulfide, ammonia, phenolic compounds, and indoles (Willis and Gibson 2000).

The gastrointestinal tract is connected with an extensive immune response system (Mayer 1998s), gut-associated lymphoid tissue (GALT), and diffuse lymphoid cells distributed along the intestine included. This system generally assures tolerance to indigenous microbes and general foods and defends against pathogens. The extent to which microorganisms, whether conventional or GMMs, connected with fermented foods can affect the intestinal microbiota lies in their ability to survive in the gastrointestinal tract or even to colonize it. While, different stress aspects, for example, the presence of bile acids, acidic conditions in the stomach, anaerobic conditions and effectively select against many types of microorganisms, human body temperature and food-related bacteria with fermentative metabolism can survive and even thrive in the intestinal tract in theory. Clearly there have been no human experiments on the colonizing potential of GMMs. Researches have concentrated on different traditional probiotic strains, as verification of their intestinal physiological and survival effects is asked for the documentation process of functional foods. The colonization of the human intestine by certain probiotic strains such as *Lactobacillus rhamnosus* GG (Alander et al. 1999; Saxelin et al. 1991), *Lactobacillus casei* F19 (Mattila-Sandholm et al. 1999), *Lactobacillus salivarius* UCC118 (Dunne et al. 2001), *Bifidobacterium infantis* UCC35624 (von Wright et al. 2002), and *Bifidobacterium lactis* Bb12 (Satokari et al. 2001) has been definitely detected. In some cases the administered strain has been detected in the feces some weeks after the stop of administration (Dunne et al. 2001), colonization in most cases is transient; after the end of feeding, the strain gradually disappears (Alander et al. 1999). Instead of colonization, the term "persistence" has been advised to represent this type of behavior (ILSI 1999).

Researches on probiotic microorganisms also offer the best documented proof that an external microorganism can change their interactions with the host and the functions of intestinal microbiota. The observed results include prevention and alleviation of different types of diarrhea, alleviation of the symptoms of lactose malabsorption, decreases in fecal mutagenicity, and modification of fecal enzymatic activities (Salminen et al. 1998). Probiotic intervention also affects immunological functions and enhances the products of IgA and IgM antibodies in individuals challenged with vaccines or bacterial or viral infections (He et al. 2000). It is worth noting that these effects have been reported with kinds of bacteria that in fact represent a small number of the intestinal microbiota. This suggests that the number of bacteria is not necessarily the decisive elements in crucial to their physiological activity. The viability of the strains does not even seem to be a complete demand on probiotic-related effects (cell adhesion, immunomodulating activities) to be clear (Ouwehand and Salminen 1998).

7.4.2 Gene Transfer Between Food-Related and Intestinal Microbes

Yeasts of the genus *Saccharomyces* are absolutely fungi, with an integrated sexual life cycle; therefore, they have the ability to exchange gene through mating. Many other filamentous fungi and yeasts that exist in fermented foodstuffs are partial, shorting of the sexual phase of the life cycle; therefore their ability to be contained exchange of gene is limited (Sutton 2000). For bacteria, the known mechanisms of natural gene transfer are phage-mediated DNA transfer or transduction from an infected cell to another host cell, then transformation or direct uptake of DNA from solution through the cell wall, and followed by conjugation or the transfer of DNA from the donor to the recipient by direct cell to cell contact. All these mechanisms may be involved in the genetic exchange between food-related microbes and gut microorganisms or among food-related microbes.

Except streptococci, which natural transformation is apparently less frequent, transduction and conjugation occur frequently in most of LAB (von Wright and Sibakov 1998). No researches about the gene transfer mechanisms of other food-associated bacterial groups appear to be reported. Including intergeneric and interspecific conjugation, conjugation could especially be a potentially essential mechanism of gene transfer between intestinal strains and food associated. Studies on conjugative transfer of broad host-range plasmids and transposable elements between lactococci, lactobacilli, *Listeria*, and enterococci have been reported, some of them are in vivo animal experiments (Morelli et al. 1988).

For the intestinal bacteria, conjugative transposable factors seem to be very general in *Clostridium* and *Bacteroides*. There are several evidences that multitude transfer of antibiotic resistance determinants has occurred both between *Bacteroides* and other intestinal genera and among the genus *Bacteroides* (Shoemaker et al. 2001).

Among streptococcal strains, transformation happens naturally. It has been probable to transform an oral streptococcus, since plasmid DNA is exposed to human saliva (Mercer et al. 1999). But the role of transformation in intestinal situations is generally difficult to estimate; it is interesting to note that experimental transformation of *Bacillus subtilis* and *Escherichia coli* in different food matrices has been observed (Bauer et al. 1999), suggesting that there could be genetic flow between diverse unrelated contaminants and food-related microorganisms. In a word, it seems that the intestinal microbial community shares a large amount of gene pool, likely mainly by conjugative DNA-transfer mechanisms. What's more, from this side, the food-related microorganisms may constitute a continuum of the intestinal microbiota, and this should be considered when evaluating the safety of food GMMs.

7.5 Regulatory Aspects

A variety of microorganisms used in food production have a long history of safe use that predates scientific assessment. However, the recipient strains do not have a history of safe use, and their safety will have to be established.

Based on the present European Union (EU) regulatory system, foods which consist of, contain, or derived from GMMs, besides complying with directive 2001/18/EC (previously 90/220/EC), fall automatically within the range of the Novel Food Regulation 258/97 EU (Hugget and Conzelmann 1997). So the safety should be firstly evaluated before final introduction to the market. But it should be noteworthy that food flavorings and additives are beyond the regulation. The final regulations on traceability and labeling of GMOs and products stemmed from GMOs remain still under consideration in the EU. Propositions which are proposed by the Commission and the Parliament are being discussed between two committees of the Parliament. They are not possible to enter into legislation until 2004.

7.6 Methods of Safety Evaluation of Food-Related GMMs

7.6.1 General Principles to Determine Risks and Benefits

But so far, there have been very few experiences of the toxicological evaluation both of complicated foods and even of individual microbial strains. Thus so as to facilitate safety assessment, the World Health Organization/Food and Agriculture Organization introduces the concept of substantial equivalence (see Chapter 8) (WHO/FAO; WHO 1995). If food component or a new food is proved to be substantially equivalent to an existing food component or food, it can be treated with the same manner for the purpose of safety. Then the toxicological evaluation can be concentrated on those sides in which the novel food or food component apparently differs from the conventional one. This concept has accidentally received criticism, but the recent WHO/FAO consultation about the safety assessment of foods stemmed from GMMs also recommended a case-by-case comparative way, using the concept of substantial equivalence. It should be highlighted that determination of substantial equivalence is not a substitute but an initial point for safety assessment.

In its strategies for the safety assessment of novel foods, the SAFEST concept (Safety Assessment of Food by Equivalence and Similarity Targeting) has been introduced by the International Life Science Institute (ILSI), so as to clear further the requirements and researches, the safety of a product needed to be established (ILSI 1995). If food composition or a novel food is substantially equivalent to a conventional foodstuff, it belongs to SAFEST Class 1, and that further safety studies are not needed. Foods that are "sufficiently similar" to conventional known food do not differ from a conventional counterpart excluding in a certain identifiable respects which belong to SAFEST Class 2, requiring the further researches and final testing. SAFEST Class 3 is reserved for foods for which no substantially equivalent or sufficiently similar conventional known foods can be found; therefore, the most comprehensive safety evaluation should be concentrated on foods in this class.

7.6.2 Viable GMMs in Foods

ILSI (1999) has used the SAFEST concept to particular cases of GMMs in foods and designed decision tree approaches to decide SAFEST classifications for specific GMMs.

Some levels of genetic modification maybe involved in final food applications of GMMs, each of them being different from the standpoint of substantial equivalence and therefore requiring a different way to safety assessment. At least the following cases should be deeply considered.

- The GMM strain includes DNA from the same or a closely associated species. For instance, a dairy starter strain can be transformed by a natural plasmid of a relevant strain coding for several functional properties that could make the strain more helpful in many specific application. Now the substantial equivalence of the eventual strain to the conventional one is clear, and the risks could be compared to those of any other starter strains having similar functions.
- The GMM strain has been modified to change, both quantitatively and qualitatively, the metabolic final products that exist in the food. So the safety assessment is more complicated, as both the safety of the intended metabolites and the possible unintended effects that the altered metabolic shifts might have caused have to be taken into consideration. Even when the metabolic pathways have been changed and are very well known and characterized, the controversial example of tryptophan products by a GMM (see above) reveals that unintended effects should not be ignored.
- Novel proteins or new enzymes have been introduced. In these cases the quantities and properties of the new proteins are, naturally, of essential significance. Proteins from known safe food origins and in quantities not different from the

conventional should not show a particular problem, but proteins with no previous history in food use or stemmed from sources that are known to induce problems in consumer subgroups need special concern. Allergenicity is one of the apparent risks that should be evaluated, based on the principles required by FAO/WHO. In most examples the toxicological evaluation of the purified proteins would possibly suffice, although if the novel enzymatic activities change the food matrix, these changes may deserve to concentrate in certain cases.

- The physiological properties of the microorganisms have been modified in a way that affects its colonization potency. In theory, for instance, modifying thermotolerance or the pH of a strain, consequently made better survival in the gastrointestinal tract, may result in disturbances in the functions and composition of the intestinal microbiota. Such conditions are hard to predict and that gastrointestinal models, animal experiments, and final human trials would possibly be demanded for the safety assessment.
- The GMM represents species with no long history in food industry use and species with known pathogenic strains. Under the circumstances the non-pathogenicity and non-toxicogenicity of the microorganism should be verified individually, besides evaluating the safety of the real genetic modification.

7.7 Researches on Food Safety of Transgenic Microorganism

7.7.1 Food Safety of Transgenic Microorganism in Some Laboratories

Existing researches on food safety of transgenic microorganism are rarely, which evaluated conventional indicators.

Different health parameters of transgenic microorganism *Lactococcus lactis* strains were evaluated, with proven in vivo beneficial effects, which are just as safe as the progenitor strain from which they were derived. Mice received an elevated concentration of different GM-LAB or the native parental strain from which they were derived during 30 days, and different health parameters were evaluated. Similar growth rates, hematological values, and other physiological parameters were obtained in the animals that received the transgenic LAB compared to those that were fed with the native strain (LeBlance et al. 2010)

The interaction between *Lactococcus lactis* NZ9000/pNPNK and intestinal microflora was evaluated as a method to assess safety of transgenic microorganism. *L. lactis* NZ9000/pNPNK is one kind of transgenic microorganism. Wistar rats were orally administered with transgenic microorganism *L. lactis* NZ9000/pNPNK and the host strain *L. lactis* NZ9000 for 6 weeks. Fecal and cecal contents were collected to determine the number of *L. lactis* NZ9000, *L. lactis* NZ9000/pNPNK, *Lactobacillus*, coliform bacteria, beneficial bacteria *Bifidobacterium*, and harmful bacteria *Clostridium perfringens*. The kidney, spleen, liver, and blood were evaluated for the bacterial translocation. No adverse effects were observed on the rat's body weight, serum biochemical or hematological parameters, or gut microflora. *L. lactis* NZ9000/pNPNK did not translocate to any organ or blood by plate cultivation method (Lee et al. 2010).

As a realistic model, genetically modified lactic acid bacteria were studied in meat fermentation. Model organisms were constructed by cloning into the competitive meat starter organisms *Lactobacillus curvatus* LTH1432 the lysostaphin gene lys of *Staphylococcus simulans* (model I) and the catalase gene katA of *Lactobacillus sake* LTH677 (model II). Investigation of the stability of the recombinant plasmids in the host revealed that the constructs were structurally stable, but the segregational stability was decreased, however, not beyond those numbers of generations required to ensure maintenance of the new property during the fermentation process. In fermenting sausages the constructs retained their full activity. Horizontal gene transfer was studied in vitro and during fermentation of sausages. No transfer of the recombinant marker genes was observed. On the other hand, conjugal transfer of pAMβ1 between strains of *L. curvatus* was detected, however, without mobilization of the recombinant plasmids. It is remarkable that under optimal laboratory conditions, no transfer of pAMβ1 between *L. curvatus* and *Staphylococcus carnosus* was observed, whereas the conditions prevailing in the meat matrix permitted an efficient transfer. The models used provide realistic examples of transgenic microorganism with the potential to improve food safety and/or quality (Hertel et al. 1995).

7.7.2 *Food Safety of Transgenic Microorganism in Our Laboratory*

To establish thorough food safety assessment of transgenic microorganism, *Lactobacillus plantarum* (Lp590) has been carried out a systematic study in our laboratory.

7.7.2.1 Assessment of Tolerance to Multistresses and In Vitro Cell Adhesion in Transgenic Microorganism Lp590

LAB have many excellent traits of industrial importance, which have been exploited for many industrial applications because of their physiological features, including their stress response.

Many transgenic organisms, including transgenic microorganism, have been engineered to meet various needs. Because LAB have many excellent features of industrial importance, they are the most common expression hosts for many good food-grade expression vectors that were developed for use in food, biology, and medical science (Mierau and Kleerebezem 2005; Simoes-Barbosa et al. 2004; Sorensen 2000).

Lp was isolated from health products. The gene transfer of Lp590 was completed in our laboratory. Lp590 derived from electroporation of pLEB590 as a food-grade vector (Takala et al. 2003) into *L. plantarum* showed that nisI functions as an immunity protein in *Lactobacilli*, enabling nisin resistant (Liu et al. 2011; Federal Register 1988). So Lp590 can continue to survive in the presence of the antibacterial peptide nisin. Nisin is produced by some *Lactococcus lactis* ssp. strains and shows an inhibitory effect on food spoilage and food-borne microorganisms. It is a food-grade bacteriocin that is widely used in the food industry as a bio-preservative because it usually inhibits the growth of most Gram-positive bacteria (Takala et al. 2002).

To fully assess the safety of transgenic microorganism, many tests can be performed, both in vitro and in vivo. The in vivo systems are used to investigate the effects of transgenic microorganism on health at the organismal level. The in vitro systems can provide in vitro insight into the effectiveness of these organisms for application purposes, and they are a prerequisite and the basis for the in vivo safety assessments. However, an in vitro assessment of the overall safety of transgenic microorganism is still lacking.

To establish a new in vitro evaluation model for transgenic microorganism, the effect of nisin resistance, five different environmental stresses (hot temperature, cold temperature, ethanol, H_2O_2, and NaCl), and upper gastrointestinal transit stress on Lp590 and Lp and their adherence properties were assessed.

The two strains of Lp and Lp590 were cultured and washed bacterial cell suspensions and colony counting. To determine the resistance of Lp590 to nisin, Lp and Lp590 of overnight culture were added to different concentrations of nisin in microtiter plates. The optical density was measured at different time.

To test environmental stress (52 °C, 4 °C, 20% ethanol (v/v), 15 mM H_2O_2, 4.0 M NaCl), the total viable counts of the two strains treated by above environmental stress conditions were determined anaerobically after 48 h of incubation at 37 °C.

To test gastrointestinal stress tolerance, two strains were cultured in simulated gastric (pH 2.0, 3.0, or 4.0) or small intestinal juices (pH 8.0). After different treatment time, total viable cell was counted.

For hydrophobicity assays, the bacterial suspensions were mixed with xylene by vortexing. The phases were allowed to separate by decantation. The aqueous phase was carefully removed, and the *A*600 was measured.

The washed bacteria and baker's yeast were incubated overnight at 4 °C. Agglutination was determined by bright-light microscopy. The reciprocal of the highest dilution that resulted in visible agglutination was recorded as the agglutination titer. Mucus from the rat's small intestine was prepared and coated in microtiter.

Wells coated with bovine serum albumin were used as the control. The bacterial suspension was added to each well and incubated at room temperature. The wells were washed and the binding was examined with an inverted microscope. The buffer was poured off, and after the wells had dried, the OD405 was measured in an ELISA plate reader.

The monolayers of Caco-2 cells, which were used in the adherence assays, were prepared by inoculating six-well tissue culture dishes. The 15-day-old Caco-2 monolayers were washed and the bacterial suspensions were added to each well. After incubation in a 5% CO_2 incubator and fixed at room, the Caco-2 monolayers of each well were washed, air-dried, and Gram stained. The adherent bacteria were detected microscopically by counting 15 randomized fields per study. The above incubation samples were centrifuged to obtain the supernatant. The lactate dehydrogenase (LDH) activity was measured based on the specific activity in the medium using a spectrophotometer. The protein content in the medium after the insult was measured by the Lowry method.

The host strain, Lp, displays the sensitivity of a nisin-negative strain. Lp590 grew well with 250 IU nisin/ml. Its growth was clearly weakened at concentrations of 250–350 IU/ml, and no growth was observed at 400 IU/ml. Hot (55 °C) and cold (4 °C) temperatures had a significant effect on the growth of both strains. When compared with Lp590, the hot temperature had a more significant effect on Lp. After 240 min at 4 °C, the viability counts of both Lp and Lp590 had been reduced to about 50% of the initial count.

Ethanol (20%), H_2O_2 (2%), and NaCl (4.0 M) also had an influence on the growth of the two strains. Exposure to ethanol (20%) for 240 min resulted in drastically decreased viability counts of Lp and Lp590. Cell counts of Lp and Lp590 exposed to H_2O_2 (2%), and 4 M NaCl for 240 min was above 50% of the initial count. Thus, Lp590 can tolerate H_2O_2 (2%) and NaCl (4.0 M) more than Lp. After 240 min, the cell counts of Lp590 exposed to 4.0 M NaCl only slightly decreased when compared with the counts of Lp cells.

Each strain had a lower viability in the simulated gastric juice at pH 2.0 than in simulated gastric juices at pH 3.0 or 4.0. The two strains maintained the same viability during the 240 min of simulated small intestinal transit in the absence of bile salts. The two strains showed a slight reduction in viable cell counts.

The two strains showed quite similar values of hydrophobicity. Lp590 had a moderate mucus-binding ability. The two strains adhered to the Caco-2 cell line poorly.

The results show that Lp and Lp590 had a moderate hydrophobicity, yeast agglutination, and adhesion capacity in vitro. The values of the hydrophobicity and agglutination had a positive correlation with the adhesion capacity, consistent with previous studies (Lindgren et al. 1990). The results of the activity of LDH revealed that the genetic transformation of Lp does not affect gene regulation or protein expression, nor does it produce potential toxic proteins, which could cause cell damage.

In this study, the resistance assay shows that Lp590 can grow well with 250 IU nisin/ml. Although the concentration of nisin that is required during food additive processing is different among industries in different countries and with different food products, 250 IU nisin/ml is generally sufficient to meet the various processing requirements, according to the GB2760-2003.

The results obtained in this study indicate that the ability of Lp590 to grow under the conditions of cold temperature (4 °C) and ethanol (20%), which are necessary in the vine and dairy industries, was equal to the ability of Lp, and both had a final viability of count of above 40%. A resistance to heat, H_2O_2, and osmotic variations are desirable properties of LAB, which are used in food fermentation, the brewing industry, and biopharmaceutical manufacturing. Compared with Lp, Lp590 was significantly more resistant to heat (52 °C), H_2O_2 (15 mM), and osmotic variation (4.0 M NaCl), and it shows an improvement on the properties shown in other lactic acid bacterium.

An ideal LAB strain with industrial potential should resist harsh conditions. In order to be used as potential probiotics, the strains need to be screened for their capacity to tolerate the conditions of the upper gastrointestinal tract. The pH of the stomach can reach as low as pH 1.5. Normally, food remains in the stomach between 2 and 4 h (Smith T 1995). A range of pH values, from pH 2.0 to 4.0, has been used to screen in vitro the acid tolerance of *Lactobacilli*, *Bifidobacteria*, and some dairy propionibacterial trains (Charteris et al. 1998; Chou and Weimer 1999; Chung et al. 1999; Conway et al. 1987; Lankaputhra and Shah 1995; Zarate et al. 2000). This study compared the effect of different pH of simulated gastric juices on the viability of the two strains (Lp and Lp590) during 240 min of simulated gastric transit. The two strains showed good trait of acid tolerance.

The transit time of food through the small intestine is generally between 1 and 4 h (Smith 1995), and the pH of the small intestine is around pH 8.0 (Keele and Neil 1965). The two strains (Lp and Lp590) demonstrated high levels of small intestinal transit tolerance.

Probiotic bacteria must survive in the small intestine. The adverse conditions of the small intestine include the presence of bile salts and pancreatin (Floch et al. 1972; Le Vay 1998). Adhesion capacity is also necessary for probiotic strains to colonize the small intestinal tract. The results show that Lp and Lp590 had only a small reduction in their viability.

For probiotic bacteria, the in vitro adherence assay is also important. The yeast cell is usually used as the host cell in an in vitro test. Bacterial adhesion to the intestinal mucosa includes a mucus-binding ability and an intestinal epithelial cell adhesion (Bernet et al. 1994). LDH is a macromolecular protein that leaks out of damaged cells. The assessment of cell damage was performed by measuring the release of LDH into the medium. Measuring the activity of LDH cell damage provided an objective evaluation. From the results of the tolerance tests in vitro, Lp has the tolerance that is necessary in the food industry. Compared with Lp, Lp590 has many of the same characteristics as Lp, but in some aspects, it also shows improved properties.

Lp590 can resist nisin and could play a role as a bio-preservative to meet the industrial requirements. This current study has demonstrated that Lp590 has a high tolerance of upper gastrointestinal transit and improved properties of environmental tolerance in vitro. It can adhere to Caco-2 cells and shows no damage to cell membranes in an in vitro cell model. This research has established a new in vitro evaluation model for transgenic microorganism.

7.7.2.2 The Effect of Transgenic Microorganism Lp590 on the Gut Health of Sprague-Dawley Rats

Whereafter, gut health of Lp590 was assessed in our laboratory.

As gene technology progresses from modifications aimed at relatively simple traits to addressing more complex attributes, the challenges of evaluating the unintended effects of transgenic microorganism on human health will increase (Lee et al. 2010). However, for issues concerning the safety, no specific instructions on how to perform the requested investigations are given.

Gastrointestinal (GI) microbiota acts as an essential "organ" and comprises a complex microecological system (Huang and Adams 2004). There is a slight difference of GI microbiota according to their histories and lifestyles, including diet. Gut health is inseparable from intestinal microbiota and their metabolites. Gut permeability and mucosal immunity can reflect the function of the intestinal barrier, further present intestine, and body health (Cani et al. 2009; Ley et al. 2008; Wang et al. 2004).

Microorganism intake can more directly effect on GI microbiota, their metabolites (Trebichavsky et al. 2009; Wang et al. 2004), gut permeability, and mucosal immunity. However, it is unfortunate that little has been done to establish a specific uniform concept of gut health with respect to transgenic microorganism safety.

In this study, we observed intestinal bacteria, their main metabolites, intestinal permeability, and intestinal mucosal immunity by administration of Lp and Lp590 to Sprague-Dawley (SD) rats for 4 weeks. We attempted to establish a transgenic microorganism food safety assessment gut health model.

One milliliter of bacterial suspension or PBS was fed to each rat using a sterile pipette once a day for 4 weeks. The general behavior signs of observation included changes in the skin, fur, eyes, and membranes, as well as respiratory, circulatory, autonomic and central nervous system function, and behavioral patterns. On day 0, 7, 14, 21, and 28, fresh fecal samples were taken from each animal for microbial and metabolic analysis. Rats were killed by exsanguinations for organ weighs, gross, and histopathological examinations. Blood biochemistry and hematology assay were also measured. The intestinal segments (jejunum and ileum) were immersed in liquid nitrogen and stored at −80 °C for further analysis.

The analysis of fecal microbiota was used by real-time quantification method. Microbiota analysis of fecal samples was used by denaturing gradient gel electrophoresis method. Fecal enzyme activity was analyzed by reference methods. Fecal SCFAs were analyzed by gas chromatograph. The DX-4000-FITC concentration was analyzed with a fluorescence spectrophotometer. The analysis of ZO-1 and occludin was used by immunohistochemistry localization and RT-PCR. And the analysis of SIgA was also used by RT-PCR.

During the course of the experiment, no treatment-related adverse effects in the clinical appearance of the animals were observed. There were no statistically significant differences between male and female microorganism-fed groups and the control group for all the parameters measured. There were no group-related alterations or statistically identified differences in the organ/body weights of male and female rats given Lp or Lp590, when compared with the control groups. No group-related histopathological observations were found. The few spontaneous findings observed were generally minimal to slight/mild in severity, were randomly distributed among all groups, and were the type commonly observed in control rats of this age and strain. In this study, there was no mortality or toxic effect observed in rats after subchronic oral administration of Lp590.

Bifidobacterium and *C. perfringens* are the most important bacteria in the GI tract, and some researchers considered them to be beneficial and harmful bacteria, respectively (de Vrese and Schrezenmeir 2008). At all five time points (day 0, 7, 14, 21, and 28), no statistically significant differences in the numbers of specific bacteria were found between the Lp590 group and its parental group (Lp). Compared with the PBS-fed group, significant changes of the two bacteria were observed in microorganism-fed groups, with higher levels of *Bifidobacterium* and lower levels of *C. perfringens* ($P < 0.05$) at day 7 and day 14. After 14 days, there was no notable difference in cell numbers of the two bacteria in all groups. Lp has long been considered as a probiotic and been used for food fermentation. Probiotic bacteria increased the number of beneficial bacteria *Bifidobacterium* and reduced the number of harmful bacteria *C. perfringens* in the early phase of treatment which is an indisputable fact. The results indicated that Lp do play a role of improvement health. The similar result can be found in many references about assessment of *Lactobacillus* strains. From the result of DGGE at the time point of day 28, the bacterial composition in Lp- and Lp590-fed rats shows high similarities to those in PBS-fed rats. Consistency of data changes in Lp and Lp590 groups demonstrated that both transgenic microorganism and nontransgenic microorganism Lp affected intestinal microbiota in a similar way.

In contrast to the control group, Lp- and Lp590-fed rats exhibited a statistically significant decrease in the activities of β-glucuronidase, β-glucosidase, and nitroreductase, which are known to be associated with the conversion of procarcinogens into proximal carcinogens (Freeman 1986) before 14 days. Then, the three enzyme activities showed no remarkable change in all groups at day 21 and day 28. According to previous studies, a decrease in β-glucuronidase, β-glucosidase, and nitroreductase, which had positive correlation with beneficial *Bifidobacterium* (Freeman 1986), can promote intestinal health (Kumar et al. 2009). Our results just proved the above correlation. And lower three enzyme activities in Lp- and Lp590-fed groups at day 7 and day 14 ($P < 0.05$) further suggested that there is no different effect of nontransgenic microorganism and transgenic microorganism *L. plantarum* on intestinal microbial metabolites.

In contrast to the control group, the level of butyric acid of Lp- and Lp590-fed groups significantly increased at day 7 and day 14 ($P < 0.05$). At other time points, for the level of butyric acid, there was no prominent change in all groups. It is known that *Bifidobacterium* produce more SCFAs especially butyric acid (Wang et al. 2004), which was in agreement with our results. Compared with control group, the level of both acetic acid and propionic acid is not significant increase in microorganism-fed groups. However, in view of no statistically difference in nontransgenic microorganism or transgenic microorganism groups, it cannot be regarded as being associated with Lp590 consumption.

Relatively large changes of the fecal bacteria and their metabolites occurred in the first 2 weeks, which were both related to the fast growing of the rats during this period and the different diets. And gender cannot affect any test result. These results were consistent with the study of Wang et al. (Wang et al. 2004). Fecal bacteria and metabolites play a crucial part in gut health. As collection of fecal samples is easy and related tests are nondestructive and can be monitored continuously, feces could be proposed to be a vital target for evaluation of genetically modified food gut health.

Both Lp590- and Lp-fed rats showed a twofold lower plasma DX-4000-FITC area under the curve compared with the PBS-fed rats. Gut permeability is controlled by several specific tight junction proteins (Cani et al. 2009). Among these, ZO-1 and occludin have been proposed as key markers of tight junction integrity [48]. Lp and Lp590 treatment increased ZO-1 and occludin mRNA in the jejunum segment. ZO-1 and occludin immunohistochemical staining appeared to be increased uniformly and continuously distributed in Lp- and Lp590-fed rat tissues. In accordance with the mRNA analysis, compared with the control group, Lp- and Lp590-fed improved the tight junctions of rats ($P < 0.05$), because the immunohistochemical staining of both proteins localized along the apical cellular border was greater. Previous studies on the relationship between intestinal beneficial bacteria and permeability illustrated that *Bifidobacteria* can augment two proteins of ZO-1 and occludin and then promote gut permeability (Cani and Delzenne 2007). The tendency also appeared in this study. To identify whether the changes in the gut microbiota and the expression of tight-junction proteins were associated with in vivo gut permeability, multiple correlation analysis between these parameters was performed. Indeed, a negative correlation between tight-junction protein mRNA level and plasma DX-4000-FITC was found in the jejunum segment.

The results of the RT-PCR analysis showed that the level of SIgA in the ileum was higher in the Lp and Lp590 groups than in the control group ($P < 0.05$). The results of previous studies with animals and volunteers as object all suggested that live bacteria are especially efficient mucosal immunogens (Muir et al. 2000). Moreover, the synthesis and excretion of SIgA depend mainly on healthy and stable beneficial intestinal bacteria especially *Bifidobacteria* (Pierce et al. 1987). It was once again verified from our results that the number of *Bifidobacteria* on day 28 and SIgA mRNA level had positive correlation.

In this study, based on subchronic test, intestinal microbiota, their metabolites, gut permeability, and gut mucosal immunity of the rats fed with Lp590 and Lp were tested, analyzed, and discussed. And eventually we concluded that there are no significant differences between Lp590- and Lp-fed groups. The conclusion of gut health was comparable with that from the traditional subchronic test. Evaluation of gut health including intestinal microbiota, their metabolites, gut permeability, and gut mucosal immunity will be a new approach of assessing the safety of transgenic microorganism.

7.7.2.3 Safety Evaluation of Subchronic Feeding of Transgenic Microorganism Lp590

The safety assessment studies of transgenic microorganism are rarely reported. Most of existing safety evaluations are based on short-term feeding trial (Lee et al. 2010; Liu et al. 2011). Previous safety tests focused on the effect of transgenic microorganism on gut immunology, intestinal microflora, or colonization. Four weeks feeding trial showed that the gut health of SD rats was not affected by Lp590 compared with Lp (Liu et al. 2011). To assess the potential adverse effects of long-term exposure of transgenic microorganism, 90-day rodent feeding studies were recommended (CAC/GL 2003; EFSA 2008; FAO/WHO 2000). A 90-day toxicity study was conducted to evaluate the subchronic safety of Lp590.

Lp and Lp590 were provided by the Chinese Academy of Microbiology Culture Collection. Lp and Lp590 were anaerobically cultured in broth (Hispanlab, Madrid, Spain) at 37 °C for 16 h. Bacteria solution was centrifuged at 6000 × g for 15 min, washed three times with phosphate buffer solution, and resuspended at 2.5×10^9 CFU/ml for subsequent study.

The feeding trial was conducted at the specific pathogen-free animal laboratory of Supervision and Testing Center for Transgenic Organisms Food Safety, Ministry of Agriculture (Beijing, China). The experiment was in compliance with the OECD Good Laboratory Practice guidelines (OECD 1997). Male and female SD rats, a 4-week-old with an average body weight of 40–60 g, were purchased from Vital River Laboratory Animal Technology Co. Ltd. (Beijing, China).

Rats were housed as 5 per cage. The animal room was maintained at 23 °C ± 2 °C with relative humidity 50% ± 20%. During two-week acclimation, all of the rats were fed with the rodent basic diet. A 12 h light/dark cycle was provided by automated fluorescent illumination. The rats had ad libitum access to their food and water. This study complied with CAC/GL 46-2003 safety assessment of transgenic microorganism and derived products: 90-day feeding tests in rats.

After acclimation, rats were randomly divided into six experimental groups (ten/sex/group) based on body weight. 1 ml bacterial suspensions (2.5×10^9 CFU/ml) or physiological saline was fed orally to each rat using a sterile pipette once a day for 90 days. The bacterial suspensions were freshly prepared every day. During the exposure period, body weight and food consumption were determined weekly. Food efficiency was calculated based on the food consumption and body weight.

$$\text{Mean weekly feed utilization}(\%) = \left(\begin{array}{l}\text{weekly body weight gain } \textit{per} \text{ cage /} \\ \text{weekly feed consumption } \textit{per} \text{ cage}\end{array}\right) \times 100.$$

Cage-side observation including the coat condition, skin, eyes, and excretion was conducted at least every 2 days throughout the study.

Hematology and serum chemistry were conducted on all rats on the 90th day (the end of the trial). Rats were fasted for 12 h before blood collection. Blood samples were collected from the inner canthus under anesthesia.

Whole blood from the inner canthus was collected with anticoagulant and analyzed for white blood cell count (WBC), red blood cell count (RBC), hemoglobin (HGB), hematocrit (HCT), mean corpuscular volume (MCV), mean corpuscular hemoglobin (MCH), mean corpuscular hemoglobin concentration (MCHC), platelet count (PLT), mean platelet volume (MPV) with Blood Cell Counter HEMAVET950 (Drew Scientific, Dallas, USA).

Whole blood from the inner canthus was collected without anticoagulant. Sera obtained following centrifugation was assayed for lactic dehydrogenase (LDH), alkaline phosphatase (ALP), alanine aminotransferase (ALT), aspartate aminotransferase (AST), total protein (TP), albumin (ALB), urea nitrogen (BUN), creatinine (CREA), glucose (GLU), calcium (Ca), phosphorus (P), triglyceride (TG), cholesterol (CHO), high-density lipoprotein (HDL), and low-density lipoprotein (LDL) with Biochemical Analyzer HITACHI 7020 (HITACHI, Tokyo, Japan).

Rats were killed by decapitation. The following organs were weighed after extraneous fat was removed (paired organs weighed together): the heart, liver, spleen, lungs, kidneys, adrenal glands, thymus, brain, ovaries and uterus (female), or testes and epididymides (male). Relative organ weights (percent of final body weight) were calculated. The stomach and intestine (duodenum, jejunum, ileum) from animals in all groups were fixed in 10% neutral buffered formalin. All processed tissues including the heart, liver, spleen, lungs, kidneys, adrenal glands, thymus, brain, ovaries and uterus (female), testes and epididymides, stomach, and intestine (duodenum, jejunum, ileum) were embedded in paraffin, sectioned approximately 5–6 μm thick, stained with hematoxylin and eosin (H&E) and examined microscopically by the pathologists at the Institute of Medicinal Plant Development, Chinese Academy of Medical Sciences (Beijing, China).

All rats survived in the 90-day study, and no signs of adverse clinical effects were observed in any rat (data not shown). No differences were observed in body weight, food consumption, weight gains, or food efficiency (data not shown) between rats fed with Lp590 and Lp. No statistically significant difference ($P > 0.05$) in mean relative organ weight was observed in the test group compared with control group.

Certain hematology variables were significantly different between CK group and the treatment groups (see RBC, MPV of male rats, and HCT of female rats) but were not different between Lp and Lp590 group. This suggests that these differences are likely attributable to probiotics and therefore were not related to exposure to Lp590.

As with the hematology variables, a number of statistically significant differences were observed between the CK group and the treatment groups. Differences observed in serum chemistry were attributed to the background variability and sporadic deviation. They were not considered to be related with the Lp590.

A complete gross necropsy and microscopic anatomic pathological analysis were conducted on all animals after a 90-day feeding study. Incidental tumid uterus (swollen and red) was observed randomly among all of treatment groups. But this phenomenon was commonly observed in rats of this species and age (He et al. 2009) and was not considered to be test substance related. Lung inflammatory cell infiltrates were also observed in a number of rats, and this is another common finding among experimental rats of this strain and age (Appenzeller et al. 2008; Delaney et al. 2008; McNaughton et al. 2008).

Therefore, this histopathological lesion was neither interpreted as adverse nor was considered related to Lp590.

LAB are generally regarded as safe organisms and therefore widely used in the production and preservation of fermented products by the food industry. There is an increasing interest in the biotechnological manipulation of LAB to improve food industrial processes. LP is an important LAB model and tools are currently available for the genetic improvement of its natural properties (Takala et al. 2003). Food-grade bacteriocin-resistance markers are closely related to the antibiotic resistance selection markers. Nisin is the most studied among bacteriocins. Nisin has been accepted in more than 50 countries (Delves-Broughton et al. 1996) as a food preservative and commercially available. Thus, the usage of a nisin resistance selection marker for developing bacteriocin-resistant food-grade starters for the food industry is favorable (Froseth and McKay 1991).

The expression of *nisI* gene can resist nisin Lp590 with *nisI* gene and can still exist under the circumstances of high concentration of nisin. As a new food source, the food safety for Lp590 should be evaluated properly. The most commonly toxicity study on transgenic food is a 90-day (i.e., subchronic) feeding study in rats (Dryzga et al. 2007; Hammond et al. 2006; Schroder et al. 2007; Seralini et al. 2007; Tang et al. 2007). So the same method was used in our study.

Appetite and activity are the most general and sensitive indicators of health status for animals. In accordance with this, the behavior and condition was one of the indexes to evaluate the subchronic toxicity of Lp590 (Gunning and Moores 2010; He et al. 2008; Liu et al. 2012; Yimam et al. 2010; Zhu et al. 2013). All animals were healthy and survived after orally administered for 90 days. No bowel disease such as diarrhea was observed. No body weight loss indicated that growth rates of the rats by administration of high dose of Lp or Lp590 were unaffected. No adverse effects on feed consumption, body weight, and appearance were found, demonstrating that Lp590 does not exhibit subchronic oral toxicity effects on the animals' growth and development.

Hematological assays are typically used to examine the adverse effects of the experimental material on blood system. The composition and changes in the number of blood cells might imply pathological changes in bone marrow or the immune system (Cao et al. 2012; Yakubu et al. 2007; Zhou et al. 2012). Serum biochemical examination is important to detect nutritional deficiencies or nutrient metabolism imbalances. If some serum biochemical parameters were altered, the corresponding tissues or organs might be affected (Liu et al. 2012; Swendseid 1987). In the present study, some mean values of hematology and serum biochemical parameters were slightly higher or lower in treatment groups compared to control group; however, these differences were considered to be randomly changed, which possibly resulted from experimental methods or the individual differences between rats.

In conclusion, administration of Lp590 showed no adverse effects on rats' physiology during our 90-day feeding study. The results suggested that genetically modified strain Lp590 can be considered as safe as the Lp.

7.8 Prospects

The use of gene technologies in the industrial manufacture is at a mature stage and has profoundly impacted the development of this process. Therefore, it is likely that the number of transgenic microorganism production strains will continue to increase. Transgenic microorganism used for production of food and food ingredients may be consumed alive or killed. Relevant criteria for their safety assessment should also include the potential for colonization of the gastrointestinal ecosystem and horizontal stability of vectors, regulation elements and foreign genes including target and marker genes, metabolites, immunological effects, viability, and residence of microorganisms in the human gastrointestinal tract, nutritional modification, allergenicity, and toxicology. In most cases, food flavors, food enzymes, food additives, and other food ingredients produced by transgenic microorganism do not include either modified DNA or the production organism. Therefore, the general principles of the safety evaluation of chemical mixtures could be used to these substances. However, the probability of unintended contaminants lying in the production microorganism or process should be considered. As a supplement of existing analyses, various omics technologies including transcriptomics, proteomics, and metabolomics will help to assess the potential for any unintended effects.

References

(JECFA), J.E.C.o.F.A. Ochratoxin A. Joint FAO/WHO Expert Committee of Food Additives. In: 37th meeting, 10–12 June 1990. Geneva: World Health Organization (WHO); 1991. p. 365–417.

(JECFA), J.E.C.o.F.A. Ochratoxin A. Joint FAO/WHO Expert Committee of Food Additives. In: 44th meeting, Technical Report Series Number 859. Geneva: World Health Organization (WHO); 1995.

†, M.L.R, Seymour RE, Whiteley LO, Sundberg JP, Hogenesch H. Anti-IL5 decreases the number of eosinophils but not the severity of dermatitis in Sharpin-deficient mice. Exp Dermatol. 2010;19:252–8.

AACC. Approved methods of the AACC (8th edn), St Paul: American Association of Cereal Chemists; 1983. Method 08–01, 44–15A, 46–13, 54–20.

Abdeen A, Miki B. The pleiotropic effects of the bar gene and glufosinate on the Arabidopsis transcriptome. Plant Biotechnol J. 2009;7:266–82.

Abdeen A, Schnell J, Miki B. Transcriptome analysis reveals absence of unintended effects in drought-tolerant transgenic plants overexpressing the transcription factor ABF3. BMC Genomics. 2010;11:69.

Abraham WM. Animal models of asthma. Curr Drug Targets. 2008;9:436–7.

Adams MR. Safety of industrial lactic acid bacteria. J Biotechnol. 1999;68(2-3):171–8.

ADAS. Strategy support for the post-market monitoring (PMM) of GM plants: Review of existing PPM strategies developed for the safety assessment of human and animal health. EFSA supporting publication. 2014:EN-739.

Ahuja V, Quatchadze M, Ahuja V, Stelter D, Albrecht A, Stahlmann R. Evaluation of biotechnology-derived novel proteins for the risk of food-allergic potential: advances in the development of animal models and future challenges. Arch Toxicol. 2010;84:909–17.

Al-Anati L, Petzinger E. Immunotoxic activity of ochratoxin A. J Vet Pharmacol Ther. 2006;29:79–90.

Alander M, Satokari R, Korpela R, Saxelin M, Vilpponen-Salmela T, Mattila-Sandholm T, von Wright A. Persistence of colonization of human colonic mucosa by a probiotic strain Lactobacillus rhamnosus GG, after oral consumption. Appl Environ Microbiol. 1999;65:351–4.

Alcantara EP, Aguda RM, Curtiss A, Dean DH, †, M.B.C. Bacillus thuringiensis δ-endotoxin binding to brush border membrane vesicles of rice stem borers. Arch Insect Biochem Physiol. 2004;55:169–77.

Aldemir H, Bars R, Herouet-Guicheney C. Murine models for evaluating the allergenicity of novel proteins and foods. Regul Toxicol Pharmacol. 2009;54:52–7.

K. Huang, *Safety Assessment of Genetically Modified Foods*,
DOI 10.1007/978-981-10-3488-6

Alderborn A, Sundstrom J, Soeria-Atmadja D, Sandberg M, Andersson HC, Hammerling U. Genetically modified plants for non-food or non-feed purposes: straightforward screening for their appearance in food and feed. Food Chem Toxicol. 2010;48:453–64.

Alexander J, Autrup H, Bard D, Benford D, Carere A, Costa L, Cravedi J, di Domenico A, Fanelli R, Fink-Gremmels J. Opinion of the scientific panel on contaminants in the food chain on a request from the commission related to ochratoxin A in food. EFSA J. 2006;365:1–56.

Altenbach SB, Kuo CC, Staraci LC, Pearson KW, Wainwright C, Georgescu A, Townsend J. Accumulation of a Brazil nut albumin in seeds of transgenic canola results in enhanced levels of seed protein methionine. Plant Mol Biol. 1992;18:235–45.

Anderson NL, Anderson NG. The human plasma proteome history, character, and diagnostic prospects. Mol Cell Proteomics. 2002;1:845–67.

Anderson FH, Zeng L, Rock NR, Yoshida EM. An assessment of the clinical utility of serum ALT and AST in chronic hepatitis C. Hepatol Res. 2000;18:63–71.

Anonymous. The safety of genetically modified foods produced through biotechnology. Toxicol Sci. 2003;71:2–8.

AOAC. Determination of lead, calcium, copper, iron, zinc, sodium, magnesium, phosphorus, potassium, manganese in foods, AOAC official methods 999.11. Association of Official Agricultural Chemists; 2000.

Appelgren L-E, Arora R. Distribution of 14 C-labelled ochratoxin A in pregnant mice. Food Chem Toxicol. 1983;21:563–8.

Appenzeller LM, Munley SM, Hoban D, Sykes GP, Malley LA, Delaney B. Subchronic feeding study of herbicide-tolerant soybean DP-356O43-5 in Sprague-Dawley rats. Food Chem Toxicol. 2008;46:2201–13.

Appenzeller LM, Malley L, Mackenzie SA, Hoban D, Delaney B. Subchronic feeding study with genetically modified stacked trait lepidopteran and coleopteran resistant (DAS-Ø15Ø7-1xDAS-59122-7) maize grain in Sprague-Dawley rats. Food Chem Toxicol. 2009;47:1512–20.

Aris A, Leblanc S. Maternal and fetal exposure to pesticides associated to genetically modified foods in Eastern Townships of Quebec. Canada Reprod Toxicol. 2011;31:528–33.

Armstrong CL, Mikel WB, Cromwell GL. Sensory evaluation of pork longissimus muscle from swine fed soybean meal from Roundup Ready or conventional soybean. J Anim Sci. 2001;79(Suppl 1):374. (Abstr 1550)

Arshad SH, Malmberg E, Krapf K, Hide DW. Clinical and immunological characteristics of Brazil nut allergy. Clin Exp Allergy. 1991;21:373–6.

Astwood JD, Leach JN, Fuchs RL. Stability of food allergens to digestion in vitro. Nat Biotechnol. 1996;14:1269–73.

Aulrich K, Böhme H, Daenicke R, Halle I, Flachowsky G. Genetically modified feeds in animal nutrition. 1st communication: Bacillus thuringiensis (Bt) corn in poultry, pig and ruminant nutrition. Arch Anim Nutr. 2001;54:183–95.

Axelsson A. Lactic acid bacteria: classification and physiology. Lactic acid bacteria: microbiology and functional aspects. New York: Marcel Dekker Inc.; 1998. p. 1–72.

Babu E, Takeda M, Narikawa S, Kobayashi Y, Enomoto A, Tojo A, Cha SH, Sekine T, Sakthisekaran D, Endou H. Role of human organic anion transporter 4 in the transport of ochratoxin A. Biochim Biophys Acta. 2002;1590:64–75.

Bajaj S, Mohanty A. Recent advances in rice biotechnology—towards genetically superior transgenic rice. Plant Biotechnol J. 2005;3:275–307.

Bakan B, Melcion D, Richard-Molard D, Cahagnier B. Fungal growth and fusarium mycotoxin content in isogenic traditional maize and genetically modified maize grown in France and Spain. J Agric Food Chem. 2002;50:728–31.

Baker JM, Hawkins ND, Ward JL, Lovegrove A, Napier JA, Shewry PR, Beale MH. A metabolomic study of substantial equivalence of field-grown genetically modified wheat. Plant Biotechnol J. 2006;4:381–92.

Ballongue J. Bifidobacteria and probiotic action. Lactic acid cacteria: microbiology and functional aspects. New York: Marcel Dekker Inc.; 1998. p. 519–87.

Barbosa HS, Arruda SCC, Azevedo RA, Arruda MAZ. New insights on proteomics of transgenic soybean seeds: evaluation of differential expressions of enzymes and proteins. Anal Bioanal Chem. 2012;402:299–314.

Barlow SM, Greig JB, Bridges JW, Carere A, Carpy AJ, Galli CL, Kleiner J, Knudsen I, Koeter HB, Levy LS, Madsen C, Mayer S, Narbonne JF, Pfannkuch F, Prodanchuk MG, Smith MR, Steinberg P. Hazard identification by methods of animal-based toxicology. Food Chem Toxicol. 2002;40:145–91.

Barreau C, Iskandar M, Turcq B, Javerzat JP. Use of linar plasmid containing telomeres as an efficient vector for direct cloning in the filamentous fungus Podospora anserine. Fungal Genet Biol. 1998;25:22–30.

Barrière Y, Vérité R, Brunschwig P, Surault F, Emile JC. Feeding value of corn silage estimated with sheep and dairy cows is not altered by genetic incorporation of Bt176 resistance to Ostrinia nubilalis. J Dairy Sci. 2001;84:1863–71.

Barros E, Lezar S, Anttonen MJ, Dijk JPV, Röhlig RM, Kok EJ, Engel KH. Comparison of two GM maize varieties with a near-isogenic non-GM variety using transcriptomics, proteomics and metabolomics. Plant Biotechnol J. 2010;8:436–51.

Bates EEM, Gilbert HJ, Hazlewood GP, Huckle J, Laurie JI, Mann SP. Expression of a Clostridium thermocellum endoglucanase gene in Lactobacillus plantarum. Appl Environ Microbiol. 1989;55:2095–7.

Bates JH, Rincon M, Irvin CG. Animal models of asthma. Am J Phys. 2009;297:L401–10.

Batt CA. Encyclopedia of food microbiology. London: Academic Press; 2000. p. 1115–8.

Baudo MM, Lyons R, Powers S, Pastori GM, Edwards KJ, Holdsworth MJ, Shewry PR. Transgenesis has less impact on the transcriptome of wheat grain than conventional breeding. Plant Biotechnol J. 2006;4:369–80.

Bauer F, Hertel C, Hammes WP. Transformation of Escherichia coli in foodstuffs. Syst Appl Microbiol. 1999;22:161–8.

Beever DE, Kemp CF. Nutrition abstracts and reviews series B: livestock feeds and feeding safety issues associated with the DNA in animal feed derived from genetically modified crops. A review of scientific and regulatory procedures. Am J Med Sci. 2000;184:113.

Behrens MR, Mutlu N, Chakraborty S, Dumitru R, Jiang WZ, Lavallee BJ, Herman PL, Clemente TE, Weeks DP. Dicamba resistance: enlarging and preserving biotechnology-based weed management strategies. Science. 2007;316:1185–8.

Belgacem O, Stübiger G, Allmaier G, Buchacher A, Pock K. Isolation of esterified fatty acids bound to serum albumin purified from human plasma and characterised by MALDI mass spectrometry. Biologicals. 2007;35:43–9.

Bendele AM, Carlton WW, Krogh P, Lillehoj EB. Ochratoxin A carcinogenesis in the (C57BL/6J X C3H)F1 mouse. J Natl Cancer Inst. 1985;75:733–42.

Benford D, Boyle C, Dekant W, Fuchs R, Gaylor D, Hard G. Ochratoxin A. International Programme on Chemical Safety (INCHEM), Joint FAO/WHO Expert Committee on Food Additives (JECFA). Monographs & Evaluations. WHO Food Addit, serie 41; 2001.

Berman KH, Harrigan GG, Riordan SG, Nemeth MA, Hanson C, Smith M, Sorbet R, Zhu E, Ridley WP. Compositions of seed, forage, and processed fractions from insect-protected soybean Mon 87701 are equivalent to those of conventional soybean. J Agric Food Chem. 2009;57:11360–9.

Berman KH, Harrigan GG, Riordan SG, Nemeth MA, Hanson C, Smith M, Sorbet R, Zhu E, Ridley WP. Compositions of forage and seed from second-generation glyphosate-tolerant soybean Mon 89788 and insect-protected soybean Mon 87701 from Brazil are equivalent to those of conventional soybean (Glycine Max). J Agric Food Chem. 2010;58:6270–6.

Berman KH, Harrigan GG, Nemeth MA, Oliveira WS, Berger GU, Tagliaferro FS. Compositional equivalence of insect-protected glyphosate-tolerant soybean MON 87701 × MON 89788 to

conventional soybean extends across different world regions and multiple growing seasons. J Agric Food Chem. 2011;59:11643–51.
Bernal-Santos G, O'Donnell AM, Vicini JL, Hartnell GF, Bauman DE. Hot topic: enhancing omega-3 fatty acids in milk fat of dairy cows by using stearidonic acid-enriched soybean oil from genetically modified soybeans. J Dairy Sci. 2010;93:32–7.
Bernet MF, Brassart D, Neeser JR, Servin AL. Lactobacillus acidophilus LA 1 binds to cultured human intestinal cell lines and inhibits cell attachment and cell invasion by enterovirulent bacteria. Gut. 1994;35:483–9.
Bernstein JA, Bernstein IL, Bucchini L, Goldman LR, Hamilton RG, Lehrer S, Rubin C, Sampson HA. Clinical and laboratory investigation of allergy to genetically modified foods. Environ Health Perspect. 2003;111:1114–21.
Betz FS, Hammond BG, Fuchs RL. Safety and advantages of Bacillus thuringiensis-protected plants to control insect pests. Regul Toxicol Pharmacol. 2000;32:156–73.
Beute MK, Porter DM, Hadley BA. Sclerotinia blight of peanut in North Carolina and Virginia and Its chemical control. Plant Dis Rep. 1975;59:697–701.
Beyer K, Ellman-Grunther L, Järvinen KM, Wood RA, Hourihane J, Sampson HA. Measurement of peptide-specific IgE as an additional tool in identifying patients with clinical reactivity to peanuts. J Allergy Clin Immunol. 2003;112:202–7.
Bhandari MR, Kawabata J. Cooking effects on Oxalate, Phytate, Trypsin and A -Amylase inhibitors of wild yam tubers of Nepal. J Food Compos Anal. 2006;19:524–30.
Bhowmik T, Fernandez L, Steele J. Gene replacement in lactobacillus helveticus CNRZ32. J Bacteriol. 1993;175:6341–4.
Blanco C, Ortega A, Dominguez C. Comparison of skin-prick test and specific serum IgE determination for the diagnosis of latex allergy. Clin Exp Allergy. 1998;28:971–6.
Blank G. Encyclopedia of food microbiology. London: Academic Press; 2000. p. 1655–62.
Bøgh KL, Kroghsbo S, †, L.D, Rigby NM, Barkholt V, Mills ENC, Madsen CB. Digested Ara h 1 has sensitizing capacity in Brown Norway rats. Clin Exp Allergy. 2009;39:1611–21.
Böhme H, Aulrich K, Daenicke R, Flachowsky G. Genetically modified feeds in animal nutrition 2nd communication: Glufosinate tolerant sugar beets (roots and silage) and maize grains for ruminants and pigs. Arch Anim Nutr. 2001;54:197–207.
Bolotin A, Wincker P, Mauger S, Jaillon O, Malarme K, Weissenbach J, Ehrlich SD, Sorokin A. The complete genome sequence of the lactic acid bacterium Lactococcus lactis ssp. lactis IL1403. Genome Res. 2001;11:731–53.
Bondzio A, Stumpff F, Schön J, Martens H, Einspanier R. Impact of Bacillus thuringiensis toxin Cry1Ab on rumen epithelial cells (REC) – a new in vitro model for safety assessment of recombinant food compounds. Food Chem Toxicol. 2008;46:1976–84.
Bonnassie S, Burini JF, Oreglia J, Trautwetter A, Patte JC, Sicard AM. Transfer of plasmid DNA to Brevibacterium lactofermentum by electrotransformation. J Gen Microbiol. 1990;136:2107–12.
Bowers E, Hellmich R, Munkvold G. Comparison of fumonisin contamination using HPLC and ELISA methods in bt and near-isogenic maize hybrids infested with European corn borer or western bean cutworm. J Agric Food Chem. 2014;62:6463–72.
Bowyer P, Osbourn AE, Daniels MJ. An "instant gene bank" method for heterologous cloning: complementation of two Aspergillus nidulans mutants with Gaeumannomyces graminis DNA. Mol Gen Genet. 1994;242:448–54.
Boyce JA, Burks AW, Jones SM, Sampson HA, Wood RA, Plaut M, Cooper SF, Fenton MJ, Arshad SH, Bahna SL. Guidelines for the diagnosis and management of food allergy in the United States: summary of the NIAID-Sponsored expert panel report. J Allergy Clin Immunol. 2010;126:e2–e17.
Brake DG, Evenson DP. A generational study of glyphosate-tolerant soybeans on mouse fetal, postnatal, pubertal and adult testicular development. Food Chem Toxicol. 2004;42:29–36.
Brake J, Vlachos D. Evaluation of transgenic event 176 "Bt" corn in broiler chickens. Poult Sci. 1998;77:648–53.

Brandt DA, Brand TS, Cruywagen CW. The use of crude protein content to predict concentrations of lysine and methionine in grain harvested from selected cultivars of wheat, barley and triticale grown in the Western Cape region of South Africa. SA J Anim Sci. 2000;30:22–5.

Bravo A, Gill SS, Soberon M. Mode of action of Bacillus thuringiensis Cry and Cyt toxins and their potential for insect control. Toxicon. 2007;49:423–35.

Breiteneder H, Ebner C. Molecular and biochemical classification of plant-derived food allergens. J Allergy Clin Immunol. 2000;106:27–36.

Breiteneder H, Radauer C. A classification of plant food allergens ☆. J Allergy Clin Immunol. 2004;113:821–30.

Brune PD, Culler AH, Ridley WP, Walker K. Safety of GM crops: compositional analysis. J Agric Food Chem. 2013;61:8243–7.

Bucchini L, Goldman LR. Starlink corn: a risk analysis. Environ Health Perspect. 2002;110:5–13.

Burlingame B. Fostering quality data in food composition databases: visions for the future. J Food Compos Anal. 2004;17:251–8.

CAC/GL 46-2003 Guideline for the conduct of food safety assessment of foods produced using recombian.

CAC/GL. Guideline for the conduct of food safety assessment of foods produced using recombinant-DNA microorganisms. Codex: Alimentarius; 2003.

Cani PD, Delzenne NM. Gut microflora as a target for energy and metabolic homeostasis. Curr Opin Clin Nutr Metab Care. 2007;10:729–34.

Cani PD, Possemiers S, Van de Wiele T. Changes in gut microbiota control inflammation in obese mice through a mechanism involving GLP-2-driven improvement of gut permeability. Gut. 2009;58:1091–103.

Cao SH, Xu WT, Luo YB, He XY, Guo X, Yuan YF, Yuan CM, Huang KL. Expression, purification and refolding of recombinant Cry1Ab/Ac obtained in Escherichia coli as inclusion bodies. J Sci Food Agric. 2009;89:1438.

Cao S, He X, Xu W, Ran W, Liang L, Luo Y, Yuan Y, Zhang N, Zhou X, Huang K. Safety assessment of Cry1C protein from genetically modified rice according to the national standards of PR China for a new food resource. Regul Toxicol Pharmacol. 2010;58:474–81.

Cao SS, Xu WT, Luo YB, He XY, Yuan YF, Ran WJ, Liang LX, Huang KL. Metabonomics study of transgenic Bacillus thuringiensis rice (T2A-1) meal in a 90-day dietary toxicity study in rats. Mol BioSyst. 2011;7:2304–10.

Cao SS, He XY, Xu WT, Luo YB, Ran WJ, Liang LX, Dai YQ, Huang K. Potential allergenicity research of Cry1C protein from genetically modified rice. Regul Toxicol Pharmacol. 2012;63(2):181–7.

Cao G, Liu Y, Zhang S, Yang X, Chen R, Zhang Y, Lu W, Liu Y, Wang J, Lin M, Wang G. A novel 5-enolpyruvylshikimate-3-phosphate synthase shows high glyphosate tolerance in Escherichia coli and tobacco plants. PLoS One. 2012a;7:e38718.

Cao S, He X, Xu W, Luo Y, Ran W, Liang L, Dai Y, Huang K. Potential allergenicity research of Cry1C protein from genetically modified rice. Regul Toxicol Pharmacol. 2012b;63:181–7.

Cao SS, He XY, Xu WT, Luo YB, Yuan YF, Liu PF, Cao B, Shi H, Huang KL. Safety assessment of transgenic Bacillus thuringiensis rice T1c-19 in Sprague-Dawley rats from metabonomics and bacterial profile perspectives. IUBMB Life. 2012c;64:242–50.

Carvalho JL, De OSJ, Conte C, Pacheco S, Nogueira EO, Souza TL, Faria JS, Aragão FG. Comparative analysis of nutritional compositions of transgenic Rnai-Mediated Virus-Resistant Bean (Event Emb-Pv051-1) with Its non-transgenic counterpart. Transgenic Res. 2015;24:1–7.

Catchpole GS, Beckmann M, Enot DP, Mondhe M, Zywicki B, Taylor J, Hardy N, Smith A, King RD, Kell DB, Fiehn O, Draper J. Hierarchical metabolomics demonstrates substantial compositional similarity between genetically modified and conventional potato crops. Proc Natl Acad Sci U S A. 2005;102:14458–62.

Caubet JC, Bencharitiwong R, Moshier E, Godbold JH, Sampson HA, Nowak-Węgrzyn A. Significance of ovomucoid- and ovalbumin-specific IgE/IgG 4 ratios in egg allergy. J Allergy Clin Immunol. 2012;129:739–47.

CDC. Investigation of human health effects associated with potential exposure to genetically modified corn: a report to the U.S. food and drug administration from the centers for disease control and prevention. Atlanta: Centers for Disease Control and Prevention; 2001.

Cellini F, Chesson A, Colquhoun IJ, Constable A, Davies HV, Engel KH, Gatehouse AMR, Kärenlampi S, Kok EJ, Leguay J-J. Unintended effects and their detection in genetically modified crops. Food Chem Toxicol. 2004;42:1089–125.

CERA. GM Crop Database, MON-Ø4Ø32-6 (GTS 40-3-2) product description. Center for environmental risk assessment, pp. GM Crop Database product description of GTS 40-43-42; 2009.

Chamberland ME, Alter HJ, Busch MP, Nemo G, Ricketts M. Emerging infectious disease issues in blood safety. Emerg Infect Dis. 2001;7:552–3.

Charteris WP, Kelly PM, Morelli L, Collins JK. Development and application of an in vitro methodology to determine the transit tolerance of potentially probiotic Lactobacillus and Bifidobacterium species in the upper human gastrointestinal tract. J Appl Microbiol. 1998;84:759–68.

Chen H, Tang W, Xu C, Li X, Lin Y, Zhang Q. Transgenic indica rice plants harboring a synthetic cry2A* gene of Bacillus thuringiensis exhibit enhanced resistance against lepidopteran rice pests. Theor Appl Genet. 2005;111:1330–7.

Cheng KC, Beaulieu J, Iquira E, Belzile FJ, Fortin MG, Strömvik MV. Effect of transgenes on global gene expression in soybean is within the natural range of variation of conventional cultivars. J Agric Food Chem. 2008;56:3057–67.

Choi H, Moon JK, Park BS, Park HW, Park SY, Kim TS, Kim DH, Ryu TH, Kweon SJ, Kim JH. Comparative nutritional analysis for genetically modified rice, iksan483 and milyang204, and nontransgenic counterparts. Appl Biol Chem. 2012;55(1):19–26.

Chong DK, Langridge WH. Expression of full-length bioactive antimicrobial human lactoferrin in potato plants. Transgenic Res. 2000;9:71–8.

Chou LS, Weimer B. Isolation and characterization of acid- and bile-tolerant isolates strains of Lactobacillus acidophilus. J Dairy Sci. 1999;82:23–31.

Chow BF, Lee C-J. Effect of dietary restriction of pregnant rats on body weight gain of the offspring. J Nutr. 1964;82:10–8.

Chrenková M, Sommer A, Ceresnáková Z, Nitrayová S, Prostredná M. Nutritional evaluation of genetically modified maize corn performed on rats. Arch Anim Nutr. 2002;56:229–35.

Chukwudebe A, Privalle L, Reed A, Wandelt C, Contri D, Dammann M, Groeters S, Kaspers U, Strauss V, van Ravenzwaay B. Health and nutritional status of Wistar rats following subchronic exposure to CV127 soybeans. Food Chem Toxicol. 2012;50:956–71.

Chung HS, Kim YB, Chun SL, Ji GE. Screening and selection of acid and bile bifidobacteria. Int J Food Microbiol. 1999;47:25–32.

Clark JH, Ipharraguerre IR. Livestock performance: feeding biotech crops. J Dairy Sci. 2001;84:E9–E18.

Clive J. Global status of commercialized Biotech/GM crops:2015. ISAAA brief No. 51. ISAAA: Ithaca; 2015. p. 210.

Codex Alimentarius Commission. Alinorm 03/34: Joint FAO/WHO Food Standard Programme, Codex Alimentarius Commission, Twenty-Fifth Session, Rome, Italy, 30 June–5 July, 2003. Appendix III, Guideline for the conduct of food safety assessment of foods derived from recombinant-DNA plants, and Appendix IV, Annex on the assessment of possible allergenicity. pp. 2003; pp. 47–60.

Cohen M, Gould F, Bentur J. Bt rice: practical steps to sustainable use. Int Rice Res Notes. 2000;25:4–10.

Coll A, Nadal A, Rossignol M, Puigdomenech P, Pla M. Proteomic analysis of MON810 and comparable non-GM maize varieties grown in agricultural fields. Transgenic Res. 2011;20:939–49.

Conner AJ, Jacobs JM. Food risks from transgenic crops in perspective. Nutrition. 2000;16:709–11.

Conway PL, Gorbach SL, Goldin BR. Survival of lactic acid bacteria in the human stomach and adhesion to intestinal cells. J Dairy Sci. 1987;70:1–12.

Cooke RD, Twiddy DR, Reilly PJA. Lactic acid fermentation as a low cost means of food preservation in tropical countries. FEMS Microbiol Rev. 1987;46:369–79.

Corpillo D, Gardini G, Vaira AM, Basso M, Aime S, Accotto GP, Fasano M. Proteomics as a tool to improve investigation of substantial equivalence in genetically modified organisms: the case of a virus-resistant tomato. Proteomics. 2004;4:193–200.

Crevel RW, Cooper KJ, Poulsen LK, Hummelshoj L, Bindslev-Jensen C, Burks AW, Sampson HA. Lack of immunogenicity of ice structuring protein type III HPLC12 preparation administered by the oral route to human volunteers. Food Chem Toxicol. 2007;45:79–87.

Cuevas-Rodríguez EO, Verdugo-Montoya NM, Angulo-Bejarano PI, Milán-Carrillo J, Mora-Escobedo R, Bello-Pérez LA, Garzón-Tiznado JA, Reyes-Moreno C. Nutritional properties of tempeh flour from quality protein maize (Zea mays L.). LWT Food Sci Technol. 2006;39:1072–9.

Curwin BD, Hein MJ, Sanderson WT, Striley C, Heederik D, Kromhout H, Reynolds SJ, Alavanja MC. Urinary pesticide concentrations among children, mothers and fathers living in farm and non-farm households in iowa. Ann Occup Hyg. 2007;51:53–65.

Daenicke R, Aulrich K, Flachowsky G. GVO in der Fu¨ tterung. Mais. 1999; 27:135–7. -293.

Dahlmann A, Dantzler WH, Silbernagl S, Gekle M. Detailed mapping of ochratoxin A reabsorption along the rat nephron in vivo: the nephrotoxin can be reabsorbed in all nephron segments by different mechanisms. J Pharmacol Exp Ther. 1998;286:157–62.

Daleprane JB, Chagas MA, Vellarde GC, Ramos CF, Boaventura GT. The impact of non- and genetically modified soybean diets in aorta wall remodeling. J Food Sci. 2010;75:T126–31.

Daniell H. Production of human serum albumin in transgenic crops without interfering with food or feed production;2003. Access entire news report at: http://www.isb.vt.edu

Daniell H, Singh ND, Mason H, Streatfield SJ. Plant-made vaccine antigens and biopharmaceuticals. Trends Plant Sci. 2009;14:669–79.

Datta SK, Datta K, Parkhi V, Rai M, Baisakh N, Sahoo G, Rehana S, Bandyopadhyay A, Alamgir M, Ali MS. Golden rice: introgression, breeding, and field evaluation. Euphytica. 2007;154:271–8.

Davies H. A role for "omics" technologies in food safety assessment. Food Control. 2010;21:1601–10.

De Steur H, Blancquaert D, Strobbe S, Lambert W, Gellynck X, van der Straeten D. Status and market potential of transgenic biofortified crops. Nat Biotechnol. 2015;33:25–9.

de Vrese M, Schrezenmeir J. Probiotics, prebiotics, and synbiotics. Adv Biochem Eng Biotechnol. 2008;111:1–66.

Dearman RJ, Beresford L, Foster ES, Mcclain S, Kimber I. Characterization of the allergenic potential of proteins: an assessment of the kiwifruit allergen actinidin. J Appl Toxicol. 2014;34:489–97.

Delaney B. Strategies to evaluate the safety of bioengineered foods. Int J Toxicol. 2007;26:389–99.

Delaney B, Appenzeller LM, Munley SM, Hoban D, Sykes GP, Malley LA, Sanders C. Subchronic feeding study of high oleic acid soybeans (Event DP-305423-1) in Sprague-Dawley rats. Food Chem Toxicol. 2008;46(12):3808–17.

Delaney B, Appenzeller LM, Munley SM, Hoban D, Sykes GP, Malley LA, Sanders C. Subchronic feeding study of high oleic acid soybeans (Event DP-3O5423-1) in Sprague-Dawley rats. Food Chem Toxicol. 2008a;46:3808–17.

Delaney B, Zhang J, Carlson G, Schmidt J, Stagg B, Comstock B, Babb A, Finlay C, Cressman RF, Ladics G. A gene-shuffled glyphosate acetyltransferase protein from Bacillus licheniformis (GAT4601) shows no evidence of allergenicity or toxicity. Toxicol Sci. 2008b;102:425–32.

Delhaize E, Hebb DM, Richards KD, Lin JM, Ryan PR, Gardner RC. Cloning and expression of a wheat (Triticum aestivum L.) phosphatidylserine synthase cDNA. Overexpression in plants alters the composition of phospholipids. J Biol Chem. 1999;274:7082–8.

Delves-Broughton J, Blackburn P, Evans RJ, Hugenholtz J. Applications of the bacteriocin, nisin. Antonie Van Leeuwenhoek. 1996;69(2):193–202.

Demont M, Stein AJ. Global value of GM rice: a review of expected agronomic and consumer benefits. New Biotechnol. 2013;30:426–36.

Deng CY, Song GS, Jun-Wang XU, Zhu Z. Increasing accumulation level of foreign protein in transgenic plants through protein targeting. Acta Bot Sin. 2003;45:1084–9.

Deng PJ, Zhou XY, Yang DY, Hou HL, Yang XK, Li YH, Yang YC, Wang XL, Fang SS, Wu SQ. The definition, source, manifestation and assessment of unintended effects in genetically modified plants. J Sci Food Agric. 2008;88:2401–13.

Denli M, Blandon JC, Guynot ME, Salado S, Perez JF. Efficacy of a new ochratoxin-binding agent (OcraTox) to counteract the deleterious effects of ochratoxin A in laying hens. Poult Sci. 2008;87:2266–72.

Dhaka V, Gulia N, Ahlawat KS, Khatkar BS. Trans fats-sources, health risks and alternative approach - a review. J Food Sci Technol-Mysore. 2011;48:534–41.

DiBaise JK, Zhang H, Crowell MD, Krajmalnik-Brown R, Decker GA, Rittmann BE. Gut microbiota and its possible relationship with obesity. Mayo Clin Proc. 2008;83:460–9.

Dibbern Jr DA, Palmer GW, Williams PB, Bock SA, Dreskin SC. RBL cells expressing human Fc epsilon RI are a sensitive tool for exploring functional IgE-allergen interactions: studies with sera from peanut-sensitive patients. J Immunol Methods. 2003;274:37–45.

Dickely F, Nilsson D, Hansen EB, Johansen E. Isolation of Lactococcus lactis suppressors and construction of a foodgrade cloning vector. Mol Microbiol. 1995;15:839–47.

Domingo JL. Toxicity studies of genetically modified plants: a review of the published literature. Crit Rev Food Sci Nutr. 2007;47:721–33.

Domon E, Takagi H, Hirose S, Sugita K, Kasahara S, Ebinuma H, Takaiwa F. 26-Week oral safety study in macaques for transgenic rice containing major human T-cell epitope peptides from Japanese cedar pollen allergens. J Agric Food Chem. 2009;57:5633–8.

Dowd PF. Indirect reduction of ear molds and associated mycotoxins in Bacillus thuringiensis corn under controlled and open field conditions: utility and limitations. J Econ Entomol. 2000;93:1669–79.

Drakakaki G, Marcel S, Glahn RP, Lund EK, Pariagh S, Fischer R, Christou P, Stoger E. Endosperm-specific co-expression of recombinant soybean ferritin and aspergillus phytase in maize results in significant increases in the levels of bioavailable iron. Plant Mol Biol. 2005;59:869–80.

Dryzga MD, Yano BL, Andrus AK, Mattsson JL. Evaluation of the safety and nutritional equivalence of a genetically modified cottonseed meal in a 90-day dietary toxicity study in rats. Food Chem Toxicol. 2007;45(10):1994–2004.

Dumitru R, Jiang WZ, Weeks DP, Wilson MA. Crystal structure of dicamba monooxygenase: a Rieske nonheme oxygenase that catalyzes oxidative demethylation. J Mol Biol. 2009;392:498–510.

Dun BQ, Wang XJ, Lu W, Chen M, Zhang W, Ping SZ, Wang ZX, Zhang BM, Lin M. Development of highly glyphosate-tolerant tobacco by coexpression of glyphosate acetyltransferase gat and EPSPS G2-aroA genes. Crop J. 2014;2:164–9.

Dunn CJ, Elliott GA, Oostveen JA, Richards IM. Development of a prolonged eosinophil-rich inflammatory leukocyte infiltration in the guinea-pig asthmatic response to ovalbumin inhalation. Am Rev Respir Dis. 1988;137:541–7.

Dunne C, O'Mahony L, Murphy L, Thornton G, Morrissey D, O'Halloran S, Feeney M, Flynn S, Fitzgerald G, Daly C, Kiely B, O'Sullivan G, Shanahan F, Collins JK. In vitro selection criteria for probiotic bacteria of human origin: correlation with in vivo findings. Am J Clin Nutr. 2001;73:386S–92S.

Dybing E, Doe J, Groten J, Kleiner J, O'Brien J, Renwick AG, Schlatter J, Steinberg P, Tritscher A, Walker R, Younes M. Hazard characterisation of chemicals in food and diet. dose response, mechanisms and extrapolation issues. Food Chem Toxicol. 2002;40:237–82.

Eccleston E, Leonard BJ, Lowe JS, Welford HJ. Basophilic leukaemia in the albino rat and a demonstration of the basopoietin. Nat New Biol. 1973;244:73–6.
EFSA. Guidance document of the scientific panel on genetically modified organisms for the risk assessment of genetically modified plants and derived food and feed. EFSA J. 2004;99:1–93.
EFSA. Safety and nutritional assessment of GM plants and derived food and feed: the role of animal feeding trials. Food Chem Toxicol. 2008:S2–S70.
EFSA. Guidance for risk assessment of food and feed from genetically modified plants. EFSA J. 2011; 9:n/a-n/a.
Eigenmann PA, Sampson HA. Interpreting skin prick tests in the evaluation of food allergy in children. Pediatr Allergy Immunol. 1998;9:186–91.
Eisenbrand G, Pool-Zobel B, Baker V, Balls M, Blaauboer BJ, Boobis A, Carere A, Kevekordes S, Lhuguenot JC, Pieters R, Kleiner J. Methods of in vitro toxicology. Food Chem Toxicol. 2002;40:193–236.
Elangovan AV, Tyagi PK, Shrivastav AK, Tyagi PK, Mandal AB. GMO (Bt-Cry1Ac gene) cottonseed meal is similar to non-GMO low free gossypol cottonseed meal for growth performance of broiler chickens. Anim Feed Sci Technol. 2006;129:252–63.
Elaroussi M, Mohamed F, Elgendy M, El Barkouky E, Abdou A, Hatab M. Ochratoxicosis in broiler chickens: functional and histological changes in target organs. Int J Poult Sci. 2008;7:414–22.
Elmer GW, Surawicz C, McFarland L. Biotherapeutic agents: a neglected modality for the treatment and prevention of selected intestinal and vaginal infections. J Am Med Assoc. 1996;275:70–6.
Emond E, Lavall ER, Drolet G, Moineau S, LaPointe G. Molecular characterization of theta-replication plasmid and its use for development of a two component food-grade cloning system for Lactococcus lactis. Appl Environ Microbiol. 2001;67:1700–9.
Erstad BL. Viral infectivity of albumin and plasma protein fraction. Pharmacotherapy. 1996;16:996–1001.
Estruch JJ, Carozzi NB, Desai N, Duck NB, Warren GW, Koziel MG. Transgenic plants: an emerging approach to pest control. Nat Biotechnol. 1997;15:137–41.
Ewen SW, Pusztai A. Health risks of genetically modified foods. Lancet. 1999;354:684.
Eyssen H, Caenepeel PH. Metabolism of fats, bile acids and steroids. Role of gut flora in toxicity and cancer. London: Academic Press; 1988. p. 263–86.
Fabienne Rancé LC. Improved screening for peanut allergy by the combined use of skin prick tests and specific IgE assays. J Allergy Clin Immunol. 2002;109:1027–33.
Fagan J, Schoel B, Haegert A, Moore J, Beeby J. Performance assessment under field conditions of a rapid immunological test for transgenic soybeans. Int J Food Sci Technol. 2001;36:357–67.
Fan Z, An W, Mi X, Ma H, Wang J. Forage (silage) maize breeding and commercially creatively utilization of quality protein maize (QPM) germplasm. J Maize Sci. 2004;12(4):23–5.
FAO. Evaluation of allergenicity of genetically modifed foods – report of a Joint FAO/WHO Expert Consultation on foods derived from biotechnology. Rome: FAO; 2001. p. 1–27.
FAO/WHO. Biotechnology and food safety – report of a joint FAO/WHO consultation, FAO Food and Nutrition Paper. Rome: Food and Agriculture Organization of the United Nations; 1996.
FAO/WHO. Safety Aspects of Genetically Modified Foods of Plant Origin. In: Report of a joint FAO/WHO expert consultation on foods derived from biotechnology. Rome: FAO; 2000a.
FAO/WHO. Evaluation of certain food additives and contaminants. World Health Organ Tech Rep Ser. 2000b;896:1–128.
FAO/WHO. Food and Agriculture Organization of the United Nations. Rome: FAO; 2001a. p. 22–5.
FAO/WHO. Evaluation of allergenicity of genetically modified foods; report of a joint FAO/WHO expert consultation on allergenicity of foods derived from biotechnology. Rome: FAO; 2001b.
Faust MA. Livestock products-composition and detection of transgenic DNA/proteins. In: ADSA–ASAS, editor. Proceedings of the symposium agricultural biotech market. Baltimore, 29 pp; 2000.

Faust MA, Miller L. Study finds no Bt in milk. Integrated coop management. IC-478. Fall Special Livest. Ed. Ames: IOWA State University Extension; 1997. p. 6–7., -62.

FDA. Biotechnology consultation agency response Letter BNF No.000141. 2015. Available at http://www.fda.gov/Food/FoodScienceResearch/GEPlants/Submissions/ucm436169.htm. Accessed 30 Oct 2015.

Federal Register. Nisin preparation: affirmation of GRAS status as a direct human food ingredient. Fed Regist. 1988;54:11247–51.

Ferrufino-Guardia E, Tangni E, Larondelle Y, Ponchaut S. Transfer of ochratoxin A during lactation: exposure of suckling via the milk of rabbit does fed a naturally-contaminated feed. Food Addit Contam. 2000;17:167–75.

Flachowsky G, Böhme H. Proposals for nutritional assessments of feeds from genetically modified plants. J Anim Feed Sci. 2005;14:49–70.

Floch MH, Binder HJ, Filburn B, Gershengoren W. The effect of bile acids on intestinal microflora. Am J Clin Nutr. 1972;25:1418–26.

Folmer J, Grant R, Milton C, Beck J. Effect of Bt corn silage on short-term lactational performance and ruminal fermentation in dairy cows. J Dairy Sci. 2000;83(Suppl 1):1182.

Food and Nutrition Board (FNB). Panel on Micronutrients, Subcommittees on Upper Reference Levels of Nutrients and of Interpretation and Use of Dietary Reference Intakes, and the Standing Committee on the Scientific Evaluation of Dietary Reference Intakes. Dietary Reference Intakes for vitamin A, vitamin K, arsenic, boron, chromium, copper, iodine, iron, manganese, molybdenum, nickel, silicon, vanadium, and zinc. Washington, DC: Food and Nutrition Board, Institute of Medicine, National Academy Press; 2001.

Freeman HJ. Effects of differing purified cellulose, pectin, and hemicellulose fiber diets on fecal enzymes in 1, 2-dimethylhydrazineinduced rat colon carcinogenesis. Cancer Res. 1986;46:5529–32.

Fritsché R. Utility of animal models for evaluating hypoallergenicity. Mol Nutr Food Res. 2009;53:979–83.

Froseth BR, McKay LL. Molecular characterization of the nisin resistance region of Lactococcus lactis subsp. lactis biovar diacetylactis DRC3. Appl Environ Microbiol. 1991;57(3):804–11.

Froseth BR, Herman RE, McKay LL. Cloning of nisin resistance determinant and replication origin on 7.6 kilobase EcoR1 fragment of pNP40 from Streptococcus lactis subsp. Diacetylactis DRC3. Appl Environ Microbiol. 1988;54:2136–9.

FSANZ. Safety assessment report (Approval) Application A1063;2012.

Fufa H, Akalu G, Wondimu A, Taffesse S, Gebre T, Schlosser K, Noetzold H, Henle T. Assessment of protein nutritional quality and effects of traditional processes: a comparison between Ethiopian quality protein maize and five Ethiopian adapted normal maize cultivars. Food. 2003;47:269–73.

Gandjar I. Encyclopedia of food microbiology. London: Academic Press; 2000. p. 767–73.

Gao Y, Gillen CM, Wheatly MG. Molecular characterization of the sarcoplasmic calcium-binding protein (SCP) from crayfish Procambarus clarkii. Comp Biochem Physiol B Biochem Mol Biol. 2006;144:478–87.

Gasson MJ, Anderson PH. High copy number plasmid vectors for use in lactic streptococci. FEMS Microbiol Lett. 1985;30:196.

Gaudry PA, Fernandes J, Reish R, Roux KH, Sathe SK, Teuber SS, Reese G, Lehrer SB. Murine IgE and IgG antibody responses to allergenic and non-allergenic foods: potential model for allergenicity assessment of foods in man. J Allergy Clin Immunol. 2004;113:S234.

Gayen D, Sarkar SN, Datta SK, Datta K. Comparative analysis of nutritional compositions of transgenic high iron rice with its non-transgenic counterpart. Food Chem. 2013;138:835–40.

Gayen D, Paul S, Sarkar SN, Datta SK, Datta K. Comparative nutritional compositions and proteomics analysis of transgenic Xa21 rice seeds compared to conventional rice. Food Chem. 2016;203:301–7.

Gekle M, Silbernagl S. The role of the proximal tubule in ochratoxin A nephrotoxicity in vivo: toxodynamic and toxokinetic aspects. Ren Physiol Biochem. 1994;17:40–9.

Gekle M, Sauvant C, Schwerdt G. Ochratoxin A at nanomolar concentrations: a signal modulator in renal cells. Mol Nutr Food Res. 2005;49:118–30.

Gems D, Aksenko A, Belenky L, Robertson S, Ramsden M, Vinetski Y, Clutterbuck AJ. An 'instant gene bank' method for gene cloning by mutant complementation. Mol Gen Genet. 1994;242:467–71.

Gendel SM. The use of amino acid sequence alignments to assess potential allergenicity of proteins used in genetically modified foods. Adv Food Nutr Res. 1998;42:45–62.

Gibbs R, Caroline G, Ian D. Postgraduate Symposium Long-chain n-3 PUFA: intakes in the UK and the potential of a chicken meat prototype to increase them. Proc Nutr Soc. 2010;69:144–55.

Goffeau A, Barrel BG, Bussey H, Davis RW, Dujon B, Feldmann H, Galibert F, Hoheisel JD, Jacq C, Johnston M, Louis EJ, Mewes HW, MurakamiY PP, Tettelin H, Oliver SG. Life with 6000 genes. Science. 1996;275:1051–2.

Goodman RE. Performing IgE serum testing due to bioinformatics matches in the allergenicity assessment of GM crops. Food Chem Toxicol. 2008;46(Suppl 10):S24–34.

Goodman RE, Taylor SL, Yamamura J, Kobayashi T, Kawakami H, Kruger CL, Thompson GP. Assessment of the potential allergenicity of a milk basic protein fraction. Food Chem Toxicol. 2007;45:1787–94.

Goodman RE, Vieths S, Sampson HA, Hill D, Ebisawa M, Taylor SL, Ree RV. Allergenicity assessment of genetically modified crops|[mdash]|what makes sense? Nat Biotechnol. 2008;26:73–81.

Gould F, Amasino RM, Brossard D, Buell CR, Dixon RA, Falck-Zepeda JB, Gallo MA, Giller K, Glenna L, Griffin TS. Genetically engineered crops: experiences and prospects. Washington, DC: National Academies Press; 2016.

Gracie JA, Bradley JA. Interleukin-12 induces interferon-γ-dependent switching of IgG alloantibody subclass. Eur J Immunol. 1996;26:1217–21.

Green JM. The benefits of herbicide-resistant crops. Pest Manag Sci. 2012;68:1323–31.

Griffiths CEM, Dearman RJ, Cumberbatch M, Kimber I. Cytokines and Langerhans cell mobilisation in mouse and man. Cytokine. 2005;32:67–70.

Gritz L, Davies J. Plasmid-encoded hygromycin B resistance: the sequence of hygromycin B phosphotransferase gene and its expression in Escherichia coli and Saccharomyces cerevisiae. Gene. 1983;25:179–88.

Gross-Steinmeyer K, Weymann J, Hege H-G, Metzler M. Metabolism and lack of DNA reactivity of the mycotoxin ochratoxin A in cultured rat and human primary hepatocytes. J Agric Food Chem. 2002;50:938–45.

Gunning RV, Moores GD. The effects of diet on the detection of resistance to Cry1Ac toxin in Australian Helicoverpa armigera Hübner (Lepidoptera: Noctuidae). Pestic Biochem Physiol. 2010;97(1):55–9.

Gupta R, Sheikh A, Strachan DP, Anderson HR. Time trends in allergic disorders in the UK. Thorax. 2007;62:91–6.

Gygi SP, Rist B, Gerber SA, Turecek F, Gelb MH, Aebersold R. Quantitative analysis of complex protein mixtures using isotope-coded affinity tags. Nat Biotechnol. 1999;17:994–9.

Hagenbuch B, Meier PJ. Organic anion transporting polypeptides of the OATP/SLC21 family: phylogenetic classification as OATP/SLCO superfamily, new nomenclature and molecular/functional properties. Pflugers Arch. 2004;447:653–65.

Halford NG, Shewry PR. Genetically modified crops: methodology, benefits, regulation and public concerns. Br Med Bull. 2000;56(1):62–73.

Hama H, Suzuki K, Tanaka H. Inheritance and stability of resistance to Bacillus thuringiensis formulations of the diamondback moth, Plutella xylostella(LINNAEUS) (Lepidoptera:Yponomeutidae). Appl Entomol Zool. 1992;27:355–62.

Hamerstrand GE. Trypsin inhibitors in soy products: modification of the standard analytical procedure. Cereal Chem. 1981;58(1):42–5.

Hamilton KA, Pyla PD, Breeze M, Olson T, Li MH, Robinson E, Gallagher SP, Sorbet R, Chen Y. Bollgard II cotton: Compositional analysis and feeding studies of cottonseed from insect-

protected cotton (Gossypium hirsutum L.) producing the Cry1Ac and Cry2Ab2 proteins. J Agric Food Chem. 2004;52:6969–76.
Hammer Ø, Harper D, Ryan P. Past: paleontological statistics software package for education and data analysis. Palaeontol Electron. 2001;4:1–9. http://palaeo-electronica.org/2001_1/past/issue1_01.html
Hammond BG, Vicini JL, Hartnell GF, Naylor MW, Knight CD, Robinson EH, Fuchs RL, Padgette SR. The feeding value of soybeans fed to rats, chickens, catfish and dairy cattle is not altered by genetic incorporation of glyphosate tolerance. J Nutr. 1996;126:717–27.
Hammond B, Lemen J, Dudek R, Ward D, Jiang C, Nemeth M, Burns J. Results of a 90-day safety assurance study with rats fed grain from corn rootworm-protected corn. Food Chem Toxicol. 2006;44(2):147–60.
Han J, Yang Y, Chen S, Wang Z, Yang X, Wang G, Men J. Comparison of nutrient composition of parental rice and rice genetically modified with cowpea trypsin inhibitor in China. J Food Compos Anal. 2005;18:297–302.
Hareau GG. Transgenic rice in asia: a general equilibrium assessment of potential welfare effects and regional distribution; 2006.
Harlander SK. Transformation of Streptococcus lactis by electroporation. Streptococcal genetics. Washington DC: American Society for Microbiology; 1987. p. 229–33.
Hart RW, Turturro A. Dietary restrictions and cancer. Environ Health Perspect. 1997;105(Suppl 4):989–92.
Hart RW, Turturro A, Leakey J, Allaben WT. Diet and test animals. Science. 1995;270:1419–21.
Hartnell GF. Best practices for the conduct of animal studies to evaluate crops genetically modified for output traits. International Life Science Institute (ILSI); 2007.
Hartung T. Toxicology for the twenty-first century. Nature. 2009;460:208–12.
Hastings GE, Wolf PG. The therapeutic use of albumin. Arch Fam Med. 1992;1:281–7.
Hausding M, Sauer K, Maxeiner JH, Finotto S. Transgenic models in allergic responses. Curr Drug Targets. 2008;9:503–10.
Håversen LA, Engberg I, Baltzer L, Dolphin G, Hanson LA, Mattsbybaltzer I. Human lactoferrin and peptides derived from a surface-exposed helical region reduce experimental Escherichia coli urinary tract infection in mice. Infect Immun. 2000;68:5816–23.
Håversen LA, Baltzer L, Dolphin G, Hanson LÅ, Mattsby-Baltzer I. Anti-inflammatory activities of human lactoferrin in acute dextran sulphate-induced colitis in mice. Scand J Immunol. 2003;57:2–10.
Hayes J, Doherty AT, Adkins DJ, Oldman K, O'Donovan MR. The rat bone marrow micronucleus test--study design and statistical power. Mutagenesis. 2009;24:419–24.
He F, Tuomola EM, Arvilommi H, Salminen S. Modulation of humoral immune response through probiotic intake. FEMS Immunol Med Microbiol. 2000;29:47–52.
He XY, Huang KL, Li X, Qin W, Delaney B, Luo YB. Comparison of grain from corn rootworm resistant transgenic DAS-59122-7 maize with non-transgenic maize grain in a 90-day feeding study in Sprague-Dawley rats. Food Chem Toxicol. 2008;46(6):1994–2002.
He XY, Tang MZ, Luo YB, Li X, Cao SS, Yu JJ, Delaney B, Huang KL. A 90-day toxicology study of transgenic lysine-rich maize grain (Y642) in Sprague-Dawley rats. Food Chem Toxicol. 2009;47(2):425–32.
He Y, Ning T, Xie T, Qiu Q, Zhang L, Sun Y, Jiang D, Fu K, Yin F, Zhang W, Shen L, Wang H, Li J, Lin Q, Li H, Zhu Y, Yang D. Large-scale production of functional human serum albumin from transgenic rice seeds. Proc Natl Acad Sci. 2011;108:19078–83.
Hefferon KL. Nutritionally enhanced food crops; Progress and perspectives. Int J Mol Sci. 2015;16:3895–914.
Helm RM. Food allergy animal models: an overview. Ann N Y Acad Sci. 2002;964:139–50.
Henderson RCA, Cox BS, Tubb RS. The transformation of brewing yeasts with a plasmid containing the gene for copper resistance. Curr Genet. 1985;9:133–8.
Hendrix KS, Petty AT, Lofgren DL. Feeding value of whole plant silage and crop residues from Bt or normal corns. J Anim Sci. 2000; 78(Suppl. 1):273(Abstr 1146). -305.

Herman RA, Phillips AM, Collins RA, Tagliani LA, Claussen FA, Graham CD, Bickers BL, Harris TA, Prochaska LM. Compositional equivalency of Cry1F corn event TC6275 and conventional corn (Zea mays L.). J Agric Food Chem. 2004;52:2726–34.

Herman RA, Storer NP, Phillips AM, Prochaska LM, Windels P. Compositional assessment of event DAS-59122-7 maize using substantial equivalence. Regul Toxicol Pharmacol. 2007;47:37–47.

Herman RA, Chassy BM, Parrott W. Compositional assessment of transgenic crops: an idea whose time has passed. Trends Biotechnol. 2009;27:555–7.

Herman RA, Phillips AM, Lepping MD, Fast BJ, Sabbatini J. Compositional safety of event DAS-40278-9 (AAD-1) herbicide-tolerant maize. GM Crops. 2010;1:294–311.

Hérouet C, Esdaile DJ, Mallyon BA, Debruyne E, Schulz A, Currier T, Hendrickx K, Klis RJVD, Rouan D. Safety evaluation of the phosphinothricin acetyltransferase proteins encoded by the pat and bar sequences that confer tolerance to glufosinate-ammonium herbicide in transgenic plants. Regul Toxicol Pharmacol. 2005;41:134–49.

Hertel C, Probst AJ, Cavadini C, Meding E, Hammes WP. Safety assessment of genetically modified microorganisms applied in meat fermentations. Syst Appl Microbiol. 1995;4(18):469–76.

Heyman MB. Lactose intolerance in infants, children, and adolescents. Pediatrics. 2006;118:1279–86.

Hileman RE, Silvanovich A, Goodman RE, Rice EA, Holleschak G, Astwood JD, Hefle SL. Bioinformatic methods for allergenicity assessment using a comprehensive allergen database. Int Arch Allergy Immunol. 2002;128:280–91.

Hiratsuka S, Nakamura K, Iwai S, Murakami M, Itoh T, Kijima H, Shipley JM, Senior RM, Shibuya M. Selection and genetic analysis of a Heliothis virescens (Lepidoptera: Noctuidae) strain with high levels of resistance to Bacillus thuringiensis toxins. J Econ Entomol. 1995;88:1545–59.

Holo H, Nes IF. High-frequency transformation by electroporation of Lactococccus lactis subsp. cremoris grown with glycine in osmotically stabilized media. Appl Environ Microbiol. 1989;55:3119–23.

Holst-Jensen A. GMO testing—trade, labeling or safety first? Nat Biotechnol. 2008;26:858–9.

Holzhauser T, Wackermann O, Ballmer-Weber BK, Bindslev-Jensen C, Scibilia J, Perono-Garoffo L, Utsumi S, Poulsen LK, Vieths S. Soybean (Glycine max) allergy in Europe: Gly m 5 (β-conglycinin) and Gly m 6 (glycinin) are potential diagnostic markers for severe allergic reactions to soy. J Allergy Clin Immunol. 2009;123:452–8.

Hommel RK, Ahnert P. Encyclopedia of food microbiology. London: Academic Press; 2000. p. 352–9.

Honeycutt Z, Rowlands H, Grace L. Glyphosate testing full report: findings in American mother's breast milk, urine and water. 2014, http://www.momsacrossamerica.com/glyphosate_testing_results. Accessed 7 Dec 2015.

Hori Y, Fujisawa M, Shimada K, Hirose Y. Determination of the herbicide glyphosate and its metabolite in biological specimens by gas chromatography–mass spectrometry. A case of poisoning by roundup herbicide. J Anal Toxicol. 2003;27:162–6.

Hothorn LA, Oberdoerfer R. Statistical analysis used in the nutritional assessment of novel food using the proof of safety. Regul Toxicol Pharmacol. 2006;44:125–35.

Huang Y, Adams MC. In vitro assessment of the upper gastrointestinal tolerance of potential probiotic dairy propionibacteria. Int J Food Microbiol. 2004;91:253–60.

Huff JE. Carcinogenicity of ochratoxin A in experimental animals. IARC Sci Publ. 1991:229–44.

Hugget AC, Conzelmann C. EU regulation on novel foods: consequences for the food industry. Trends Food Sci Technol. 1997;8:133–9.

Hyun Y, Bressner GE, Ellis M, Lewis AJ, Fischer R, Stanisiewski EP, Hartnell GF. Performance of growing-finishing pigs fed diets containing Roundup Ready corn (event nk603), a nontransgenic genetically similar corn, or conventional corn lines. J Anim Sci. 2004;82:571–80.

ILSI. The safety assessment of novel foods. Brussels: International Life Science Institute; 1995.

ILSI. Safety assessment of genetically modified microorganisms used in food: consensus guidelines reached at a workshop held in April 1999. Brussels: International Life Sciences Institute Novel Food Task Force; 1999.
ILSI. Best practices for the conduct of animal studies to evaluate crops genetically modified for input traits. Washington, DC: International Life Sciences Institute; 2003.
ILSI. Nutritional and safety assessments of foods and feeds nutritionally improved through biotechnology. Compr Rev Food Sci Food Saf. 2004;3:36–104.
Ivanciuc O, Garcia T, Torres M, Schein CH, Braun W. Characteristic motifs for families of allergenic proteins. Mol Immunol. 2008;46:559–68.
Jackson KD, Howie LD, Akinbami LJ. Trends in allergic conditions among children: United States, 1997-2011. Nchs Data Brief. 2013;121:1–8.
James C. 20th Anniversary (1996 to 2015) of the global commercialization of biotech crops and biotech crop highlights in 2015. ISAAA Brief No. 51. Ithaca: ISAAA; 2015a.
James C. Global status of commercialized biotech/GM crops: 2015. ISAAA Brief, No. 51. Ithaca: ISAAA; 2015b.
Janczyk P, Wolf C, Hartmann A, Junghans H, Schwerin M, Souffrant W. Modification of potatoes, by either recombinant DNA technology or conventional breeding, affects their nutritional value for the rat. Archiva Zootechnica. 2007;10:8–21.
Janssen PT, Bijsterveld OPV. A simple test for lacrimal gland function: a tear lactoferrin assay by radial immunodiffusion. Graefes Arch Clin Exp Ophthalmol. 1983;220:171–4.
Jenkins ZA, Montgomery GW. Genetic mapping of the ovine homologue of the mouse Hacl1 gene to sheep chromosome 1. Anim Genet. 1999;30:74.
Jenkins N, Parekh RB, James DC. Getting the glycosylation right: implications for the biotechnology industry. Nat Biotechnol. 1996;14:975–81.
Jenkins JA, Griffiths-Jones S, Shewry PR, Breiteneder H, Mills ENC. Structural relatedness of plant food allergens with specific reference to cross-reactive allergens: an in silico analysis. J Allergy Clin Immunol. 2005;115:163–70.
Jenkins JA, Breiteneder H, Mills ENC. Evolutionary distance from human homologs reflects allergenicity of animal food proteins. J Allergy Clin Immunol. 2007;120:1399–405.
Jiao Z, Deng J, Li G, Zhang Z, Cai Z. Study on the compositional differences between transgenic and non-transgenic papaya (Carica papaya L.). J Food Compos Anal. 2010;23:640–7.
Jiménez JJ, Bernal JL, Nozal MJ, Toribio L, Bernal J. Profile and relative concentrations of fatty acids in corn and soybean seeds from transgenic and isogenic crops. J Chromatogr A. 2009;1216:7288–95.
Jiménezsaiz R, Belloque J, Molina E, Lópezfandiño R. Human immunoglobulin E (IgE) binding to heated and glycated Ovalbumin and Ovomucoid before and after in Vitro Digestion. J Agric Food Chem. 2011;59:10044–51.
Jin Z, Cai GL, Li XM, Gao F, Yang JJ, Lu J, Dong JJ. Comparative proteomic analysis of green malts between barley (Hordeum vulgare) cultivars. Food Chem. 2014;151:266–70.
Jonge JD, Knippels L, Ezendam J. The importance of dietary control in the development of a peanut allergy model in Brown Norway rats. Methods. 2007;41:99–111.
Jonnala RS, Dunford NT, Chenault K. Nutritional composition of genetically modified peanut varieties. J Food Sci. 2005;70(4):S254–6.
Jore KP, van Luijk N, Luiten RG, van der Werf MJ, Pouwels PH. Efficient transformation of Propionibacterium freundenreischii based on a novel vector. Appl Environ Microbiol. 2001;67:499–503.
Jung KY, Takeda M, Kim DK, Tojo A, Narikawa S, Yoo BS, Hosoyamada M, Cha SH, Sekine T, Endou H. Characterization of ochratoxin A transport by human organic anion transporters. Life Sci. 2001;69:2123–35.
Jung SH, Ahn HH, Nam HJ. Comparison of nutritional composition in Korean Rices. J Korean Soc Food Sci Nutr. 2002;31(5):885–92.

Junhua H, Yuexin Y, Shurong C, Zhu W, Xiaoli Y, Guodong W, Jianhua M. Comparison of nutrient composition of parental rice and rice genetically modified with cowpea trypsin inhibitor in China. J Food Compos Anal. 2005;18:297–302.

Just F, Sanchez S, et al. Generation and assembly of secretory antibodies in plants. Science. 1995;268:716–9.

Kahala K, Palva A. The expression signals of the Lactococcus brevis slpA gene direct efficient heterologous protein production in lactic acid bacteria. Appl Microbiol Biotechnol. 1999;51:71–8.

Karim S, Dean DH. Toxicity and receptor binding properties of Bacillus thuringiensis delta-endotoxins to the midgut brush border membrane vesicles of the rice leaf folders, Cnaphalocrocis medinalis and Marasmia patnalis. Curr Microbiol. 2000;41:276–83.

Kaul S, Lüttkopf D, Kastner B, Vogel L, Höltz G, Vieths S, Hoffmann A. Mediator release assays based on human or murine immunoglobulin E in allergen standardization. Clin Exp Allergy. 2007;37:141–50.

Kawahigashi H, Hirose S, Ohkawa H, Ohkawa Y. Herbicide resistance of transgenic rice plants expressing human CYP1A1. Biotechnol Adv. 2007;25:75–84.

Keele CA, Neil E. Secretion of digestive juices. In: Samson, editors Wright's applied physiology. Oxford University 11th edn, chap. 60. London: Oxford University Press; (1965). pp. 353–363.

Keese P. Risks from GMOs due to horizontal gene transfer. Environ Biosaf Res. 2008;7:123–49.

Khoudi H, Laberge S, Ferullo JM, Bazin R, Darveau A, Castonguay Y, Allard G, Lemieux R, Vézina LP. Production of a diagnostic monoclonal antibody in perennial alfalfa plants. Biotechnol Bioeng. 1999;64:135–43.

Kim MK, Kim JH, Nam SJ, Ryu S, Kong G. Dietary intake of soy protein and tofu in association with breast cancer risk based on a case-control study. Nutr Cancer. 2008;60:568–76.

Kim SJ, Rim KT, Kang MG, Kim JK, Chung YH, Yang JS. A study of micronucleus induction with methyl formate and 2-Methylbutane in bone marrow cells of male ICR mice. Saf Health Work. 2010;1:80–6.

Kimber I, Helm RM. Assessment of protein allergenicity on the basis of immune reactivity: animal models. Environ Health Perspect. 2003;111:1125–30.

Kitessa SM, Young P. Enriching milk fat with n−3 polyunsaturated fatty acids by supplementing grazing dairy cows with ruminally protected Echium oil - Animal feed science and technology. Anim Feed Sci Technol. 2011;170:35–44.

Klein TM, Wolf ED, Wu R, Sanford JC. High-velocity microprojectiles for delivering nucleic acids into living cells. 1987. Biotechnology. 1987;24:384–6.

Klurfeld DM, Kritchevsky D. Isolation and quantitation of lectins from vegetable oils. Lipids. 1987;22:667–8.

Knight AK, Shreffler WG, Sampson HA, Sicherer SH, Noone S, Mofidi S, Nowak-Wegrzyn A. Skin prick test to egg white provides additional diagnostic utility to serum egg white–specific IgE antibody concentration in children. J Allergy Clin Immunol. 2006;117:842–7.

Knippels LM, Penninks AH. Assessment of the allergic potential of food protein extracts and proteins on oral application using the brown Norway rat model. Environ Health Perspect. 2003;111:233–8.

Knippels L, Houben G, Spanhaak S, Penninks A. An oral sensitization model in Brown Norway rats to screen for potential allergenicity of food proteins. Methods. 1999a;19:78–82.

Knippels LMJ, Penninks AH, Smit JJ, Houben GF. Immune-mediated effects upon oral challenge of ovalbumin-sensitized Brown Norway rats: further characterization of a rat food allergy model. Toxicol Appl Pharmacol. 1999b;156:161–9.

Knippels LMJ, Kleij HPMVD, Koppelman SJ, Houben GF, Penninks AH, Knippels LMJ, Penninks AH, Felius AA. Comparison of antibody responses to hen's egg and cow's milk proteins in orally sensitized rats and food-allergic patients. Allergy. 2000;55:251–8.

Kok J, Van Der Vossen JMBM, Venema G. Construction of plasmid cloning vectors for lactic streptococci which also replicate in Bacillus subtilis and Escherichia coli. Appl Environ Microbiol. 1984;48:726–31.

Kok EJ, Kuiper HA. Comparative safety assessment for biotech crops. Trends Biotechnol. 2003;21:439–44.

Kok EJ, Keijer J, Kleter GA, Kuiper HA. Comparative safety assessment of plant-derived foods. Regul Toxicol Pharmacol. 2008;50:98–113.

Kondo JK, McKay LL. Transformation of Streptococcus lactis protoplasts by plasmid DNA. Appl Environ Microbiol. 1982;43:1213–5.

Kreis M, Forde BG, Rahman S, Miflin BJ, Shewry PR. Molecular evolution of the seed storage proteins of barley, rye and wheat. J Mol Biol. 1985;183:499–502.

Kroghsbo S, Madsen C, Poulsen M, Schrøder M, Kvist PH, Taylor M, Gatehouse A, Shu Q, Knudsen I. Immunotoxicological studies of genetically modified rice expressing PHA-E lectin or Bt toxin in Wistar rats. Toxicology. 2008;245:24–34.

Krüger M, Schledorn P, Schrödl W, Hoppe HW, Lutz W. Detection of glyphosate residues in animals and humans. J Environ Anal Toxicol. 2014;4:1–5.

Kruse TA, Bolund L, Grzeschik KH, Ropers HH, Sjostrom H, Noren O, Mantei N, Semenza G. The human lactase-phlorizin hydrolase gene is located on chromosome 2. FEBS Lett. 1988;240:123–6.

Krzanowska H. Inheritance of sperm head abnormality types in mice – the role of the Y chromosome. Genet Res. 1976;28:189–98.

Kuiper HA, Noteborn HP, Peijnenburg AA. Adequacy of methods for testing the safety of genetically modified foods. Lancet. 1999;354:1315–6.

Kuiper HA, Kleter GA, Noteborn HPJM, Kok EJ. Assessment of the food safety issues related to genetically modified foods. Plant J. 2001;27:503–28.

Kuiper-Goodman T, Scott PM. Risk assessment of the mycotoxin ochratoxin A. Biomed Environ Sci. 1989;2:179–248.

Kullen MJ, Klaenhammer TR. Genetic modification of intestinal lactobacilli and bifidobacteria. Curr Issues Mol Biol. 2000;2:41–50.

Kumar VK, Vennila S, Nalini N. Modifying effects of morin on the development of aberrant crypt foci and bacterial enzymes in experimental colon cancer. Food Chem Toxicol. 2009;47:309–15.

Kumar A, Gul MZ, Zeeshan A, Bimolata W, Qureshi IA, Ghazi IA. Differential antioxidative responses of three different rice genotypes during bacterial blight infection. Aust J Crop Sci. 2013;7:1893–900.

Ladics GS, Selgrade MJK. Identifying food proteins with allergenic potential: evolution of approaches to safety assessment and research to provide additional tools. Regul Toxicol Pharmacol. 2009;54:2–6.

Ladics GS, Knippels LMJ, Penninks AH, Bannon GA, Goodman RE, Herouet-Guicheney C. Review of animal models designed to predict the potential allergenicity of novel proteins in genetically modified crops. Regul Toxicol Pharmacol RTP. 2010;56:212–24.

Lankaputhra WEV, Shah NP. Survival of Lactobacillus acidophilus and Bifidobacterium spp. in the presence of acid and bile salts. Cult Dairy Prod J. 1995;30:2–7.

Latta M, Eskin M. A simple and rapid colorimetric method for phytate determination. J Agric Food Chem. 1980;28:1313–5.

Le Vay D. 13 Digestion. Human anatomy physiology, 3rd edn. Sevenoaks, Kent: Hodder & Stoughton; 1998. pp. 231–233.

LeBlance JG, Van Sindere D, Hugenholtz J, Piard JC, Sesma F, de Giori GS. Risk assessment of genetically modified Lactic Acid Bacteria using the concept of substantial equivalence. Curr Microbiol. 2010;61:590–5.

Lee HU, Bae EA, Han MJ, Kim NJ, Kim DH. Hepatoprotective effect of ginsenoside Rb1 and compound K on tert-butyl hydroperoxide-induced liver injury. Liver Int. 2005;25:1069–73.

Lee KC, Liu CF, Lin TH, Pan TM (2010). Safety and risk assessment of the genetically modified Lactococci on rats intestinal bacterial flora. Int J Food Microbiol 142: 164-169.

Lehesranta SJ, Davies HV, Shepherd LV, Nunan N, McNicol JW, Auriola S, Koistinen KM, Suomalainen S, Kokko HI, Kärenlampi SO. Comparison of tuber proteomes of potato varieties, landraces, and genetically modified lines. Plant Physiol. 2005;138:1690–9.

Leier I, Hummel-Eisenbeiss J, Cui Y, Keppler D. ATP-dependent para-aminohippurate transport by apical multidrug resistance protein MRP2. Kidney Int. 2000;57:1636–42.

Lemee R, Lortal S, Casselin B, van Heijenoort J. Involvement of an N-acetylglucosamine in autolysis of Propionibacterium freudenreichii CNRZ. Appl Environ Microbiol. 1994;60:4351–8.

Lepping MD, Herman RA, Potts BL. Compositional Equivalence of DAS-444empty set6-6 (AAD-12+2mEPSPS+PAT);Herbicide-Tolerant Soybean and Nontransgenic Soybean. J Agric Food Chem. 2013;61:11180–90.

Ley RE, Lozupone CA, Hamady M. Worlds within worlds: evolution of the vertebrate gut microbiota. Nat Rev Microbiol. 2008;6:776–88.

Li Y. Liberty link rice 601 contamination and the impact on US rice exports. ProQuest; 2008.

Li Z, Upadhyaya NM, Meena S, Gibbs AJ, Waterhouse PM. Comparison of promoters and selectable marker genes for use in Indica rice transformation. Mol Breed. 1997;3:1–14.

Li X, Huang K, He X, Zhu B, Liang Z, Li H, Luo Y. Comparison of nutritional quality between Chinese Indica rice with sck and cry1Ac genes and its nontransgenic counterpart. J Food Sci. 2007;72:420–4.

Li X, He XLY, Xiao G, Jiang X, Huang K. Comparative analysis of nutritional composition between herbicide-tolerant rice with bar gene and its non-transgenic counterpart. J Food Compos Anal. 2008;21:535–9.

Li SS, Nugroho A, Rocheford T, White WS. Vitamin A equivalence of the β-carotene in β-carotene-biofortified maize porridge consumed by women. Am J Clin Nutr. 2010;92:1105–12.

Li X, Li S, Lang Z, Zhang J, Zhu L, Huang D. Chloroplast-targeted expression of the codon-optimized truncated cry1Ah gene in transgenic tobacco confers a high level of protection against insects. Plant Cell Rep. 2013;32:1299–308.

Liang Q, Richardson T. Expression and characterization of human lactoferrin in yeast Saccharomyces cerevisiae. J Agric Food Chem. 1993;41:1800–7.

Lichtenstein AH, Ausman LM, Carrasco W, Jenner JL, Ordovas JM, Schaefer EJ. Hydrogenation impairs the hypolipidemic effect of corn oil in humans. Hydrogenation, trans fatty acids, and plasma lipids. Arterioscler Thromb. 1993;13:154–61.

Lichtenstein AH, Matthan NR, Jalbert SM, Resteghini NA, Schaefer EJ, Ausman LM. Novel soybean oils with different fatty acid profiles alter cardiovascular disease risk factors in moderately hyperlipidemic subjects. Am J Clin Nutr. 2006;84:497–504.

Liénard D, Sourrouille C, Gomord V, Faye L. Pharming and transgenic plants. Biotechnol Annu Rev. 2007;13:115–47.

Lindgren S, Dobrogosz WJ. Antagonistic activities of lactic acid bacteria in food and feed fermentation. FEMS Microbiol Rev. 1990;87:149–64.

Lipp M, Anklam E, Stave JW. Validation of an immunoassay for detection and quantitation of a genetically modified soybean in food and food fractions using reference materials: interlaboratory study. J AOAC Int. 2000;83:919–27.

Liscombe DK, Macleod BP, Loukanina N, Nandi OI, Facchini PJ. Evidence for the monophyletic evolution of benzylisoquinoline alkaloid biosynthesis in angiosperms. Phytochemistry. 2005;66:2501–20.

Liu WH, Stephen Inbaraj B, Chen BH. Analysis and formation of trans fatty acids in hydrogenated soybean oil during heating. Food Chem. 2007;104:1740–9.

Liu HY, Xu WT, Luo YB, Tian HT, Wang HX, Wang HX, Guo X, Yuan YF, Huang KL. Assessment of tolerance to multistresses and in vitro cell adhesion in genetically modified Lactobacillus plantarum 590. Anton Leeuw. 2011;99:759–69.

Liu PF, He XY, Chen DL, Luo YB, Cao SS, Song H, Liu T, Huang KL, Xu WT. A 90-day subchronic feeding study of genetically modified maize expressing Cry1Ac-M protein in Sprague-Dawley rats. Food Chem Toxicol. 2012;50(9):3215–21.

Liu H-Y, Xu W-T, Yuan Y-F, Cao S-S, He X-Y, Li S-Y, Huang K-L, Luo Y-B. The effect of genetically modified Lactobacillus plantarum 590 on the gut health of sprague-dawley rats. IUBMB Life. 2012a;64:617–27.

Liu M, Li D, Wang Z, Meng F, Li Y, Wu X, Teng W, Han Y, Li W. Transgenic expression of ThIPK2 gene in soybean improves stress tolerance, oleic acid content and seed size. Plant Cell Tissue Org Cult. 2012b;111:277–89.

Livingstone DM, Hampton JL, Phipps PM, Grabau EA. Enhancing resistance to sclerotinia minor in peanut by expressing a barley oxalate oxidase gene. Plant Physiol. 2005;137:1354–62.

Llorente B, Alonso GD, Bravo-Almonacid F, Rodríguez V, López MG, Carrari F, Torres HN, Flawiá MM. Safety assessment of nonbrowning potatoes: opening the discussion about the relevance of substantial equivalence on next generation biotech crops. Plant Biotechnol J. 2011;9:136–50.

Lönnerdal B, Iyer S. Lactoferrin: molecular structure and biological function. Annu Rev Nutr. 1995;15:93–110.

Lu Y, Xu W, Kang A, Luo Y, Guo F, Yang R, Zhang J, Huang K. Prokaryotic expression and allergenicity assessment of Hygromycin B phosphotransferase protein derived from genetically modified plants. J Food Sci. 2007;72:228–32.

Lu B-R. Transgene escape from GM crops and potential biosafety consequences: an environmental perspective. International Centre for Genetic Engineering and Biotechnology (ICGEB). Collection Biosafety Rev. 2008;4:66–141.

Lucca P, Hurrell R, Potrykus I. Fighting iron deficiency anemia with iron-rich rice. J Am Coll Nutr. 2002;21:184S–90S.

Maaroufi K, Achour A, Hammami M, el May M, Betbeder AM, Ellouz F, Creppy EE, Bacha H. Ochratoxin A in human blood in relation to nephropathy in Tunisia. Hum Exp Toxicol. 1995;14:609–14.

Magana-Gomez JA, de la Barca AM. Risk assessment of genetically modified crops for nutrition and health. Nutr Rev. 2009;67:1–16.

Mallet AK, Rowland IR. Bacterial enzymes: their role in the formation of mutagens and carcinogens in the intestine. Dig Dis. 1990;8:71–6.

Mally A, Zepnik H, Wanek P, Eder E, Dingley K, Ihmels H, Völkel W, Dekant W. Ochratoxin A: lack of formation of covalent DNA adducts. Chem Res Toxicol. 2004;17:234–42.

Manimaran P, Ramkumar G, Mohan M, Mangrauthia SK, Padmakumari AP, Muthuraman P, Bentur JS, Viraktamath BC, Balachandran SM. Bt rice evaluation and deployment strategies. GM Crops. 2011;2:135–7.

Marchand F, Mecheri S, Guilloux L, Iannascoli B, Weyer A, Blank U. Human serum IgE-mediated mast cell degranulation shows poor correlation to allergen-specific IgE content. Allergy. 2003;58:1037–43.

Mari A. Multiple pollen sensitization: a molecular approach to the diagnosis. Int Arch Allergy Immunol. 2001;125:57–65.

Marsh WR, Irvin CG, Murphy KR, Behrens BL, Larsen GL. Increases in airway reactivity to histamine and inflammatory cells in bronchoalveolar lavage after the late asthmatic response in an animal model. Am Rev Respir Dis. 1985;131:875–9.

Marth E, Kleinhappl B. Albumin is a necessary stabilizer of TBE-vaccine to avoid fever in children after vaccination. Vaccine. 2001;20:532–7.

Martín-Orúe SM, O'Donnell AG, Ariño J, Netherwood T, Gilbert HJ, Mathers JC. Degradation of transgenic DNA from genetically modified soya and maize in human intestinal simulations. Br J Nutr. 2002;87:533–42.

Marzok MA. Detection of genetically modified soybeans and maize in Egypt as well as comparative nutritional safety investigations of isogenic and transgenic (Bt) maize in broiler nutrition: broiler performance, degradation and metabolic fate of maize DNA in some tissues and organs. Berlin: Mensch & Buch Verlag; 2004.

Matulka RA, Lyon MR, Wood S, Ann Marone P, Merkel DJ, Burdock GA. The safety of PolyGlycopleX (PGX) as shown in a 90-day rodent feeding study. Nutr J. 2009;8:1.

Mayer L. Current concepts in mucosal immunity. I. Antigen presentation in the intestine: new rules and regulations. Am J Physiol. 1998;274:G7–9.

Mayo B, Kok J, Venema K, Bockelmann W, Teuber M, Reinke H, Venema G. Molecular cloning and sequence analysis of the X-prolyl dipeptidyl aminopeptidase gene from Lactococcus lactis subsp. cremoris. Appl Environ Microbiol. 1991;57:38–44.

Mayo-Wilson E, Imdad A, Herzer K, Yakoob MY, Bhutta Z. Vitamin A supplements for preventing mortality, illness and blindness in children aged under 5: systematic review and meta-analysis. BMJ. 2011;343:d5094.

McComas KA, Besley JC, Steinhardt J. Factors influencing U.S. consumer support for genetic modification to prevent crop disease. Appetite. 2014;78:8–14.

Mcgaughey WH. Insect resistance to the biological insecticide Bacillus thuringiensis. Science. 1985;229:193–5.

McGregor D, Anderson D. DNA damage and repair in mammalian cells in vitro and in vivo as indicators of exposure to carcinogens. IARC Sci Publ. 1999:309–54.

McGuire MK, McGuire MA, Price WJ, Shafii B, Carrothers JM, Lackey KA, Goldstein DA, Jensen PK, Vicini JL. Glyphosate and aminomethylphosphonic acid are not detectable in human milk. Am J Clin Nutr. 2016;103:1285–90.

McKay LL. Functional properties of plasmids in lactic streptococci. Antonie Van Leeuwenhoek. 1983;49:259–74.

Mckay LL, Baldwin KA. Applications for biotechnology-present and future improvements in lactic acid bacteria. FEMS Microbiol Rev. 1990;87:3–14.

McNaughton J, Roberts M, Smith B, Rice D, Hinds M, Sanders C, Layton R, Lamb I, Delaney B. Comparison of broiler performance when fed diets containing event DP-305423-1, Nontransgenic Near-Isoline control, or commercial reference soybean meal, hulls, and oil. Poultry Sci. 2008;87(12):2549–61.

Melo VMM, Xavier-Filho J, Lima MS, Prouvost-Danon A. Allergenicity and tolerance to proteins from brazil nut H. B. K. Food Agric Immunol. 1994;6:185–95.

Mendoza C, Viteri FE, Lönnerdal B, Young KA, Raboy V, Brown KH. Effect of genetically modified, low-phytic acid maize on absorption of iron from tortillas. Am J Clin Nutr. 1998;68:1123–7.

Mendoza C, Viteri F, Lonnerdal B, Raboy V, Young K, Brown K. Absorption of iron from unmodified maize and genetically altered, low-phytate maize fortified with ferrous sulfate or sodium iron EDTA. Am J Clin Nutr. 2001;73:80–5.

Mercenier A. Molecular genetics of streptococcus thermophilus. FEMS Microbiol Rev. 1990;7:61–77.

Mercer DK, Scott KP, Bruce-Johnson A, Glover LA, Flint H. Fate of free DNA and transformation of the oral bacterium Streptococcus gordonii DL1 by plasmid DNA in human saliva. Appl Environ Microbiol. 1999;65:6–10.

Mesnage R, Clair E, Gress S, Then C, Székács A, Séralini GE. Cytotoxicity on human cells of Cry1Ab and Cry1Ac Bt insecticidal toxins alone or with a glyphosate-based herbicide. J Appl Toxicol. 2013;33:695.

Metcalfe DD, Astwood JD, Townsend R, Sampson HA, Taylor SL, Fuchs RL. Assessment of the allergenic potential of foods derived from genetically engineered crop plants. Crit Rev Food Sci Nutr. 1996;36:S165–86.

Mierau I, Kleerebezem M. 10 years of the nisin-controlled gene expression system (NICE) in Lactococcus lactis. Appl Microbiol Biotechnol. 2005;68:705–17.

Mikkelsaar M, Mandar R, Sepp E. Lactic acid microflora in the human microbial ecosystem and its development. Lactic acid bacteria: microbiology and functional aspects. New York: Marcel Dekker Inc.; 1998. p. 279–342.

Miller GW. The safety of genetically modified foods produced through biotechnology. Toxicol Sci. 2003;71:2–8.

Mills ENC, Jenkins JA, Alcocer MJC, Shewry PR. Structural, biological, and evolutionary relationships of plant food allergens sensitizing via the gastrointestinal tract. Crit Rev Food Sci Nutr. 2004;44:379–407.

Mj VDV, Van RR, Aalberse RC, Akkerdaas J, Koppelman SJ, Jansen HM, Js VDZ. Poor biologic activity of cross-reactive IgE directed to carbohydrate determinants of glycoproteins. J Allergy Clin Immunol. 1997;100:327–34.

Modirroosta BH, Tohidfar M, Saba J, Moradi F. The substantive equivalence of transgenic (Bt and Chi) and non-transgenic cotton based on metabolite profiles. Funct Integr Genomics. 2013;14:1–8.

Momma K, Hashimoto W, Ozawa S, Kawai S, Katsube T, Takaiwa F, Kito M, Utsumi S, Murata K. Quality and safety evaluation of genetically engineered rice with soybean glycinin: analyses of the grain composition and digestibility of glycinin in transgenic rice. Biosci Biotechnol Biochem. 1999;63:314–8.

Montel MC. Fermented meat products. Encyclopedia of food microbiology. London: Academic Press; 2000. p. 744–53.

Montero M, Coll A, Nadal A, Messeguer J, Pla M. Only half the transcriptomic differences between resistant genetically modified and conventional rice are associated with the transgene. Plant Biotechnol J. 2011;9:693–702.

Morelli L, Sarra PG, Bottazzi V. In vivo transfer of pAMb from Lactobacillus reuteri to Enterococcus faecalis. J Appl Bacteriol. 1988;65:371–5.

Mosmann TR, Cherwinski HM. Two types of murine T-cell clone. J Immunol. 1986;136:2348.

Munkvold GP, Hellmich RL, Rice LG. Comparison of fumonisin concentrations in kernels of transgenic Bt maize hybrids and nontransgenic hybrids. Plant Dis. 1999;83:130–8.

Nandi S, Yalda D, Lu S, Nikolov Z, Misaki R, Fujiyama K, Huang N. Process development and economic evaluation of recombinant human lactoferrin expressed in rice grain. Transgenic Res. 2005;14:237–49.

National Academies of Sciences, Engineering, and Medicine. Genetically engineered crops: experiences and prospects. Washington, DC: The National Academies Press; 2016. doi:10.17226/23395.

Netherwood T, Martín-Orúe SM, O'Donnell AG, Gockling S, Graham J, Mathers JC, Gilbert HJ. Assessing the survival of transgenic plant DNA in the human gastrointestinal tract. Nat Biotechnol. 2004;22:204–9.

Nielsen J, Oliver S. The next wave in metabolome analysis. Trends Biotechnol. 2005;23:544–6.

Niemann L, Sieke C, Pfeil R, Solecki R. A critical review of glyphosate findings in human urine samples and comparison with the exposure of operators and consumers. J Verbr Lebensm. 2015;10:3–12. doi:10.1007/s00003-014-0927-3.

Nordlee JA, Taylor SL, Townsend JA, Thomas LA, Bush RK. Identification of a Brazil-nut allergen in transgenic soybeans. N Engl J Med. 1996;334:688–92.

Noteborn HPJM, Lommen A, Jagt RCVD, Weseman JM. Chemical fingerprinting for the evaluation of unintended secondary metabolic changes in transgenic food crops. J Biotechnol. 2000;77:103–14.

Nout RMJ. Fermented foods and their production. Fermentation and food safety. Gaithersburg: Aspen Publishers Inc.; 2001. p. 1–38.

NRC (National Research Council). Safety of genetically engineered foods: approaches to assessing unintended health effects. Washington, DC: National Academies Press; 2004.

Nutrition WHO. Micronutrient deficiencies. Available from: http://www.who.int/nutrition/topics/vad/en (cited 21 March2012).

Nwaru BI, Hickstein L, Panesar SS, Muraro A, Werfel T, Cardona V, Dubois AEJ, Halken S, Hoffmann-Sommergruber K, Poulsen LK. The epidemiology of food allergy in Europe: a systematic review and meta-analysis. Allergy. 2014;69:62–75.

NY/T1102-2006, C.S. Safety assessment of genetically modified plant and derived products 90-day feeding test in rats. Beijing: Standards Press of China; 2006.

Oberdoerfer RB, Shillito RD, De BM, Mitten DH. Rice (oryza sativa l.) containing the bar gene is compositionally equivalent to the nontransgenic counterpart. J Agric Food Chem. 2005;53(5):1457–65.

Ocmant A, Mulier S, Hanssens L, Goldman M, Casimir G, Mascart F, Schandené L. Basophil activation tests for the diagnosis of food allergy in children. Clin Exp Allergy. 2009;39:1234–45.

OECD. Safety evaluation of foods derived by modern biotechnology: concepts and principles. Paris: Organisation for Economic Co-operation and Development; 1993a.

OECD. Safety evaluation of foods derived by modern biotechnology.Paris: Organisation for Economic Coorperation and Development; 1993b. http://oecd.org/dsti/sti/s_t/biotech/prod/modernhtm

OECD. Safety evaluation of foods derived through modern biotechnology: concepts and principles. Paris: OECD; 1993c.

OECD. Report of the OECD workshop on the toxicological and nutritional testing of novel foods. Paris: France; 1997.

OECD. Series on the Safety of Novel Foods and Feeds No. 1. Consensus document on key nutrients and key toxicants in low erucic acid rapeseed (canola). Paris: Organization for Economic Co-operation and Development; 2001.

OECD. Report of the OECD workshop on nutritional assessment of novel foods and feeds. 5–7 February 2001. Ottawa/Paris: Organization for Economic Co-operation and Development; 2002.

OECD. Consensus document on compositional considerations for new varieties of rice (Oryza sativa): key food and feed nutrients and antinutrients. Organisation for Economic Cooperation and Development; 2004.

Oh SW, Park SY, Lee SM, Oh SD, Cho HS, Park SK, Lee HJ, Kim HY, Yeo YS. Multivariate analysis for the safety assessment of genetically modified rices in the anti-nutrients and phenolic compounds. Int J Food Sci Technol. 2016;51:765–76.

Ohtani T, Galili G, Wallace JC, Thompson GA, Larkins BA. Normal and lysine-containing zeins are unstable in transgenic tobacco seeds. Plant Mol Biol. 1991;16:117–28.

Onose J, Imai T, Hasumura M, Ueda M, Ozeki Y, Hirose M. Evaluation of subchronic toxicity of dietary administered Cry1Ab protein from Bacillus thuringiensis var. Kurustaki HD-1 in F344 male rats with chemically induced gastrointestinal impairment. Food Chem Toxicol. 2008;46:2184–9.

Ouwehand A, Salminen S. The health effects of cultured milk products with viable and non-viable bacteria. Int Dairy J. 1998;8:749–58.

Özer D, Özer BH. Encyclopedia of food microbiology. London: Academic Press; 2000. p. 798–805.

Padgette SR, Taylor NB, Nida DL, Bailey MR. The composition of glyphosate-tolerant soybean seeds is equivalent to that of conventional soybeans. J Nutr. 1996;126(3):702–16.

Paine JA, Shipton CA, Chaggar S, Howells RM, Kennedy MJ, Vernon G, Wright SY, Hinchliffe E, Adams JL, Silverstone AL, Drake R. Improving the nutritional value of golden rice through increased pro-vitamin A content. Nat Biotechnol. 2005;23:482–7.

Pandey A, Kamle M, Yadava L, Muthukumar M, Kumar P, Gupta V, Ashfaque M, Pandey B. Genetically modified food: Its uses, future prospects and safety assessments. Biotechnology. 2010;9:444–58.

Pariza M, Foster E. Determining the safety of enzymes used in food processing. J Food Prot. 1983;46:453–68.

Pariza MW, Johnson EA. Evaluating the safety of microbial enzyme preparations used in food processing: update for a new century. Regul Toxicol Pharmacol. 2001;33:173–86.

Park SY, Si ML, Lee JH, Ko HS, Kweon SJ, Suh SC, Shin KS, Kim JK. Compositional comparative analysis between insect-resistant rice (Oryza Sativa L.) with a Synthetic Cry1ac gene and its non-transgenic counterpart. Plant Biotechnol Rep. 2012;6:29–37.

Park SY, Suwon RDA, Republic of Korea. Comparison of the nutritional compositions of insect-resistant and glufosinate-tolerant rice and conventional rice. J Appl Biol Chem. 2013;56(1):431–7.

Partridge-Telenko DE, Hu J, Livingstone DM, Shew BB, Phipps PM, Grabau EA. Sclerotinia blight resistance in virginia-type peanut transformed with a barley oxalate oxidase gene. Phytopathology. 2011;101:786–93.

Party BW, Lack G, Chapman M, Kalsheker N, King V, Robinson C, Venables K. Report on the potential allergenicity of genetically modified organisms and their products. Clin Exp Allergy. 2002;32:1131–43.

Passante E, Ehrhardt C, Sheridan H, Frankish N. RBL-2H3 cells are an imprecise model for mast cell mediator release. Agents Actions. 2009;58:611–8.

Pastorello EA, Farioli L, Pravettoni V, Ispano M, Scibola E, Trambaioli C, Giuffrida MG, Ansaloni R, Godovac-Zimmermann J, Conti A. The maize major allergen, which is responsible for food-induced allergic reactions, is a lipid transfer protein. J Allergy Clin Immunol. 2000;106:744–51.

Paul M, Ma JK. Plant-made pharmaceuticals: leading products and production platforms. Biotechnol Appl Biochem. 2011;58:58–67.

Penninks AH, Knippels LMJ. Determination of protein allergenicity: studies in rats. Toxicol Lett. 2001;120:171–80.

Pfohl-Leszkowicz A, Manderville RA. Ochratoxin A: an overview on toxicity and carcinogenicity in animals and humans. Mol Nutr Food Res. 2007;51:61–99.

Piacentini GL, Vicentini L, Bodini A, Mazzi P, Peroni DG, Maffeis C, Boner AL. Allergenicity of a hydrolyzed rice infant formula in a guinea pig model. Ann Allergy Asthma Immunol. 2003;91:61–4.

Pierce NF, Kaper JB, Mekalanos JJ. Determinants of the immunogenicity of live virulent and mutant vibrio cholerae in rabbit intestine. Infect Immun. 1987;55:477–81.

Pilacinski W, Crawford A, Downey R, Harvey B, Huber S, Hunst P, Lahman LK, MacIntosh S, Pohl M, Rickard C, Tagliani L, Weber N. Plants with genetically modified events combined by conventional breeding: an assessment of the need for additional regulatory data. Food Chem Toxicol. 2011;49:1–7.

Piva G, Morlacchini M, Pietri A, Rossi F, Prandini A. Growth performance of broilers fed insect protected (MON810) or near isogenic control corn. Poult Sci. 2001;80(Suppl. 1):320(Abstr 1324). -8.

Poulsen LK. Allergy assessment of foods or ingredients derived from biotechnology, gene-modified organisms, or novel foods. Mol Nutr Food Res. 2004;48:413–23.

Poulter M, Hollox E, Harvey CB, Mulcare C, Peuhkuri K, Kajander K, Sarner M, Korpela R, Swallow DM. The causal element for the lactase persistence/non-persistence polymorphism is located in a 1 Mb region of linkage disequilibrium in Europeans. Ann Hum Genet. 2003;67:298–311.

Pray CE, Huang J, Hu R, Rozelle S. Five years of Bt cotton in China – the benefits continue. Plant J Cell Mol Biol. 2002;31:423–30.

Prescott VE, Hogan SP. Genetically modified plants and food hypersensitivity diseases: usage and implications of experimental models for risk assessment. Pharmacol Ther. 2006;111:374–83.

Qaim M. The economics of genetically modified crops. Ann Rev Resour Econ. 2009;1:665–94.

Qi XZ, He XY, Luo YB, Li SY, Zou SY, Cao SS, Tang MZ, Delaney B, Xu WT, Huang KL. Subchronic feeding study of stacked trait genetically-modified soybean (3O5423 x 40-3-2) in Sprague-Dawley rats. Food Chem Toxicol. 2012;50:3256–63.

Qi X, Chen S, Sheng Y, Guo M, Liu Y, He X, Huang K, Xu W. Safety assessment of genetically modified rice expressing human serum albumin from urine metabonomics and fecal bacterial profile. Food Chem Toxicol. 2014;76:1.

Raskin I, Ribnicky DM, Komarnytsky S, Ilic N, Poulev A, Borisjuk N, Brinker A, Moreno DA, Ripoll C, Yakoby N. Plants and human health in the twenty-first century. Trends Biotechnol. 2002;20:522–31.

Rayan AM, Abbott LC. Compositional analysis of genetically modified corn events (NK603, MON88017 × MON810 and MON89034 × MON88017) compared to conventional corn. Food Chem. 2015;176:99–105.

Raybourne RB, Williams KM, Vogt R, Reissman DB, Winterton BS, Rubin C. Development and use of an ELISA test to detect IgE antibody to Cry9C following possible exposure to bioengineered corn. Int Arch Allergy Immunol. 2003;132:322–8.

Remig V, Franklin B, Margolis S, Kostas G, Nece T, Street JC. Trans Fats in America: a review of their use, consumption, health implications, and regulation. J Am Diet Assoc. 2010;110:585–92.

Reuter T, Aulrich K, Berk A. Investigations on genetically modified maize (Bt-maize) in pig nutrition: fattening performance and slaughtering results. Arch Anim Nutr. 2002a;56:319–26.

Reuter T, Aulrich K, Berk A, Flachowsky G. Investigations on genetically modified maize (Bt-maize) in pig nutrition: chemical composition and nutritional evaluation. Arch Anim Nutr. 2002b;56:23–31.

Ridley WP, Sidhu RS, Pyla PD, Nemeth MA, Breeze ML, Astwood JD. Comparison of the nutritional profile of glyphosate-tolerant corn event NK603 with that of conventional corn (Zea mays L.). J Agric Food Chem. 2002;50:7235–43.

Ridley WP, Shillito RD, Coats I, Steiner HY, Shawgo M, Phillips A, Dussold P, Kurtyka L. Development of the international life sciences institute crop composition database. J Food Compos Anal. 2004;17:423–38.

Robert SS. Production of eicosapentaenoic and docosahexaenoic acid-containing oils in transgenic land plants for human and aquaculture nutrition. Mar Biotechnol. 2006;8:103–9.

Rogan GJ, Dudin YA, Lee TC, Magin KM, Astwood JD, Bhakta NS, Leach JN, Sanders PR, Fuchs RL. Immunodiagnostic methods for detection of 5-enolpyruvylshikimate-3-phosphate synthase in Roundup Ready®; soybeans. Food Control. 1999;10:407–14.

Rommens CM, Yan H, Swords K, Richael C, Ye J. Low-acrylamide French fries and potato chips. Plant Biotechnol J. 2008;6:843–53.

Ross P, O'Gara F, Condon S. Thymidylate synthase gene from Lactococcus lactis as a genetic marker: an alternative to antibiotic resistance genes. Appl Environ Microbiol. 1990;56:2164–9.

Rothbard JB, Gefter ML. Interactions between immunogenic peptides and MHC proteins. Annu Rev Immunol. 1991;9:527–65.

Ruebelt MC, Lipp M, Reynolds TL, Astwood JD, Engel K, Jany K. Application of two-dimensional gel electrophoresis to interrogate alterations in the proteome of genetically modified crops. 2. Assessing natural variability. J Agric Food Chem. 2006;54:2162–8.

Ruiter B, Knol EF, Neerven RJJV, Garssen J, Bruijnzeel-Koomen CAFM, Knulst AC, Hoffen EV. Maintenance of tolerance to cow's milk in atopic individuals is characterized by high levels of specific immunoglobulin G4. Clin Exp Allergy. 2007;37:1103–10.

Rymer C, Hartnell GF, Givens DI. The effect of feeding modified soyabean oil enriched with C18: 4 n-3 to broilers on the deposition of n-3 fatty acids in chicken meat. Br J Nutr. 2010;105:866–78.

Salminen S, Deighton MA, Benno Y, Gorbach SL. Lactic acid bacteria: microbiology and functional Aspects. New York: Marcel Dekker Inc.; 1998. p. 211–53.

Salovaara H. Lactic acid bacteria: microbiology nd functional aspects. New York: Marcel Dekker Inc; 1998.

Sandhir R, Garg SK, Modi DR. Genetic engineering: modification of yeast and moulds. Encyclopedia of food microbiology. London: Academic Press; 2000. p. 907–16.

Saoudi A, Kuhn J, Huygen K, De Kozak Y, Velu T, Goldman M, Druet P, Bellon B. TH2 activated cells prevent experimental autoimmune uveoretinitis, a TH1-dependent autoimmune disease. Eur J Immunol. 1993;23:3096–103.

Satokari R, Vaughan EE, Akkermans ADL, Saarela M, De Vos W. Polymerase chain reaction and denaturing gradient gel electrophoresis monitoring of fecal Bifidobacterium populations in a prebiotic and probiotic feeding trial. Syst Appl Microbiol. 2001;24:227–31.

Sava V, Velasquez A, Song S, Sanchez-Ramos J. Adult hippocampal neural stem/progenitor cells in vitro are vulnerable to the mycotoxin ochratoxin-A. Toxicol Sci. 2007;98:187–97.
Sawahel W. Improved performance of transgenic Glycinebetaine-accumulating rice plants under drought stress. Biol Plant. 2003;47:39–44.
Scheirlinck T, Mahillon J, Joos H, Dhaese P, Michielis F. Integration and expression of a-amylase and endoglucanase genes in the Lactobacillus plantarum chromosome. Appl Environ Microbiol. 1989;55:2130–7.
Schrickx J, Lektarau Y, Fink-Gremmels J. Ochratoxin A secretion by ATP-dependent membrane transporters in Caco-2 cells. Arch Toxicol. 2006;80:243–9.
Schroder M, Poulsen M, Wilcks A, Kroghsbo S, Miller A, Frenzel T, Danier J, Rychlik M, Emami K, Gatehouse A. A 90-day safety study of genetically modified rice expressing Cry1Ab protein (Bacillus thuringiensis toxin) in Wistar rats. Food Chem Toxicol. 2007;45(3):339–49.
Schwerdt G, Gekle M, Freudinger R, Mildenberger S, Silbernagl S. Apical-to-basolateral transepithelial transport of Ochratoxin A by two subtypes of Madin-Darby canine kidney cells. Biochim Biophys Acta. 1997;1324:191–9.
Scott KP, Mercer DK, Richardson AJ, Melville CM, Glover LA, Flint HJ. Chromosomal integration of the green fluorescent protein gene in lactic acid bacteria and the survival of marked strains in human gut simulations. FEMS Microbiol Lett. 2000;182:23–7.
Selgrade MK, Bowman CC, Ladics GS, Privalle L, Laessig SA. Safety assessment of biotechnology products for potential risk of food allergy: implications of new research. Toxicol Sci. 2009;110:31–9.
Seralini GE, Cellier D, de Vendomois JS. New analysis of a rat feeding study with a genetically modified maize reveals signs of hepatorenal toxicity. Arch Environ Contam Toxicol. 2007;52(4):596–602.
Seralini GE, Clair E, Mesnage R, Gress S, Defarge N, Malatesta M, Hennequin D, de Vendomois JS. Long term toxicity of a Roundup herbicide and a Roundup-tolerant genetically modified maize. Food Chem Toxicol. 2012;50:4221–31.
Serror P, Sasaki T, Ehrlich S, Maguin E. Electrotransformation of Lactobacillus delbrueckii subsp. bulgaricus and L. delbrueckii subsp. Lactis with various plasmids. Appl Environ Microbiol. 2002;68:46–52.
Serup J, Staberg B. Quantification of weal reactions with laser doppler flowmetry. Allergy. 1985;40:233–7.
Sharma P, Singh AK, Singh BP, Gaur SN, Arora N. Allergenicity assessment of osmotin, a pathogenesis-related protein, used for transgenic crops. J Agric Food Chem. 2011;59:9990–5.
Sharman J, Kumar L, Singh S. Comparison of results of skin prick tests, enzyme-linked immunosorbent assays and food challenges in children with respiratory allergy. J Trop Pediatr. 2001;47:367–8.
Shaw RC, Sherman JM. The production of volatile fatty acids and carbon dioxide by propionic bacteria with special reference to their action in cheese. J Dairy Sci. 1923;6:303–9.
Sheng Y, Qi X, Liu Y, Guo M, Chen S, He X, Huang K, Xu W. Subchronic toxicity study in vivo and allergenicity study in vitro for genetically modified rice that expresses pharmaceutical protein (human serum albumin). Food Chem Toxicol. 2014;72:242–6.
Shireen KF, Pace RD, Egnin M, Prakash CS. Bioavailability of calcium from sweetpotato and soy flour supplemented diets in hamsters. J Environ Sci Health B. 2002;37:637–45.
Shoemaker NB, Vlamakis H, Hayes K, Salyers AA. Evidence for extensive resistance gene transfer among Bacteroides ssp. and among Bacteroides and other genera in human colon. Appl Environ Microbiol. 2001;67:561–8.
Sicherer SH, Sampson HA. Food allergy. J Allergy Clin Immunol. 2010;125:116–125.
Simoes-Barbosa A, Abreu H, Silva Neto A, Gruss A, Langella P. A food-grade delivery system for Lactococcus lactis and evaluation of inducible gene expression. Appl Microbiol Biotechnol. 2004;65:61–7.
Sjoblad RD, McClintock JT, Engler R. Toxicological considerations for protein components of biological pesticide products. Regul Toxicol Pharmacol. 1992;15:3–9.

Smith T (ed). The digestive system. In: The human body. Collingwood: Ken Fin Books; 1995. pp 150–73.

Soares-Weiser K, Takwoingi Y, Panesar SS, Muraro A, Werfel T, Hoffmann-Sommergruber K, Roberts G, Halken S, Poulsen L, van Ree R, Vlieg-Boerstra BJ, Sheikh A, Allergy EF, Anaphylaxis Guidelines G. The diagnosis of food allergy: a systematic review and meta-analysis. Allergy. 2014;69:76–86.

Sokol PP, Ripich G, Holohan PD, Ross CR. Mechanism of ochratoxin A transport in kidney. J Pharmacol Exp Ther. 1988;246:460–5.

Song Y, Liang C, Wang W, Fang J, Sun N, Jia X, Li N. Immunotoxicological evaluation of corn genetically modified with Bacillus thuringiensis Cry1Ah gene by a 30-day feeding study in BALB/c mice. PLoS One. 2014;9:e78566.

Sorensen KI, Larsen R, Kibenich A, Junge MP, Johansen E. A food-grade cloning system for industrial strains of Lactococcus lactis. Appl Environ Microbiol. 2000;66:1253–8.

Sørensen IK. Guidance document of the scientific panel on genetically modified organisms for the risk assessment of genetically modified plants and derived food and feed:Question No EFSA-Q-2003-005. Appl Mech Mater. 2012;174–177:1698–701.

Spencer JD, Allee GL, Sauber TE. Growing-finishing performance and carcass characteristics of pigs fed normal and genetically modified low-phytate corn. J Anim Sci. 2000a;78:1529–36.

Spencer JD, Allee GL, Sauber TE. Phosphorus bioavailability and digestibility of normal and genetically modified low-phytate corn for pigs. J Anim Sci. 2000b;78:675–81.

Spök A, Gaugitsch H, Laffer S, Pauli G, Saito H, Sampson H, Sibanda E, Thomas W, van Hage W, Valenta R. Suggestions for the assessment of the allergenic potential of genetically modified organisms. Int Arch Allergy Immunol. 2005;137:167–80.

Stave JW. Protein immunoassay methods for detection of biotech crops: applications, limitations, and practical considerations. J AOAC Int. 2002;85:780–6.

Stein AJ, Rodriguez-Cerezo E. The global pipeline of new GM crops: implications of asynchronous approval for international trade. Luxembourg: European Commission Joint Research Centre; 2009.

Steinborn A, Alder L, Michalski B, Zomer P, Bendig P, Martinez SA, Mol HG, Class TJ, Costa Pinheiro N. Determination of glyphosate levels in breast milk samples from Germany by LC-MS/MS and GC-MS/MS. J Agric Food Chem. 2016;64:1414–21.

Strathmann M, Wingender J, Flemming HC. Application of fluorescently labelled lectins for the visualization and biochemical characterization of polysaccharides in biofilms of pseudomonas aeruginosa. J Microbiol Methods. 2002;50:237–48.

Styrna J. Linkage between sperm abnormality level and major urinary protein phenotype in mice. Genet Res. 1991;57:135–8.

Sun N, Zhou C, Pu QK, Wang J, Huang KL, Che HL. Allergic reactions compared between BN and Wistar rats after oral exposure to ovalbumin. J Immunotoxicol. 2013;10:67–74.

Sun N, Wang J, Zhou C, Wang CY, Wang SP, Che HL. Cell-based immunological assay: complementary applications in evaluating the allergenicity of foods with FAO/WHO guidelines. Food Res Int. 2014;62:735–45.

Sun N, Tekutyeva L, Wang SP, Pu QK, Zhou C, Wang J, Che HL. A modified weight-of-evidence approach to evaluate the allergenic potential of food proteins. Toxicol Res. 2015a;4:476–85.

Sun N, Zhou C, Zhou X, Sun L, Che H. Use of a rat basophil leukemia (RBL) cell-based immunological assay for allergen identification, clinical diagnosis of allergy, and identification of anti-allergy agents for use in immunotherapy. J Immunotoxicol. 2015b;12:1–7.

Sussman GL, Harvey RP, Schocket AL. Evaluation of skin test response using two techniques of measurement. Ann Allergy. 1982;48:75–7.

Sussulini A, Garcia JS, Mesko MF, Moraes DP, Flores ÉMM, Pérez CA, Arruda MAZ. Evaluation of soybean seed protein extraction focusing on metalloprotein analysis. Microchim Acta. 2007;158:173–80.

Sutton BC. Classification of the Deuteromycetes. Encyclopedia of food microbiology. London: Academic Press; 2000. p. 893–901.

Swallow DM. Genetics of lactase persistence and lactose intolerance. Annu Rev Genet. 2003;37:197–219.

Swendseid ME. Biochemical assessment of protein and amino acid status. In: Livingston GE, editor. Proceedings of a council of the American Health Foundation. Nutritional Status Assessment of the Individual. New York: Food and Nutrition Press; 1987.

Takala TM, Saris PE. A food-grade cloning vector for actic acid bacteria based on the nisin immunity gene nisI. Appl Microbiol Biotechnol. 2002;59:467–71.

Takala TM, Saris PE, Tynkkynen SSH. Food-grade host/vector expression system for Lactobacillus casei based on complementation of plasmid-associated phospho- β -galactosidase gene lacG. Appl Microbiol Biotechnol. 2003;60:564–70.

Takeuchi A, Masuda S, Saito H, Abe T, Inui K. Multispecific substrate recognition of kidney-specific organic anion transporters OAT-K1 and OAT-K2. J Pharmacol Exp Ther. 2001;299:261–7.

Tan ASL. An approach to building the case for nutrition policies to limit trans-fat intake - a Singapore case study. Health Policy. 2011;100:264–72.

Tang M, Huang K, Zhou XLK, He X, Luo Y. Absence of effect after introducing Bacillus thuringiensis gene on nutritional composition in cottonseed. J Food Sci. 2006;71:S38–41.

Tang G, Qin J, Dolnikowski GG, Russell RM, Grusak MA. Golden rice is an effective source of vitamin A. Am J Clin Nutr. 2009;89:1776–83.

Tang M, Xie T, Cheng W, Qian L, Yang S, Yang D, Cui W, Li K. A 90-day safety study of genetically modified rice expressing rhIGF-1 protein in C57BL/6J rats. Transgenic Res. 2012a;21(3):499–510.

Tang G, Hu Y, Yin SA, Wang Y, Dallal GE, Grusak MA, Russell RM. β-Carotene in golden rice is as good as β-carotene in oil at providing vitamin A to children. Am J Clin Nutr. 2012;96:658–64.

Tang M, He X, Luo Y, Ma L, Tang X, Huang K. Nutritional assessment of transgenic lysine-rich maize compared with conventional quality protein maize. J Sci Food Agric. 2013;93:1049–54.

Tashiro M, Maki Z. Purification and characterization of a trypsin inhibitor from rice bran. J Nutr Sci Vitaminol. 1979;25(3):255–64.

Taylor SL. Chemistry and detection of food allergens. Food Technol. 1992;46:146–52.

Taylor ML, Stanisiewski EP, Riordan SG, Nemeth MA, George B, Hartnell GF. Comparison of broiler performance when fed diets containing roundup ready (event RT73), nontransgenic control, or commercial canola meal. Poult Sci. 2004;83:456–61.

Tenovuo J. Antimicrobial agents in saliva – protection for the whole body. J Dent Res. 2002;81:807–9.

Tenovuo J, Lehtonen OP, Aaltonen AS, Vilja P, Tuohimaa P. Antimicrobial factors in whole saliva of human infants. Infect Immun. 1986;51:49–53.

Tohidfar M, Ghareyazie B, Mosavi M, Yazdani S, Golabchian R. Agrobacterium-mediated transformation of cotton (Gossypium hirsutum) using a synthetic cry1Ab gene for enhanced resistance against Heliothis armigera. Iran J Biotechnol. 2008;6:164–73.

Tohidfar M, Hossaini R, Bashir NS, Meisam T. Enhanced resistance to Verticillium dahliae in transgenic cotton expressing an endochitinase gene from Phaseolus vulgaris. Czech J Genet Plant Breed. 2012;48:33–41.

Tong Jen Fu, †, Upasana RA, ‡, §, Hatzos‡ C Digestibility of food allergens and nonallergenic proteins in simulated gastric fluid and simulated intestinal fluida comparative study. J Agric Food Chem. 2002;50:7154–60.

Torrent M, Alvarez I, Geli MI, Dalcol I, Ludevid D. Lysine-rich modified gamma-zeins accumulate in protein bodies of transiently transformed maize endosperms. Plant Mol Biol. 1997;34:139–49.

Travis RC, Appleby PN, Siddiq A, Allen NE, Kaaks R, Canzian F, Feller S, Tjonneland A, Fons Johnsen N, Overvad K, Ramon Quiros J, Gonzalez CA, Sanchez MJ, Larranaga N, Chirlaque MD, Barricarte A, Khaw KT, Wareham N, Trichopoulou A, Valanou E, Oustoglou E, Palli D,

Sieri S, Tumino R, Sacerdote C, Bueno-de-Mesquita HB, Stattin P, Ferrari P, Johansson M, Norat T, Riboli E, Key TJ. Genetic variation in the lactase gene, dairy product intake and risk for prostate cancer in the European prospective investigation into cancer and nutrition. Int J Cancer. 2013;132:1901–10.

Trebichavsky I, Rada V, Splichalova A, Splichal I. Cross-talk of human gut with Bifidobacteria. Nutr Rev. 2009;67:77–82.

Trials, E.G.P.W.G.o.A.F. Safety and nutritional assessment of GM plants and derived food and feed: the role of animal feeding trials. Food Chem Toxicol. 2008;46(Suppl 1):S2–70.

Tsuda M, Sekine T, Takeda M, Cha SH, Kanai Y, Kimura M, Endou H. Transport of ochratoxin A by renal multispecific organic anion transporter 1. J Pharmacol Exp Ther. 1999;289:1301–5.

Twyman RM, Stoger E, Schillberg S, Christou P, Fischer R. Molecular farming in plants: host systems and expression technology. Trends Biotechnol. 2003;21:570–8.

Tyagi AK, Mohanty A. Rice transformation for crop improvement and functional genomics. Plant Sci. 2000;158:1–18.

Udell T, Gibson RA, Makrides M. The effect of alpha-linolenic acid and linoleic acid on the growth and development of formula-fed infants: a systematic review and meta-analysis of randomized controlled trials. Lipids. 2005;40:1–11.

USDA. Agricultural research service, USDA nutrient database for standard reference, Release 16-1. Nutrient Data Laboratory;2004. http://www.nal.usda.gov/fnic/foodcomp

Utsumi S. Plant food protein engineering. Adv Food Nutr Res. 1992;36:89–208.

Valenta R, Duchene M, Ebner C, Valent P, Sillaber C, Deviller P, Ferreira F, Tejkl M, Edelmann H, Kraft D. Profilins constitute a novel family of functional plant pan-allergens. J Exp Med. 1992;175:377–85.

Valenta H, Dänicke S, Flachowsky G, Böhme T. Comparative study on concentrations of deoxynivalenol and zearalenone in kernels of transgenic Bt maize hybrids and nontransgenic maize hybrids. Mycotoxin Res. 2001;17(suppl 1):15–8.

Van RR, Aalberse RC. Specific IgE without clinical allergy. J Allergy Clin Immunol. 1999;103:1000–1.

van der Werf J, Venema K. Bifidobacteria: genetic modification and the study of their role in the colon. J Agric Food Chem. 2001;49:378–83.

van Jaarsveld PJ, Faber M, Tanumihardjo SA, Nestel P, Lombard CJ, Benadé AJ. Beta-carotene-rich orange-fleshed sweet potato improves the vitamin A status of primary school children assessed with the modified-relative-dose-response test. Am J Clin Nutr. 2005;81:1080–7.

Van Linda BM, Cohn RA, Hajjar JJ. Percutaneous endoscopic gastrostomy. Report of two cases and a review of the literature. Conn Med. 1983;47:136–8.

Vasconcelos M, Datta K, Oliva N, Khalekuzzaman M, Torrizo L, Krishnan S, Oliveira M, Goto F, Datta SK. Enhanced iron and zinc accumulation in transgenic rice with the ferritin gene. Plant Sci. 2003;164:371–8.

Vaz NM, Vaz EM, Levine BB. Relationship between histocompatibility (H-2) genotype and immune responsiveness to low doses of ovalbumin in the mouse. J Immunol. 1970;104:1572–4.

Vaz NM, Phillipsquagliata JM, Levine BB, Vaz EM. H-2-linked genetic control of immune responsiveness to ovalbumin and ovomucoid. J Exp Med. 1971;134:1335–48.

Vazquez-Padron RI, Moreno-Fierros L, Neri-Bazan L, Martinez-Gil AF, de la Riva GA, Lopez-Revilla R. Intragastric and intraperitoneal administration of Cry1Ac protoxin from Bacillus thuringiensis induces systemic and mucosal antibody responses in mice. Life Sci. 1999;64:1897–912.

Vazquez-Padron RI, Moreno-Fierros L, Neri-Bazan L, Martinez-Gil AF, de la Riva GA, Lopez-Revilla R. Characterization of the mucosal and systemic immune response induced by Cry1Ac protein from Bacillus thuringiensis HD 73 in mice. Braz J Med Biol Res. 2000;33:147–55.

Vidal N, Barbosa H, Jacob S, Arruda M. Comparative study of transgenic and non-transgenic maize (Zea mays) flours commercialized in Brazil, focussing on proteomic analyses. Food Chem. 2015;180:288–94.

Vinje NE, Larsen S, Løvik M. A mouse model of lupin allergy. Clin Exp Allergy. 2009;39:1255–66.

Vinje NE, Namork E, Løvik M. Anaphylactic reactions in mice with Fenugreek allergy. Scand J Immunol. 2011;74:342–53.

von Wright A, Taimisto AM, Sivela¨ S (1985). Effect of Ca2+ ions on the plasmid transformation of Streptococcus lactis protoplasts. Appl Environ Microbiol, 50: 1100-1102.

von Wright A, Wessels S, Tynkkynen S, Saarela M. Isolation of a replication region of a large lactococcal plasmid and use in cloning of a nisin resistance determinant. Appl Environ Microbiol. 1990;56:2029–35.

von Wright A, Sibakov M. Genetic modification of lactic acid bacteria. In: Salminen S, von Wright A, editors. Lactic acid bacteria*microbiology and functional aspects. 2nd ed. New York: Marcel Dekker; 1998. p. 161–95.

von Wright A, Salminen S. Probiotics: established effects and open questions. Eur J Gastroenterol Hepatol. 1999;11:1195–8.

Wakasa Y, Takaiwa F. The use of rice seeds to produce human pharmaceuticals for oral therapy. Biotechnol J. 2013;8:1133–43.

Wal JM. 8–Assessing and managing allergenicity of genetically modified (GM) foods; 2015.

Waldron C, Murphy EB, Roberts JL, Gustafson GD, Armour SL, Malcolm SK. Resistance to hygromycin B: a new marker for plant transformation studies. Plant Mol Biol. 1985;5:103–8.

Wang Y, Harvey CB, Hollox EJ, Phillips AD, Poulter M, Clay P, Walker-Smith JA, Swallow DM. The genetically programmed down-regulation of lactase in children. Gastroenterology. 1998;114:1230–6.

Wang JF, Zhu YH, Li DF, Wang M, Jensen BB. Effect of type and level of dietary fibre and starch on ileal and faecal microbial activity and short-chain fatty acid concentrations in growing pigs. Anim Sci. 2004a;78:109–17.

Wang ZT, Yao YM, Xiao GX. Risk factors of development of gut-derived bacterial translocation in thermally injured rats. World J Gastroenterol. 2004b;11:1619–24.

Wang K, Han X, Dong K, Gao L, Li H, Ma W, Yan Y, Ye X. Characterization of seed proteome in Brachypodium distachyon. J Cerealence. 2010;52:177–86.

Wang Y, Xu WT, Zhao WW, Hao JR, Luo YB, Tang XG, Zhang Y, Huang KL. Comparative analysis of the proteomic and nutritional composition of transgenic rice seeds with Cry1ab/ac genes and their non-transgenic counterparts. J Cereal Sci. 2012;55:226–33.

Wang EH, Yu Z, Hu J, Jia XD, Xu HB. A two-generation reproduction study with transgenic Bt rice TT51 in Wistar rats. Food Chem Toxicol. 2014a;65:312–20.

Wang J, Sun N, Zhou C, Zhou X, Lu J, Wang CY, Che HL. Food proteins from different allergen families sensitize Balb/c mice to family-specific immune responses. J Immunotoxicol. 2014b;11:172–9.

Wang J, Zhou C, Tian J, Sun N, Lu J, Wang CY, Che HL. The analysis of specific allergenicity of food allergens families. Food Agric Immunol. 2014c;25:375–85.

Waters CM, Bassler BL. Quorum sensing: cell-to-cell communication in bacteria. Annu Rev Cell Dev Biol. 2005;21:319–46.

Weimer B. Brevibacterium. Encyclopedia of food microbiology. London: Academic Press; 2000. p. 308–14.

Welborn JR, Groves CE, Wright SH. Peritubular transport of ochratoxin A by single rabbit renal proximal tubules. J Am Soc Nephrol. 1998;9:1973–82.

WHO. Application of the principles of substantial equivalence to the safety evaluation of foods on food components from plants derived by modern biotechnology. Geneva: World Health Organization; 1995.

WHO. Operational guidelines for ethics committees that review biomedical research. Geneva: WHO; 2000.

Wild CP, Yun YG. Mycotoxins and human disease: a largely ignored global health issue. Carcinogenesis. 2010;31:71–82.

Williams GM, Kroes R, Munro IC. Safety evaluation and risk assessment of the herbicide Roundup and its active ingredient, glyphosate, for humans. Regul Toxicol Pharmacol. 2000;31:117–65.

Williams AL, Watson RE, DeSesso JM. Developmental and reproductive outcomes in humans and animals after glyphosate exposure: a critical analysis. J Toxicol Environ Health B Crit Rev. 2012;15:39–96.

Willis CL, Gibson GR. Microflora of the intestine. Encyclopedia of food microbiology. London: Academic Press; 2000. p. 1351–5.

Wilt TJ, Shaukat A, Shamliyan T, Taylor BC, MacDonald R, Tacklind J, Rutks I, Schwarzenberg SJ, Kane RL, Levitt M. Lactose intolerance and health, Evid Rep Technol Assess (Full Report). Rockville: Agency for Healthcare Research and Quality, U.S. Department of Health and Human Services; 2010. p. 1–410.

Wu X, Zhao R, Wang D, Bean SR, Seib PA, Tuinstra MR, Campbell M, O'Brien A. Effects of amylose, corn protein, and corn fiber contents on production of ethanol from starch-rich media. Cereal Chem. 2006;83:569–75.

Xin L, Xiaoyun H, Yunbo L, Guoying X, Xianbin J, Kunlun H. Comparative analysis of nutritional composition between herbicide-tolerant rice with bar gene and its non-transgenic counterpart. J Food Compos Anal. 2008;21:535–9.

Xu W, Cao S, He X, Luo Y, Guo X, Yuan Y, Huang K. Safety assessment of Cry1Ab/Ac fusion protein. Food Chem Toxicol. 2009;47:1459–65.

Xu W, Li L, Lu J, Luo Y, Shang Y, Huang K. Analysis of caecal microbiota in rats fed with genetically modified rice by real-time quantitative PCR. J Food Sci. 2011;76:M88–93.

Xue J, Liang G, Crickmore N, Li H, He K, Song F, Feng X, Huang D, Zhang J. Cloning and characterization of a novel Cry1A toxin from Bacillus thuringiensis with high toxicity to the Asian corn borer and other lepidopteran insects. FEMS Microbiol Lett. 2008;280:95–101.

Yakubu M, Akanji M, Oladiji A. Hematological evaluation in male albino rats following chronic administration of aqueous extract of Fadogia agrestis stem. Pharmacogn Mag. 2007;3(9):34.

Yamanishi R, Yusa I, Bando N, Terao J. Adjuvant activity of alum in inducing antigen specific IgE antibodies in BALB/c mice: a reevaluation. Biosci Biotechnol Biochem. 2003;67:166–9.

Yang D, Guo F, Liu B, Huang N, Watkins SC. Expression and localization of human lysozyme in the endosperm of transgenic rice. Planta. 2003;216:597–603.

Yang P, Wang J, Gong G, Sun X, Zhang R, Zhuo D, Liu Y, Li R, Ding F, Tang B, Dai Y, Li N. Cattle mammary bioreactor generated by a novel procedure of transgenic cloning for large-scale production of functional human lactoferrin. PLoS One. 2008;3:e3453.

Yano A, Takekoshi M. Transgenic plant-derived pharmaceuticals - the practical approach? Expert Opin Biol Ther. 2004;4:1565–8.

Ye X, Al-Babili S, Klöti A, Zhang J, Lucca P, Beyer P, Potrykus I. Engineering the provitamin A (β-carotene) biosynthetic pathway into (carotenoid-free) rice endosperm. Science. 2000;287:303–5.

Yimam M, Zhao Y, Ma W, Jia Q, Do SG, Shin JH. 90-day oral toxicity study of UP446, a combination of defined extracts of Scutellaria baicalensis and Acacia catechu, in rats. Food Chem Toxicol. 2010;48(5):1202–9.

Yokoyama H, Anzai N, Ljubojevic M, Ohtsu N, Sakata T, Miyazaki H, Nonoguchi H, Islam R, Onozato M, Tojo A, Tomita K, Kanai Y, Igarashi T, Sabolic I, Endou H. Functional and immunochemical characterization of a novel organic anion transporter Oat8 (Slc22a9) in rat renal collecting duct. Cell Physiol Biochem. 2008;21:269–78.

Yoshikawa Y, Kishimoto Y, Tagami H, Kanahori S. Assessment of the safety of hydrogenated resistant maltodextrin: reverse mutation assay, acute and 90-day subchronic repeated oral toxicity in rats, and acute no-effect level for diarrhea in humans. J Toxicol Sci. 2013;38:459–70.

Youngblood GL, Sweet DH. Identification and functional assessment of the novel murine organic anion transporter Oat5 (Slc22a19) expressed in kidney. Am J Phys Renal Phys. 2004;287:F236–44.

Yu JJ, Peng P, Zhang XJ, Zhao Q, Zhy DY, Sun XH, Liu JQ, Ao GM. Seed-specific expression of a lysine rich protein sb401 gene significantly increases both lysine and total protein content in maize seeds. Mol Breed. 2004;14:1–7.

Yuan DT. Management of lactose intolerance in infants. Am Fam Physician. 2010;81:933. author reply 933–4

Zaied C, Bouaziz C, Azizi I, Bensassi F, Chour A, Bacha H, Abid S. Presence of ochratoxin A in Tunisian blood nephropathy patients. Exposure level to OTA. Exp Toxicol Pathol. 2011;63:613–8.

Zarate G, Chaia AP, Gonzalez S, Oliver G. Viability and beta-galactosidase activity of dairy propionibacteria subjected to digestion by artificial gastric and intestinal fluids. J Food Prot. 2000;63(1214):1221.

Zhang HC, Chen XY, Wang XF, Yang GL, Jun-Feng XU, Yue-Ying LI. Unintended effects and their detection in genetically modified crops. Food Chem Toxicol. 2004a;42:1089–125.

Zhang X, Groves CE, Bahn A, Barendt WM, Prado MD, Rödiger M, Chatsudthipong V, Burckhardt G, Wright SH. Relative contribution of OAT and OCT transporters to organic electrolyte transport in rabbit proximal tubule. Am J Physiol Ren Physiol. 2004b;287:F999–F1010.

Zhang X, Boesch-Saadatmandi C, Lou Y, Wolffram S, Huebbe P, Rimbach G. Ochratoxin A induces apoptosis in neuronal cells. Genes Nutr. 2009;4:41–8.

Zhang M, Liu N, Qian C, Wang Q, Wang Q, Long Y, Huang Y, Zhou Z, Yan X. Phylogenetic and functional analysis of gut microbiota of a fungus-growing higher termite: bacteroidetes from higher termites are a rich source of β-glucosidase genes. Microb Ecol. 2014a;68:416–25.

Zhang M, Zhuo Q, Tian Y, Piao J, Yang X. Long-term toxicity study on transgenic rice with Cry1Ac and sck genes. Food Chem Toxicol. 2014b;63:76–83.

Zhang X, Zhao P, Wu K, Zhang Y, Peng M, Liu Z. Compositional equivalency of RNAi-mediated virus-resistant transgenic soybean and its nontransgenic counterpart. J Agric Food Chem. 2014c;62:4475–9.

Zhou C, Sun N. Allergenicity assessment of a genetically modified protein-recombinant human lactoferrin. J Allergy Ther. 2013;S3:002.

Zhou XH, Dong Y, Xiao X, Wang Y, Xu Y, Xu B, Shi WD, Zhang Y, Zhu LJ, Liu QQ. A 90-day toxicology study of high-amylose transgenic rice grain in Sprague-Dawley rats. Food Chem Toxicol. 2011;49:3112–8.

Zhou XH, Dong Y, Wang Y, Xiao X, Xu Y, Xu B, Li X, Wei XS, Liu QQ. A three generation study with high-lysine transgenic rice in Sprague-Dawley rats. Food Chem Toxicol. 2012;50(6):1902–10.

Zhou C, Wang JW, Sun N, Tian J, Wang J, Lv YS, Wang P, Huang KL, Che HL. Allergenicity of recombinant human lactoferrin to an animal model Brown Norway rats. Food Agric Immunol. 2014;25:34–48.

Zhu Z. Research and development of highly insert-resistant transgenic rice. Bull Chin Acad Sci. 2001;05:353–7.

Zhu Y, He X, Luo Y, Zou S, Zhou X, Huang K, Xu W. A 90-day feeding study of glyphosatetolerant maize with the G2-aroA gene in Sprague-Dawley rats. Food Chem Toxicol. 2013;51:280–7.

Zingerle M, Silbernagl S, Gekle M. Reabsorption of the nephrotoxin ochratoxin A along the rat nephron in vivo. J Pharmacol Exp Ther. 1997;280:220–4.

Zolla L, Rinalducci S, Antonioli P, Righetti PG. Proteomics as a complementary tool for identifying unintended side effects occurring in transgenic maize seeds as a result of genetic modifications. J Proteome Res. 2008;7:1850–61.

Zou S, He X, Liu Y, Chen D, Luo Y, Huang K, Zhang W, Xu W. Toxicological evaluation of lactase derived from recombinant Pichia pastoris. PLoS One. 2014;9:e106470.

http://www.gruene-bundestag.de/themen/agrar/glyphosat-in-muttermilch_ID_4396067.html (only available in German) Accessed 7 Dec 2015.

Kolibri 360 GmbH
Ferdinand-Jühlke-Straße 7
99095 Erfurt, Deutschland
produktsicherheit@kolibri360.de